# FOUNDATIONS
## of
# MOBILE RADIO
# ENGINEERING

## Michel Daoud Yacoub

University of Campinas
Sao Paulo, Brazil

**CRC Press**
**Boca Raton   Ann Arbor   London   Tokyo**

**Library of Congress Cataloging-in-Publication Data**

Yacoub, Michel Daoud.
    Foundations of mobile radio engineering / Michel Daoud Yacoub.
      p.   cm.
    Includes bibliographical references and index.
    ISBN 0-8493-8677-2
    1. Mobile radio stations.   2. Mobile communication systems.
   I. Title
  TK6570.M6Y33   1993
  621.3845—dc20
                                                92-38563
                                                   CIP

© 1993 by CRC Press, Inc.

International Standard Book Number 0-8493-8677-2

Library of Congress Card Number 92-38563

Printed in the United States of America   2 3 4 5 6 7 8 9 0

Printed on acid-free paper

# Preface

In the past few years that I have been involved with mobile radio communications I have felt the need for a textbook or a reference that would treat the various topics related to this challenging area in an accessible and comprehensive way. However, mobile radio engineering involves virtually all the areas of telecommunications, and a book covering the subject thoroughly would require many volumes and several authors.

Some of the phenomena of mobile radio communications have already been explored in various classic textbooks, where mobile radio would appear just as an advanced application case. Moreover, many techniques, initially used in other applications, are found to be perfectly suited to mobile radio needs. Therefore, writing a specialized book on this area would require a careful compilation of the various materials in the well-established literature.

This is only partially true. Since the emergence of the first cellular systems, a great deal of research has been undertaken and a significant amount of interesting results obtained. What initially used to belong to a very restricted area of research has now become the main part of most of the investigations. It seems to me that all the communications research and development activities are steering toward mobile radio applications.

To compile both the basic and the more advanced materials in a one-volume–one-author book is literally impossible. In this book I have endeavored to address what I consider to be a useful and broad subset of this vast field.

Intended for use by advanced students and professional engineers involved in mobile communications research, the book includes both basic and advanced materials covered in considerable depth. The chapters are ordered so that knowledge is acquired in a logical and progressive way. The reader will find, in sequence, an introduction to the subject, the description and analysis of the basic phenomena, some practical solutions to the problems, and more advanced materials. Each chapter starts with a preamble, abstracting the topics to be found in it, and ends with a section of summary and conclusions, where the main results and general comments are included. The mathematical derivations are meant to be clear, logical, and thorough; whenever necessary, additional information is appended at the end of the corresponding chapter, providing a self-contained book.

The book covers both analog and digital systems, making use of analytical as well as Monte Carlo simulation solutions to the various problems. The problem of cell coverage area is tackled on a deterministic and on a statistical basis, whereas adjacent-channel and cochannel interference problems take into account the system traffic load. Digital techniques include speech coding and modulation schemes, with emphasis given to those used by the American and European systems. Multiple-access architectures, such as FDMA, TDMA, and CDMA, are investigated and the ALOHA access protocol is analyzed in a mobile radio environment. The book also explores the traffic engineering aspects, studying several channel allocation techniques where some new algorithms are included.

The book is divided into six parts.

*Part I   Introduction* comprises Chapters 1 and 2.

Chapter 1 traces a brief history of mobile radio communications engineering, bringing up the main events that have contributed to its development. It also describes the various mobile radio services and explores the spectrum allocation problems.

Chapter 2 addresses the basic principles of cellular mobile radio systems, giving an overview of the main points to be considered in a system design.

*Part II   Mobile Radio Channel* comprises Chapters 3 and 4.

Chapter 3 analyzes the mobile radio propagation phenomena, such as path loss, shadowing, and multipath propagation, and their statistics. The problem of cell coverage area is tackled on a deterministic and statistical basis. The boundaries between cells are also investigated, with the aim of determining the proportion of overlapped service areas.

Chapter 4 is entirely devoted to the various topics related to multipath propagation phenomena, such as time delay, delay spread, Doppler effect, coherence bandwidth, random FM, etc. It outlines the main points to be considered in field measurements. Finally, it describes some analog and digital mobile radio channel simulators.

*Part III   Diversity-Combining Methods* comprises Chapters 5 and 6.

Chapter 5 analyzes the various diversity schemes used to combat fading. The problem is first approached from the macroscopic side and then from the microscopic side. The chapter also investigates envelope-combining methods, where they are assessed by a measure of the SNR obtained at the output of the combiners.

Chapter 6 examines the performance of digital transmission over a fading environment where several diversity techniques are used and compared. The chapter also includes an appendix on channel coding.

*Part IV   Noise, Interference, and Modulation* comprises Chapters 7, 8, and 9.

Chapter 7 is concerned with noise and with cochannel and adjacent-channel interference problems. It investigates the influence of the traffic load on the adjacent-channel and cochannel interferences. Monte Carlo simulation is used to approach the problem of cochannel interference in a more realistic way.

Chapter 8 examines the analog modulation schemes, such as AM, SSB, and FM, in a mobile radio environment.

Chapter 9 describes and analyzes some speech-coding and digital modulation techniques in a mobile radio environment. Emphasis is given to those schemes used by the American and European mobile digital systems.

*Part V Multiple Assess* comprises Chapters 10 and 11.

Chapter 10 describes multiple-access architectures such as FDMA, TDMA, and CDMA. In particular, CDMA systems are investigated in the light of the information of spread-spectrum technology.

Chapter 11 analyzes the various multiple-access protocols such as CSMA, slotted ALOHA, and PRMA. In particular, the performance of the slotted ALOHA scheme is investigated in a mobile radio environment.

*Part VI Traffic* comprises Chapter 12.

Chapter 12 examines some of the main channel allocation techniques that can be used in a mobile radio system. More specifically, it analyzes one global approach by means of Monte Carlo simulation and a local approach by means of analytical methods, using some well-known tools from queueing theory.

The tutorial value of this book would make it suitable for use both in the classroom and as a reference. The book can be used as a textbook for a two-term graduate course, with the first six chapters and the remaining six comprising each term, respectively.

# Acknowledgments

I am grateful to Professor Attílio J. Giarola for his encouragement and patient review of the manuscript. I am indebted to my colleagues of the Department of Communications, School of Electrical Engineering, at The State University of Campinas (UNICAMP) for their assistance and helpful inputs.

I have been privileged to share the wisdom and experience of the many engineers attending my courses who stimulated discussions and gave me valuable suggestions. In the same way, I have had the honor of supervising the research projects of many students who have contributed with original results, some of them partially included in this book. In particular, I thank J. C. E. Mencia, O. C. Branquinho, J. L. A. D'Annibale, G. Fernandes, E. J. Leonardo, N. F. Keffer, A. A. Shinoda, A. F. Victória, and E. Nisembaum.

I acknowledge the support of the School of Electrical Engineering at UNICAMP, the R & D Center of Telebrás (CPqD-Telebrás), Serifa Editoração e Informática, CNPq, CAPES, and FAPESP.

Among those who have lent their help in different ways I would like to mention Professor K. W. Cattermole, Professor C. J. Hughes, Dr. J. Szajner, Dr. B. S. Ramos, Dr. C. D. Yacoub, Mr. J. Claypool, Ms. Sarah W. Roesser, and Ms. Janete S. Toma.

Finally, my wholehearted gratitude to my wife, Maria Nídia, and my children, Alexandre, Helena, Carolina, Ricardo, and Vinícius, for enduring the apparently endless writing period of this book.

*To my dear parents, Daoud and Helena*

# Table of Contents

# PART I
# Introduction

# CHAPTER 1

# Mobile Radio System

This chapter traces a brief history of mobile radio communications, bringing up the main events that have contributed to the development of such an efficient and successful means of communication. Spectrum allocation problems and technological evolution—the latter addressing equipment packaging, modulation schemes, and system architecture aspects—are outlined. Some mobile radio services are then briefly described and system design considerations are examined. It is also shown that, because radio propagation does not recognize geopolitical boundaries, effective use of the spectrum is possible only if international agreement is achieved. Under such circumstances the International Telecommunication Union (ITU) has emerged to provide "worldwide harmony". The international frequency allocation for some radio services is then presented in tabular form.

## 1.1  INTRODUCTION

The first successful use of mobile radio dates from the late 1800s, when M. G. Marconi established a radio link between a land-based station and a tugboat, over an 18-mile path. Since then mobile systems have developed and spread considerably. The usefulness of mobile radio services was first recognized by the public safety services (police and fire departments and the forestry conservation, highway maintenance, and local government services), followed by the private sector (power, oil, motion picture, telephone maintenance, and transportation services, as well as taxis and lorry fleets). The growth rate of these services in the United States by the 1950s (when the number of subscribers was just a few thousands) was greater than 20% per year.[1] Consequently, by 1963 the number of users exceeded 1.3 million although only a few channels (about 12) were available.

These early systems (conventional mobile systems [CMS]) consisted of a base station whose transmitter and receiver were assembled on a hilltop. The coverage area was chosen to be large, in a way similar to the radio or television broadcast services. The transmitter usually operated with high power (200 or 250 W [Reference 3]), assuring a large coverage area (25-mi radius).

Conventional mobile systems are usually isolated from each other, with only a few of them accessing the public switched telephone network. Those having connection with the telephone network are named *mobile telephone systems*, where the communication unit is assigned to the subscriber and not to a physical location.

## 1.2  CONVENTIONAL MOBILE SYSTEMS

By the end of the nineteenth century H. G. Hertz, a German scientist, demonstrated that radio waves could propagate in a wireless medium, in fact, over a path of a few yards, between transmitter and receiver. Still before the twentieth century, Marconi showed the first wireless communication "on the move" between a land-based station and a tugboat. Thereafter, many maritime mobile services were established and operated successfully.

On land, the Detroit Police Department started its experiments with mobile radio in 1921, first operating as a dispatching system. Initially, only the base station could transmit. Later the mobile unit was also able to communicate to the base station. The frequency used was around 2 MHz. Soon, other police departments installed their own systems. As a consequence, the available frequency spectrum became congested. By the middle of the 1930s the Federal Communications Commission (FCC) authorized four more channels between 30 and 40 MHz. In 1946, six channels near 150 MHz were available for use. In fact, due to technological restraints, out of those six channels only three could be utilized because of adjacent-channel interference problems. These radio frequencies were first used by the mobile telephone services. Starting in St. Louis, this system quickly spread throughout the United States. Shortly after, frequencies around 40 MHz were available for use on highways. Again the highway mobile services grew rapidly, although this did not seem to have worked out well due to radio interference. By the middle of the 1950s, due to the reduction of the channel spacing from 60 kHz to 30 kHz, a total of 11 channels could be used at 150 MHz. Almost at the same time, the FCC released 12 channels at 450 MHz. Up to that time the mobile telephone systems were manually operated, with each call having to be handled through an operator. It was only in the 1960s that the automatic systems appeared, allowing the subscribers themselves to do the direct dialling. Automatic mobile telephone systems operated initially with frequencies around 150 MHz, later moving to 450 MHz.

Already in 1975, after a long period of negotiations involving the mobile industries and the FCC, a 40-MHz band between 800 and 900 MHz was released. The year 1978 was marked by the beginning of a new era in mobile communications history, when the first cellular system was sent into the field testing.[22]

## 1.3 TECHNOLOGICAL EVOLUTION

Before the Second World War, mobile radio systems were largely dominated by military (and paramilitary) users. Consequently, the evolution and development of such systems were supported by and closely linked to military needs, requirements, and standards.

Recently, however, this tendency has been reversing, as more and more mobile services are steering toward civil applications. Consequently, as far as system design and technology are concerned, commercial mobile systems have taken the lead.[14]

### 1.3.1 Equipment Package Evolution

The early mobile radio systems were equipped with vacuum-tube equipment requiring powerful batteries. In fact, before 1930 only the receivers were mobile,[2] implying one-way communication. Mobile transmitters soon appeared, but they were still bulky and heavy, requiring special power supplies. The total apparatus could occupy most of a vehicle's trunk space.

By the 1950s equipment was already small enough to be man-transportable, although the volume was still quite considerable and the main application was for military purposes.

Transistorization of mobile radio products started in 1957 with the power supplies. Soon after, the vacuum tubes in receivers and in some parts of transmitters were replaced by transistors, contributing to a 50% reduction of the volume, lower power consumption, and higher reliability. The packages could then be mounted in a car's dashboard or even on motorcycles. Maintenance costs diminished because of the reduced number of spare parts required and also because of the very rapid decrease of transistor prices. By the 1960s the mobile products were all being built with solid-state components. The design of the equipment introduced some new components, such as printed circuit boards, sockets, and heat sinks.

The first hand-held portable radios appeared in the 1960s, initially for commercial and later for military systems. At this point, mobile civil applications started to take the lead.

The change from transistors to integrated circuits was an obvious and natural step, occurring by the middle of the 1970s. By this time cordless telephones were already available. Today the equipment includes LSI and VLSI circuits, rendering the products smaller, lighter, and less costly.

## 1.3.2  Modulation Scheme Evolution

### 1.3.2.1 *Analog Systems*

Similar to any other early systems using voice transmission by radio, mobile radio systems also employed amplitude modulation (AM), initially in the HF band and later in the VHF band.

The introduction of frequency modulation (FM) into mobile radio services started soon after its invention in 1935. In the 1940s most military and some commercial equipment was already operating in VHF FM. However, if on the one hand FM systems had a remarkably better signal-to-noise ratio performance in the presence of fading as compared to AM systems, on the other hand they required a much wider frequency band. Although for voice transmission AM occupied around 6 kHz in the spectrum, an early FM channel needed 120 kHz. The 120-kHz bandwidth used in the 1940s was reduced to 60 kHz in the 1950s, 30 kHz in the 1960s, and 25 kHz in the 1970s. It is believed that a further reduction to 12 kHz would not substantially deteriorate the speech quality.

Except for (1) cordless telephones and citizen's band radio, which used AM in the 1970s, and (2) the military mobile systems of the 1960s, which used single-sideband modulation, FM is the modulation scheme widely used in analog mobile systems.

Nevertheless, because the radio spectrum is a scarce resource, a great deal of research has been steered toward the "old" single-sideband (SSB) modulation. Like the AM systems, one of the biggest problems of SSB is its disastrous signal-to-noise ratio performance in the presence of fading. On the other hand, it requires five to six times less spectrum than FM, rendering this scheme very attractive for mobile applications. Accordingly, many enhanced forms of SSB, such as "transparent tone in band," "feedforward signal regeneration," etc., have been investigated.

The reader is referred to Chapter 8 for a deeper study of the analog modulation schemes.

### 1.3.2.2 *Digital Systems*

Digital techniques are always seen as the natural evolution of the analog approach. This is quite true for most of the applications where there is physical connection or where bandwidth is not a limiting factor.[4] In the mobile radio environment, however, spectrum efficiency is probably one of the strongest aspects to be taken into account if a change of technology is to be considered. Digital techniques appeared before the analog techniques, but only became feasible with the advent of the transistor and, later, integrated circuits. As far as mobile radio is concerned, the word "digital" is essentially related to voice digitization and to digital transmission.

The main problem with digital transmission is spectrum utilization. The standard (conventional PAM) 64 kbit/s, used for voice communication in the fixed telephone network, if transmitted by radio, would occupy approximately

100 kHz of the spectrum, four times the 25-kHz FM currently in use. This, by itself, is such a constraint that it makes any digital mobile radio unthinkable. Efforts have been concentrated in diminishing the transmission rate to 16 kbit/s (as adopted by the pan-European GSM system) or even less (8 kbit/s as proposed by the American digital system[5]). On the other hand, digital techniques provide robustness against interference and flexibility for integration into the emerging digital network, with the capability of transmitting data, voice, etc.

Another clear advantage of the digital system is to allow other options for channel access. In analog systems only frequency division multiple access (FDMA) can be used. In digital systems, besides FDMA, other access schemes such as time division multiple access (TDMA) and code division multiple access (CDMA) can be used.

The reader is referred to Chapter 9 for the studies of digital techniques and to Chapter 10 for the various access schemes.

### 1.3.3  System Evolution

As far as access methods and operational aspects of mobile radio are concerned, there have been several stages of evolution as follows.

#### 1.3.3.1 *Simplex System (SS)*

A simplex system operates with one frequency, and only the base station can transmit. The mobile unit is just a single receiver (Figure 1.1a).

#### 1.3.3.2 *(Single) Half-Duplex System ([S]HDS)*

In this system both the mobile unit and the base station use only one frequency. The system operates on a push-to-talk basis, where the base station competes with the mobile unit (Figure 1.1b).

#### 1.3.3.3 *Double Half-Duplex System (DHDS)*

In this case the transmitter of the mobile and the receiver of the base station operate with one frequency, whereas the receiver of the mobile and the transmitter of the base station operate with another frequency. Here again, the push-to-talk scheme is used, but the base station does not compete with the mobiles (Figure 1.1c).

#### 1.3.3.4 *Duplex Base Double Half-Duplex System (DBDHDS)*

This system operates with two frequencies, and only the base station can transmit and receive simultaneously. The mobiles make use of the push-to-talk scheme (Figure 1.1d).

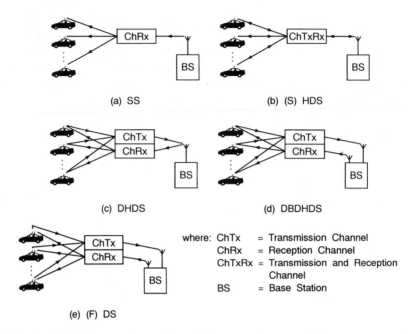

(a) SS

(b) (S) HDS

(c) DHDS

(d) DBDHDS

where:  ChTx   = Transmission Channel
        ChRx   = Reception Channel
        ChTxRx = Transmission and Reception
                 Channel
        BS     = Base Station

(e) (F) DS

**Figure 1.1.**    The various mobile systems and their operation modes (all connections are via radio).

### 1.3.3.5 *(Full) Duplex System ([F]DS)*

Both base station and mobile operate duplex. They are able to transmit on one frequency while receiving on another frequency. Two antennas in each end (mobile and base) must be provided, so that transmitter and receiver work independently with different frequencies. It is possible, however, to include filters between transmitters and receivers in order to avoid interference, and to use only one antenna at each end. Here the push-to-talk procedure is no longer used (Figure 1.1e).

### 1.3.3.6 *Mobile Telephone System (MTS)*

In general terms, mobile services can be classified in two categories, namely, nontrunked mobile system (NTMS) and trunked mobile system (TMS).

The first category—NTMS—comprises the systems where only one or just a small number of channels is available for use and the channels are allocated for special services. Nontrunked systems are characterized by the absence of privacy, where all the users can take part in the communication. Examples of such systems include the private mobile radio (PMR) networks such as the radio taxis, lorry fleet radio, citizens' band (CB) radio, etc. These systems are intended to serve a very limited number of users for a particular application.

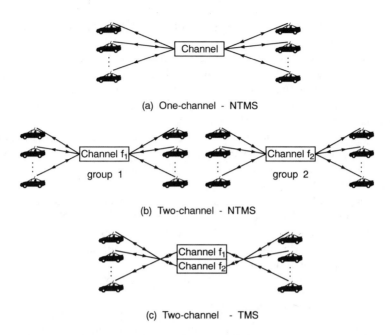

**Figure 1.2.** Nontrunked and trunked mobile systems (all connections are via radio).

The early mobile systems were limited by the available technology. The radios could only operate with one or a small subset of the already scarce set of channels, because each frequency was obtained by means of a quartz crystal. With the advent of solid-state technology and the use of frequency synthesizers, the subscribers themselves were able to select a free channel for conversation. This improvement, however, was not enough to enable the system to cope with the traffic demand.

Consider an NTMS operating with only one channel (Figure 1.2a). Suppose also that this channel is, on the average, busy 30% of the time (i.e., the traffic carried is 0.3 erl, which, in a one-channel system, corresponds to a blocking probability of 30%). Thus, the total traffic offered is $0.3/(1 - 0.3) \simeq$ 0.43 erl. If each mobile can generate 0.1 erl, then the maximum number of users this system can support is 43 (43 subscribers will experience 30% blocking probability!). In order to maintain this grade of service and increase the number of subscribers, another channel must be provided and allocated to the other group of users as shown in Figure 1.2b (now 86 subscribers will experience 30% blocking probability!).

Suppose now that this system has been doubled in size, but with all the subscribers being able to share all the channels (86 users and 2 channels), as in Figure 1.2c. The total amount of traffic is also doubled but the blocking probability will go down to approximately 16%. If the (unacceptable) figure of 30% is to be maintained, then the number of subscribers could go up to

145. This is the principle of a trunked mobile system (refer to Chapter 12 for studies of traffic).

Both the NTMS and TMS used in a private network require the subscribers to have some sort of discipline so that the channels can be efficiently used. In the case of the TMS, if there are many channels available, usually one channel is reserved for control (e.g., the conversation will always start on a specific channel and move to another channel).

Although in TMS the channel usage is substantially increased, there still remains a serious concern with respect to the bounds imposed on the subscriber's mobility. The users can only communicate among themselves within the restricted area covered by the radio transmitter.

A higher degree of flexibility can be achieved when the mobile system is integrated into the fixed telephone network, resulting in the mobile telephone system (MTS). On the other hand, the subscribers of the fixed telephone network will experience some unusual degradation in the voice signal due to fading and man-made noise.

The MTS's have gone through several stages of evolution:

1. Initially the MTSs were NTMSs, operating in half-duplex and push-to-talk mode with the calls being handled by a special mobile operator.
2. Then, with the availability of more channels, the MTS went through a TMS phase, but the channels were selected manually. The mobiles operated on a push-to-talk basis and the landline subscribers could already dial directly, without need of the operator.
3. The next step was to operate in a duplex mode with both sides (mobile and landline users) able to dial directly. Channel selection was still manual.
4. The only change at this stage was toward automatic channel selection.

Figure 1.3 gives a schematic representation of a mobile telephone system.

The MTSs reached their final stage of evolution by the mid-1960s. One decade later there were more than 1000 such systems already in operation in the United States, most of them at their limit of capacity. Just to have an idea of how overloaded the systems were,[6] the New York Telephone Company

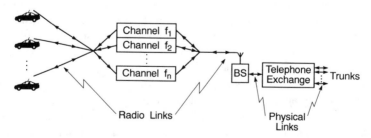

**Figure 1.3.**   Mobile telephone system.

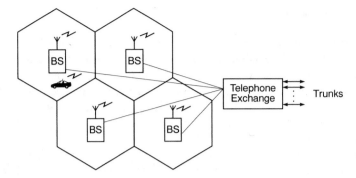

**Figure 1.4.** Basic structure of a cellular system.

operated two MTSs with six channels each in 1976 serving a total of 543 subscribers (a concentration of approximately 46 subscribers per channel) in a 50-mi area. The number of applicants on waiting-lists was about 3700.

### 1.3.3.7 *Cellular Mobile Radio System*

In a cellular system the service area is divided into regions (the cells) each containing a subset of the total number of available channels. Channels used in a given cell can be reused in another cell sufficiently far away that interference is minimized. The cells can be placed in a modular way and, in theory, the system can grow indefinitely. The structure of a cellular system is shown in Figure 1.4.

A proper treatment of the cellular concept will be given in Chapter 2. In terms of revolutionary ideas the cellular architecture is a cornerstone of this revolution. Systems now can be considered as *before* or *after* cellular. There has been some progress as far as the size of the cells are concerned, but still the architecture remains cellular.

## 1.4  MOBILE RADIO SERVICES

In the previous section we showed how the mobile systems evolved to the configuration of the present day's systems. It must be emphasized, however, that the advent of a better network, in general, did not make the earlier systems unviable. New and old systems can coexist for a long time until the gradual replacement of the latter is finally accomplished.

Mobile radio systems do vary according to the specific application for which they are designed. The most flexible network using the latest technology will be able to meet the requirements of a simpler system, but will also be much more expensive. In this sense there are many different types of radio systems, some of them using solutions considered to be obsolete for certain types of mobile services.

### 1.4.1   Radio Paging

Paging systems provide only one-way communication between base station and mobile users. In the simplest systems the base station operator selects the wanted user and sends him or her an alert tone. The user is then required to report by telephone to a fixed location for further information.[15] More-advanced systems allow voice or alphanumeric messages to be transmitted. Initially, the paging systems covered small areas (2–5 km radius), requiring small and low-consumption receivers.

In large cities, however, users of the paging services experienced the need for a wider-range operation. Accordingly, some wide-area paging systems appeared in various countries, each of which had its own standards. Later, a paging code and signalling format were proposed that gained international recognition by the CCIR. The POCSAG code,[16] as it was known, "has the capacity to cope with eight million pagers, provides the facility for alphanumeric messages and can operate in 12.5 or 25 kHz channels".[2]

### 1.4.2   Packet Radio

"Packet radio networks (PR nets) represent the extension of packet switching technology into the environment of mobile radio. They are intended to provide data communications to users located over a broad geographic region where connection between the source and destination users is not practical or cost effective".[17] The network is composed of a number of packet radio units, consisting of transmitter, receiver, antenna, and controller. Usually, full connectivity among all the radio units in the network is not provided, because not all nodes are within line-of-sight of one another. Accordingly, a packet may reach its destination after being received by and relayed to as many radio units in its path as required. Therefore, all nodes must provide for store-and-forward operation. Note that, due to the limited range of transmission and the mobility of the nodes, the connectivity (network topology) changes dynamically.

### 1.4.3   Future Public Land Mobile Telecommunication Systems (FPLMTS)

FPLMTS is a new concept, arising from the CCIR study groups, where cellular mobile technology is used for fixed-service purposes. The initial application of FPLMTS aimed at the developing countries where the lack of service in rural and remote areas or lack of capacity to offer a good grade of service in urban areas are critical. Providing rural wireline networks is usually extremely costly due to the long distances, difficult terrain, and climatic conditions. Moreover, planning network expansion in urban areas is not easy because of the high and unpredictable demand. Accordingly, FPLMTS may become an attractive alternative to the wireline systems because of its (1) flexibility, (2) modular design, (3) capability for covering wide geographical areas, and (4) cost reductions resulting from technological improvements. Industrialized countries with large territories, rough terrain, and thinly populated areas may also profit from FPLMTS. Besides voice, other services such as point-to-multipoint, short messages, paging, facsimile, text, and data are also planned to be offered by these systems.[19]

### 1.4.4  Cordless Telephone

#### 1.4.4.1 *CT1 — The First Generation*

The cordless telephone (CT) has been on the market for a long time, but with analog technology. It is, in fact, a wireless extension of the wireline telephone, with a very limited transmission range (50–200 m). Moreover, because a small number of channels is available and the apparatus uses only one fixed channel, interference (or lack of privacy) is quite common.

#### 1.4.4.2 *CT2 — The Second Generation*

The second generation of cordless telephone uses digital technology, providing "high speed quality, higher degree of security and the possibility of introducing new services".[20] It has three main applications:

- *Domestic use*—Here CT2 replaces its analog counterpart with the advantage of automatic channel selection as well as other advantages provided by digital technology.
- *Cordless PABX*—A linked network of base stations allows staff with handsets to move around inside a building, being able to receive and make calls.
- *Telepoint*—Base stations are connected to the public switched telephone network. These base stations are usually located at shops, bus stations, train stations, airports, etc. and the users within their coverage areas are able to make calls (but not to receive them).

The CT2 systems use FDMA architecture,* 4-MHz bandwidth, 1 channel per carrier, and a total of 40 carriers.

#### 1.4.4.3 *CT3 — The Third Generation*

This system uses TDMA architecture,* 8-MHz bandwidth, eight channels per carrier, and a total of eight carriers. The great difference between CT2 and CT3 is that in the former "the cordless handset is sending or receiving radio transmissions all the time while a call is in progress".[20] In the latter the

---

*Refer to Chapter 10.

handset operates for one-eighth of the time. Accordingly, the rest of the time can be used for other applications.

### 1.4.4.4  *DECT — The Third Generation*

The digital European cordless telephone (DECT) is also a third-generation cordless telephone, specified to be used throughout Europe. It is very similar to CT3, using TDMA architecture, 20-MHz bandwidth, 12 channels per cell, and a total of 12 channels per carriers.

## 1.4.5  Personal Communications

Personal communications networks (PCN) aim at providing a "go-any-where telephone that can be used at any time with a guarantee of a quality service".[21] The idea is to provide low-cost and high-quality mobile communications for the mass market. The cellular architecture is kept but the cells are much smaller, with a consequent reduction of the handset power. On the other hand, the number of base stations required for this service increases drastically. Cell size may vary from a radius of 1 km in city centers to 6 km in the country. The spectrum space planned to be allotted to PCNs is at 1.8 GHz, where the spectrum is fairly clean.

## 1.4.6  Mobile – Satellite Communications

"Mobile–Satellite Service provides communication between earth stations and one or more space stations, or between mobile earth stations via one or more space stations".[23] The earth stations can be located on board ships, aircrafts, or terrestrial vehicles—characterizing the maritime, aeronautical, or land mobile–satellite services, respectively. In addition, "this service can be used to detect and locate distress signals from survival craft stations and emergency position-indicating radio beacon stations."[23]

Among the various international systems providing telecommunications (in general) by satellite we mention INTELSAT (Internatioanl Telecommunications Satellite Organization), EUTELSAT (European Telecommunications Satellite Organization), and INMARSAT (International Maritime Satellite Organization). These organizations are financed by their members (countries), to whom the given services are hired out.

In particular, INMARSAT's primary mission is to provide data, voice, and emergency communications to ships and platforms. More recently, INMARSAT is aiming at offering its services to be used for air traffic control worldwide.

A global integration of the mobile services is likely to occur in the near future and will certainly make use of satellites. Motorola has given a step forward proposing the Iridium System, where 66 low-orbit satellites in connection with 37 cells would cover the entire globe. The advantage of having low-orbit satellites is the use of low-power handsets. Other companies (e.g., Qualcomm) followed the same steps, proposing their own systems.

## 1.5  SYSTEM DESIGN CONSIDERATIONS

Although the use of electromagnetic waves would restrict mobile communication to some specific types of services, connection with the fixed telephone network gives the flexibility that makes the mobile network able to give support to all sorts of communications services. On the other hand, the design of such a system is complicated for it may involve most of the areas in telecommunications as well as other related areas.

Let us mention some of these areas.

- *Radio system design and propagation*—For instance, the topography of the terrain is carefully studied; the results of this study constitute input data for the radio coverage planning. At this point the need for the use of *repeaters** (a very common method for range extension is most mobile radio networks) may be detected.
- *Frequency regulation and planning*—Each radio service utilizes its own portion of the frequency spectrum. Given that only a limited amount of the spectrum is available, the frequency allocation plan has to be made carefully so that interferences are minimized and the traffic demand is satisfied.
- *Modulation*—As far as voice in an analog system is concerned, very few services still use AM, whereas the majority uses narrowband FM. With the advent of digital systems, there is a variety of possibilities and the choice is not so straightforward.
- *Antenna design*—In the early systems only omnidirectional antennas were used. With the increase in traffic demand the cells were then split into sectors for which directional antennas are required. Today new cellular patterns[7] are being suggested for which more-complex antennas are required (sectorization and the use of directional antennas are covered in Chapter 2 and 7).
- *Transmission planning*—There are many aspects to be considered here, such as the structures of the channels used for signalling and voice, the characteristics of the message voice transmitted, and the performance of the components of the transmission systems (power capacity, noise, bandwidth, stability, etc.). We can also include the design of the transmitters and the receivers whose main features include robustness and portability.

---

*A repeater is an electronic assembly that processes a receiving signal for retransmission. Basically, the processing may consist of a pure amplification of the signal either keeping or changing its frequency. In the first case it is important to obtain a high isolation between receive and retransmit antennas, to avoid oscillation.

- *Switching system design*—In most of the cases this consists of adapting the existing switching network for mobile purposes.
- *Teletraffic*—One of the ways of assessing the performance of the system is to estimate the overall blocking probability. Consequently, the concepts of traffic engineering are required. For a given grade of service and number of channels available, how many subscribers can be served? What routing strategy should be able to enhance the traffic performance? How many channels should be used for voice and how many for signalling?
- *Software design*—With the use of microprocessors throughout the system, there are software applications in the mobile, in the base station, and in the exchange adapted for that service.

Other aspects, such as human factors, economics, etc., will also influence the design of a mobile radio system.

## 1.6    FREQUENCY PLANNING AND SPECTRUM ALLOCATION

Mobile radio is just one among tens of radio communication services. Like all the other systems involving radio transmission, the allocation of radio frequencies for this service is subject to a series of administrative and technical factors.

The electromagnetic spectrum is a renewable resource that, in theory, can be used indiscriminately. In practice, however, the chaotic manipulation of radio frequencies would only lead to, at least, an inefficient use of this limited resource, restricting the advent of new services and the growth of those already established. Because radio propagation does not recognize geopolitical boundaries and the political, economic, and social aspirations may vary from country to country, there is a need for international cooperation and participation, so that an orderly worldwide use of the radio spectrum can be agreed. Many other factors, such as historical developments, regional policy, and extent of current usage, also influence the selection of frequencies.

In order to accomplish effective use of the radio spectrum, the frequencies are allocated to specific services with similarity in terms of radiated power and interference features. This has been agreed under a global organization —the International Telecommunication Union (ITU).

### 1.6.1    The Worldwide Harmony

The International Telecommunication Union was founded in 1932 as a result of the fusion of the International Telegraph Union, founded in 1865, and the Radio Telegraph Union, founded in 1903, and today constitutes a specialized agency of the United Nations. Its main objective is to harmonize

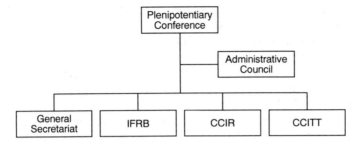

**Figure 1.5.**  General structure of the ITU.

and care for telecommunications around the world, including the efficient utilization of the radio spectrum.

The ITU is backed by an international treaty—the ITU Convention—where its structure, responsibilities, rights, and obligations are defined. Its general structure is shown in Figure 1.5.

The Plenipotentiary Conference reexamines the convention in periodic meetings (usually every five years), determines the general policies, budget, and salaries and elects the Secretary General, the Deputy Secretary General, and the members of the Administrative Council.

The Administrative Council is composed of 36 members responsible for directing the work of the ITU in the period between the plenipotentiary conferences. It usually meets every year.

The next four blocks shown in Figure 1.5 constitute the permanent organs of the ITU:

- The General Secretariat is in charge of executive management and technical cooperation. It translates, interprets, and publishes the documents of the ITU.
- The International Frequency Registration Board (IFRB) is in charge of the analysis and record-keeping of the countries, frequency assignments that may potentially cause interference with the services of other administrations. Should an interference problem be found, a solution is suggested and returned to the originating administration.
- The International Radio Consultative Committee (Comité Consultatif International de Radio [CCIR]) has the task of studying the technical and operating issues related to radio communications.
- The International Telegraph and Telephone Consultative Committee (Comité Consultatif International des Téléphonique et Télégraphique [CCITT]) has the task of studying the technical, operational, and tariff questions concerning telegraphy and telephony.

The studies by CCIR and CCITT result in decisions, resolutions, recommendations, and reports. Study groups are then set up to analyze the recommendations.

Besides these four permanent organs the ITU is also composed of the Administrative Conferences, where the technical regulations are agreed upon. These conferences, called for by the Administrative Council, are held if a sufficient number if ITU member nations agree to participate. The aim is to revise the technical regulations whenever this is required. Accordingly, there are the Administrative Radio Conference and the Administrative Telegraph and Telephone Conference.

The administrative conferences may be worldwide or just regional. The World Administrative Conference attends to regulations worldwide. The Regional Administrative Conferences deal with the telecommunication matters of a regional nature. For this the ITU has divided the world into three regions:

Region 1: Europe, including all former USSR territory outside Europe; Mongolian People's Republic; Asia Minor; and Africa

Region 2: Western hemisphere, including Hawaii

Region 3: Australia; New Zealand; Oceania; and Asia, excluding former USSR territory and Asia Minor

All the decisions taken by these conferences must be in conformity with the ITU Convention and, moreover, the outcomes of the regional conferences have to be in accordance with the world conference.

The world conference responsible for the radio communication questions is the World Administrative Radio Conference (WARC). There have been nine general WARCs: 1906 in Berlin, 1912 in London, 1927 in Washington, D.C., 1932 in Madrid, 1938 in Cairo, 1947 in Atlantic City, 1959 in Geneva, 1979 again in Geneva, and 1992 in Torremolinos. Several other small conferences have been held in order to deal with specific matters (e.g., mobile services).

### 1.6.2  Frequency Assignment and National Control

Several regions of the frequency spectrum are allocated to more than one radio service. There are, however, other regions for the exclusive use of a particular service. Again, for the sake of efficient and orderly usage of the spectrum, in one or the other case, there must be a centralized organization regulating the assignment and use of the frequencies.

In most countries this central administration is performed by a government entity such as a ministry or secretary or department of communications. A well-known exception to this is the United States, where nongovernment radio services are administered by the Federal Communications Commission (FCC) and government services are controlled by the Interdepartment Radio Advisory Committee (IRAC), which in fact acts under a government agency.

**TABLE 1.1. Allocation of Frequencies to Some Radio Services[a]**

| Bandwidth (kHz) | BRC | FIX | MOB | Bandwidth (MHz) | BRC | FIX | MOB |
|---|---|---|---|---|---|---|---|
| 130–160 | | 23 | 23 | 18.168–18.780 | | 123 | |
| 130–148.5 | | 1 | 1 | 87–100 | 3 | 3 | 3 |
| 148.5–255 | 1 | | | 87.5–100 | 1 | | |
| 160–190 | | 23 | | 88–100 | 2 | | |
| 255–283.5 | 1 | | | 100–108 | 123 | | |
| 415–435 | | | 1 | 136–137 | | | 123 |
| 415–495 | | | 23 | 150.05–153 | | 1 | 1 |
| 435–495 | | | 1 | 150.05–156.7625 | | 23 | 23 |
| 505–510 | | | 2 | 153–154 | | 1 | 1 |
| 505–526.5 | | | 13 | 156.7625–156.8365 | | | 123 |
| 510–525 | | | 2 | 156.8375–174 | | 123 | 123 |
| 525–535 | 2 | | | 174–216 | 2 | | |
| 526.5–535 | 3 | | | 174–223 | 13 | 3 | 3 |
| 526.5–1605.5 | 1 | | | 216–225 | | 2 | 2 |
| 535–1605 | 23 | | | 223–230 | 13 | 3 | 3 |
| 1605–1625 | 2 | | | 225–235 | | 2 | 2 |
| 1605.5–1800 | | 3 | 3 | 230–235 | | 13 | 13 |
| 1606.5–1625 | | 1 | 1 | 406–406.1 | | | 123[b] |
| 1625–1705 | 2 | 2 | 2 | 470–512 | 2 | | |
| 1635–1800 | | 1 | 1 | 470–585 | 3 | 3 | 3 |
| 1705–1800 | | 2 | 2 | 470–790 | 1 | | |
| 1800–2000 | | 3 | 3 | 512–608 | 2 | | |
| 1850–2000 | | 12 | 12 | 585–610 | 3 | 3 | 3 |
| | | | | 610–890 | 3 | 3 | 3 |
| | | | | 614–806 | 2 | | |
| | | | | 790–862 | 1 | 1 | |
| | | | | 806–890 | 2 | 2 | 2 |
| | | | | 862–890 | 1 | 1 | 1 |
| | | | | 890–902 | | 2 | 2 |
| | | | | 890–942 | 13 | 13 | 13 |
| | | | | 902–942 | | 2 | |
| | | | | 942–960 | 13 | 13 | 13 |
| | | | | 1530–1660.5 | Dedicated to mobile satellite communications (land-mobile, maritime-mobile, or aeronautical-mobile) | | |
| | | | | 1700–2450 | | 123 | 23 |
| | | | | 2450–2500 | | 123 | 123 |
| | | | | 2500–2535 | | 3 | 3 |
| | | | | 2500–2655 | 123[b] | 123 | 123 |

[a] BRC = broadcasting; FIX = fixed; MOB = mobile; 1 = Region 1; 2 = Region 2; 3 = Region 3.
[b] Using satellite.

The concession for the radio service operation is a rather complicated matter because all sorts of factors, such as historical, economic, political, social, etc., may be involved. Another very important issue is congestion, which may occur in many portions of the spectrum already occupied by other services. In this case there may occur a frequency reassignment procedure when the concerned service is shifted to another region of the spectrum. Again this is an extremely difficult issue where many interests are at play.

Table 1.1 shows some radio services, namely, broadcasting (BRC), fixed (FIX), and mobile (MOB), allocated on a primary basis to the various regions (1, 2, 3, or combinations of them). The mobile services may include land, maritime, and/or aeronautical mobile communications.

## 1.7  SUMMARY AND CONCLUSIONS

Mobile radio services started soon after the invention of radio. Since then they have undergone several stages of regulatory and technological evolution, starting with bulky and extremely heavy equipment and evolving to pocket-sized portable telephones.

As in the majority of communication systems, the evolution of the various services occurred independently, with each system using its own solution. Today, there has been a move toward global standardization and, perhaps, a worldwide integration of the mobile radio services. As far as frequency planning and spectrum allocation are concerned, chaotic manipulation of radio frequencies has been avoided by the creation of the ITU, a global organization where this matter is carefully discussed and controlled.

## REFERENCES

1. Shepherd, N. H., Mobile radio services, in *Communications System Engineering Handbook*, Chapter 17, D. H. Hamsher, Ed. McGraw-Hill, New York, 1967.
2. Calhoun, G., *Digital Cellular Radio*, Artech House, New Jersey, 1988.
3. Young, W. R., Advanced mobile phone service: introduction, background and objectives, *Bell Syst. Tech. J.*, 58(1), 1, 1979.
4. Lomer, G. J., Telephoning on the move—Dick Tracy to Captain Kirk, *IEE Proc.*, 134(1), Part F, 1, 1987.
5. Zysman, G., AT&T proposed digital cellular system, in *Proc. 2nd Int. Seminar on Cellular Radio Telephony* (in Portuguese), São Paulo, S.P., Brazil, 1988.
6. Lee, W. C. Y., *Mobile Communications Engineering*, McGraw-Hill, New York, 1976.
7. Heeralall, S., The applications of directional antennas in cellular mobile radio systems, Ph.D. thesis, University of Essex, England, July 1988.
8. Steile, R. and Prabru, V. K., High-user-density digital cellular mobile radio systems, *IEE Proc.*, 132, 396, 1985.

9. Hughes, C. J., Notes on Mobile Radio Systems Course, *University of Essex*, 1987.

10. Hoff, J., Mobile telephony in the next decade, in *Proc. 2nd Nordic Seminar on Digital Land Mobile Radio Communication*, Stockholm, October 1986.

11. Siling, P. H., Radio-frequency allocation and assignment, in *Communications System Engineering Handbook*, Chapter 19, D. H. Hamsher, Ed., McGraw-Hill, New York, 1967.

12. Bodson, D., Gould, R. G., Hagn, G. H., and Utlaut, W. F., Spectrum management and the 1979 World Administrative Radio Conference, *IEEE Trans. Commun.*, Com-29(8), 1085, 1981.

13. Shrum, R. E., A nontechnical overview of 1979 WARC, *IEEE Trans. Commun.*, Com-29(8), 1089, 1981.

14. Final Acts of the World Administrative Radio Conference for the Mobile Services, (MOB-87), Geneva, 1987.

15. Parsons, J. D. and Gardiner, J. G., *Mobile Communication Systems*, Blackie & Son, Ltd., Glasgow, 1989.

16. British Telecom, A Standard Code for Radio Paging. Report of the Post Office Code Standardization Advisory Group (POCSAG), June 1978 and November 1980.

17. Leimer, B. M., Nielsen, D. L., and Tobagi, F. A., Scanning the issue, *Proc. IEEE*, 75(1), 3, 1987. (Special issue on packet radio networks.)

18. Shacham, N. and Westcott, J., Future directions in packet radio architectures and protocols, *Proc. IEEE*, 75(1), 83, 1987. (Special issue on packet radio networks.)

19. CCIR Study Groups (1986–1990), Iterim Working Party 8/13 (Question 77/8), Adaptation of Mobile Radiocommunication Technology to the Needs of Developing Country.

20. Caldwell, R., Towards a European standard with a common air interface, *Telecommunications, Int. Ed.*, 24(9), 1990.

21. Potter, R., Personal communications for the mass market, *Telecommunications, Int. Ed.*, 24(9), 1990.

22. Huff, D. L., Advanced mobile phone service: the developmental system, *Bell Syst. Tech. J.*, 58(1) 249, 1979.

23. Maral, G. and Bousquet, M., *Satellite Communications Systems*, John Wiley & Sons, New York, 1986.

# CHAPTER 2

# Cellular Mobile Radio

This chapter addresses the basic principles of cellular mobile radio systems, giving an overview of the main points to be considered in a system design. It traces the outstanding differences between conventional and cellular networks, introducing the jargon commonly used in the latter. There is a brief functional description of each cellular component and of how the components can be interconnected to give different architectures. A theory of patterns and symmetry is introduced so that some of the well-accepted cell geometry assumptions, today taken for granted, can be better understood.

The cellular system has, from its inception, a modular architecture that, in theory, can be expanded indefinitely as needed. There are, however, several practical constraints that limit this modular growth, to a certain extent, above which some alternative techniques are recommended. These techniques are examined in this chapter. The issue of system performance is also considered, where three different types of efficiency measures are described. The next topic considered in the chapter is related to the data and signalling control over the radio channels. In this same topic, a successful mobile-to-fixed and fixed-to-mobile call sequence and also the hand-off process are illustrated in pictorial form.

Finally, still within this overview approach, the basic system design specifications followed by the main steps in a cellular design are given. To end the chapter, some nonconventional traffic performance enhancement techniques are shown. Again the approach given here is superficial, but a deeper study is carried out in Chapter 12.

## 2.1 INTRODUCTION

The early conventional mobile systems consisted of a base station with its transmitter and receiver assembled on a hilltop. The coverage area was chosen to be large and, generally, a very small number of channels was available. As initially those systems were manually controlled, each call had to be conducted through a special mobile operator. Automatic systems appeared by the middle of the 1960s but they were still very limited—for example, if a customer had a call already established in one area, that call would drop as the caller moved into another area, and a new call process would have to be initiated.

Notwithstanding the limitations of the systems and the high cost of the services, the growth of demand for mobile telephones was indeed remarkable. Such systems, having a small number of channels, poor signalling protocols, and limited capacity for modular growth, could not cope with such demand. New conceptual ideas for the development of a versatile system, where the subscribers could roam with their telephones in a worldwide integrated network, started to materialize. The cornerstone of this dream is the cellular idea as described in this chapter.

The expansion of a conventional mobile system relies heavily upon spectrum availability. Radio frequency planning is a complicated matter because the frequency spectrum must be shared by several other different types of services. There is plenty of spectrum available at higher frequency bands, but technology has not as yet been able to overcome the difficulties of operating in those frequency bands.

However, if on the one hand there is scarcity of frequency spectrum, strictly controlled by the regulatory agencies, on the other hand there is the pressure of the demand for mobile services. The way out is the conception of new ideas. The implementation of these new ideas may require a new technology, giving rise to a new system capable not only of meeting the initial demand, but also of offering new services. These new services can generate new demand and need to be regulated. A new cycle starts as illustrated in Figure 2.1.

In the analysis that follows we will be considering an analog system. Digital systems will be continuously mentioned and analyzed throughout the following chapters at convenient instants.

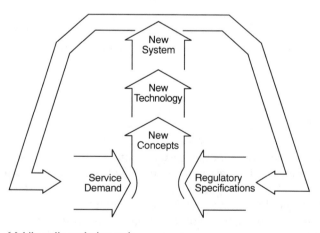

**Figure 2.1.**  Mobile radio evolution cycle.

## 2.2 THE CELLULAR JARGON

The "new" conceptual ideas implemented today in mobile systems were proposed by the Bell Telephone Laboratories during the middle of this century. The implementation of such ideas required, among others, complicated means for administering the system, which was feasible with the advent of electronic switching technology.[21]

The basic concept is *frequency reuse*: If a channel of a certain frequency covers an area of radius $R$, then the same frequency can be reused to cover another area. Each one of the areas constitutes a *cell*. Cells using the same carrier frequency are called *cocells*. They are positioned sufficiently apart from each other that *cochannel interference* can be within tolerable limits. With this new concept, a region initially served by one base station in a conventional mobile system is split into several cells, each of which has its own base station, served by its own set of channels.

"The main purpose of defining cells in a mobile radio system is to delineate areas in which either specific channels or a specific base station will be used at least preferentially, if not exclusively".[2] If omnidirectional transmitting antennas are used, ideally, the coverage area would be circular. Therefore, if for modelling purposes a cell shape is required, a circle would be the straightforward choice. However, a plane filled up with circles can exhibit overlapped areas or gaps, adding difficulties to frequency planning. Regular polygons such as equilateral triangles, squares, and hexagons do not present these constraints. The *regular hexagon* is the most convenient format, because it most closely resembles the circle.

To understand this in a quantitative way, consider three different cellular systems, the first with a triangular, the second with a square, and the third with a hexagonal array of cells. Suppose that the base station is situated at the center of the cells, i.e., equidistantly from the vertices. Let $R$ be this distance. Therefore, the coverage areas, corresponding to the area of the triangle, square, and hexagon are $3\sqrt{3}\,R^2/4$, $2R^2$, and $3\sqrt{3}\,R^2/2$, respectively. Note, for a given $R$, that the hexagon (then square, followed by the triangle) can cover a larger area. Accordingly, a hexagonal array requires fewer cells (and consequently fewer base stations) to cover the same region.[2]

The set of channels available in the system is assigned to a group of cells constituting the *cluster*. The same set can be reused only in different clusters. The number of cells per cluster determines the *repeat pattern*. Due to geometrical constraints (see Section 2.6) only certain repeat patterns can tessellate, the most common having 4, 7, and 12 cells per cluster. The smaller the repeat pattern, the higher the number of channels per cell, corresponding to a higher traffic capacity system. On the other hand, the smaller the repeat pattern, the smaller the distance between cocells, leading to a higher cochannel interference.

Channel allocation strategies play an important role as far as *adjacent-channel interference* is concerned. Although the radio equipment is designed to select only the wanted channels, cutting off adjacent frequencies, there may be situations when adjacent channels can cause interference. For instance, this may happen when two mobiles using adjacent channels transmit to the base station from a short and from a long distance, respectively. The interference problem is worsened in the presence of fading.

The sizes of the cells vary according to the network planning, made to meet a certain traffic demand. The cells are considered to be very small when compared with the conventional systems. Recent studies[1] suggest the use of microcells (90-m radius) with the frequencies allocated at a 60-GHz band.

Although the mobile can travel from cell to cell, it is required that the call already established (or in progress) must not be interrupted. Therefore, changes of channels may occur without the need of the intervention of the subscriber. This action of changing channels is known as *hand-over* or *hand-off*. Hand-off occurs whenever the signal strength measured at the base station falls below a threshold value. The process of monitoring the quality of the signal and recommending the change of channels, if necessary, is called *locating*. The process of determining a mobile's availability to receive a given incoming call is named *paging*. The function of beginning a call, performed by the mobile unit, is termed *access*.

## 2.3  ESSENTIAL CHARACTERISTICS OF THE CELLULAR SYSTEMS

The main difference between conventional and cellular systems is that the first is a noise-limited system (NLS) and the second is an interference-limited system (ILS). Let us describe some of main features of each:

*Conventional (noise-limited) systems:*
1. Serve low subscriber densities
2. Rely heavily upon spectrum availability
3. Do not take advantage of frequency reuse
4. Require high-power transmitters mounted on a hilltop
5. Have a large-coverage service area
6. Do not allow modular expansion
7. Do not allow hand-off

*Cellular (interference-limited) systems:*
1. Serve high subscriber densities
2. Consider spectrum availability as a limiting factor
3. Take advantage of frequency reuse
4. Require low-power transmitters located at low elevation
5. Divide the whole service area into small-coverage cells
6. Allow (theoretically) an endless modular expansion
7. Allow hand-off

In terms of figures, the typical characteristics of the cellular systems are as follows:

1. Around 650 channels are available.
2. Transmitters provide output power ranging from 10 to 45 W.
3. Transmitters are mounted at elevations typically 30–100 m above the service area.
4. The coverage area radius varies from 2 to 16 km.
5. Generally each cell is served by 25 to 75 channels.

## 2.4 BASIC COMPONENTS OF A CELLULAR MOBILE SYSTEM

A general overview of a cellular mobile network is illustrated in Figure 2.2. There are four main elements:

1. Mobile station
2. Base station
3. Mobile switching center
4. Public switched telephone network

We shall briefly describe each component.

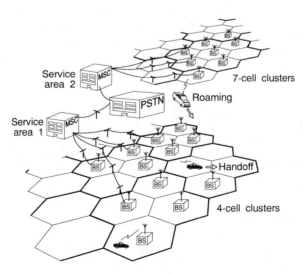

**Figure 2.2.** Overview of a cellular mobile network.

### 2.4.1  Mobile Station

The mobile station (MS) constitutes the interface between the mobile subscriber and the base station. Besides voice communication the MS also provides control and signalling functions, usually performed by a microprocessor. The mobile unit is able to tune, under the system command, to any channel in the frequency spectrum allocated to the system. Each channel comprises a pair of frequencies for a two-way conversation. The control messages are transmitted in a digital form and can be sent through voice or signalling channels, depending on the task to be performed, as seen in Section 2.10. The power levels of the transmitter can also be controlled by the system.

Every call from a mobile unit yields the subscriber's identification and the dialled digits. In order to provide a more efficient use of the channels, the digits are first held in a memory in the telephone and, when the called party number is complete, they are sent, initiating the communications with the base station.

The mobile station consists of two elements: the telephone set and the radio set.

The telephone set provides the contact between the subscriber and the system. It is mounted near the car's front seat, where the driver may have comfortable and safe access to it. It includes a handset, keypad, display, and some kind of alert signal (lamp or tone). It is also possible to have some sort of loudspeaker so that the driver does not need to use the handset while driving.

The radio set comprises two distinct pieces of equipment: the radio itself and the control. The radio deals with all the frequencies available in the system in a full duplex mode. The transmitter provides up to (approximately) 10 W output power and has its level controlled by the system. It transmits both voice and data, which, in the analog systems, experience different types of modulation. The receiver demodulates voice and data. The control is a logic unit with a microprocessor with the role of managing the control tasks within the mobile. Some examples of control messages include the following:

1. Origination request from the mobile in order to access a channel
2. Registration of the mobile in the current service area
3. Channel assignment message from the base station to the mobile
4. Hand-off message from the base station, for retuning to another channel

### 2.4.2  Base Station

The base stations are responsible for serving the calls to or from the mobile units located in their respective cells. They are connected to the mobile switching center via land links. Base stations consist of two elements:

the radio and the control. The radio comprises the transmitters, receivers, towers, and antennas. The control is a microprocessor unit responsible for the control, monitoring, and supervision of the calls. The assignment and reassignment of channels to the mobile units can be carried out by the base station. In addition, the base station monitors the signal levels to recommend the hand-off.

### 2.4.3 Mobile Switching Center

The mobile switching center (MSC) is a telephone exchange specially assembled for cellular radio services. It works as a central controller, interfacing the mobile units and the fixed telephone network. The MSC uses standard telephone signalling and is equivalent to a class 5 local exchange.

The number of cells connected to (controlled by) an MSC varies according to the needs. One MSC can be responsible for a large metropolitan area or for a small number of small neighboring cities. The area served by one MSC is known as a *service area*. The mobile subscriber within a service area is a *home* subscriber. It is possible, however, for the subscriber to move out from this area into another area, in which case he (she) is called a *roamer*. The main tasks performed by an MSC include paging, locating, and hand-off. Moreover, it performs functions of an ordinary digital switching exchange such as signalling, switching, A/D conversion of the audio circuits (in the analog systems), detection of on-hook–off-hook status, line scanning for dial pulse signalling, etc.

Communications between the base stations and the MSC are usually carried out through landlines providing speech and data paths. The MSC is also connected to the fixed network by means of land links.

### 2.4.4 Public Switched Telephone Network

The public switched telephone network (PSTN) treats the MSCs as ordinary fixed telephone exchanges.

It has not been shown in Figure 2.2, but the cellular systems have a control center where some operations and information are centralized. This control center can be located at one of the mobile switching centers. It contains the main data base for the whole system and performs the following: call record administration, traffic analysis, network management, maintenance, equipment configurations, etc.

## 2.5 SYSTEM ARCHITECTURE

A cellular mobile radio system can be built according to a centralized or a decentralized architecture. In a centralized architecture the mobile switching center is usually very large and controls many base stations. In this case the MSC includes clusters situated nearby and in remote areas as well. In a decentralized system the MSCs are smaller, controlling a smaller number of clusters.

Small systems are generally centralized and large systems tend to the decentralized approach. In fact, there are different degrees of decentralization in that there may or may not be interconnection between the MSCs. In

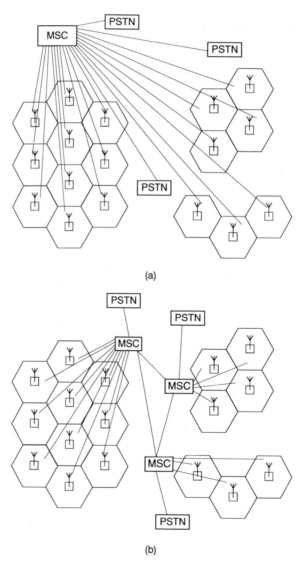

(a)

(b)

**Figure 2.3.**  Mobile system architectures: (a) centralized architecture; (b) decentralized architecture.

the first case a call from a mobile will go through the PSTN only when the called party is a fixed subscriber. In the second case, even if the called party is a mobile, but from a different service area, this call will have to pass through the PSTN (calls within the same service area are handled by the corresponding MSC without involving a PSTN). The different architectures are shown in Figure 2.3.

When a mobile moves out of its service area (roams into another area), the first task to be done is to inform the corresponding MSC that this mobile is within that new zone. The visited MSC informs the home MSC of the presence of its user. When this particular mobile is called, his (her) original MSC will direct that call to the MSC within which he (she) is registered.

## 2.6  THE THEORY OF CELLULAR PATTERNS

The applications of the theory of symmetry and patterns have usually had crystallographic studies as their main target; only recently (by the 1970s[6]) has this theory been used in engineering, mainly in cellular mobile radio (a good reference can be found in Reference 7). In this section we should first recall the main concepts for our particular purposes.

1. Any plane repetitive pattern can always be decomposed into fundamental regions, also known as areas or cells, with a general shape of a parallelogram. In fact, the sides of the cells can be of any shape, but opposite sides must have the same shape.
2. A whole pattern can be obtained by translating the fundamental region in a regular way. In a plane this means placing each cell in a position $X^m Y^n$, where $X^m$ represents $m$ steps of $X$ along the $OX$ direction and $Y^n$ represents $n$ steps of $Y$ along the $OY$ direction, where $m$ and $n$ are integers and $O$ is the origin. This is done until the whole area is filled up.
3. A *tessellation* is defined as a pattern having a polygon as a fundamental region. Regular polygons generate regular tessellation. Equilateral triangles, squares, and hexagons are the only regular polygons that can tessellate.[3]
4. *Color symmetry* is a feature presented by figures or patterns having symmetrical parts with a choice of colors. In cellular radio terms this corresponds to a set of channels (i.e., each set of channels is a color). It will be seen that such a system exhibits a polychromatic pattern.

Although the theory of symmetry and patterns is well established, the creation of a pattern is still an art and, in this respect, it is open to the imagination. Some patterns, however, due to their symmetry or regularity may be easily tessellated in a logical and sequential manner. The creation of a pattern can be carried out in different ways, a few of which (extracted from

Reference 4) are listed here:

- By breaking down a fundamental area into subareas and placing an identical design into each subarea
- By making a design across several subareas
- By distorting the boundaries of subareas to form similar interlocking shapes
- By using a basic pattern as a grid and by coloring only certain areas between the lines of the grid
- By replacing the straight lines of a basic pattern with curved ones
- By using mirrors, as in the kaleidoscope
- By using a computer to replicate a motif with any combination of symmetry operations

### 2.6.1  The Regular Patterns

In this section we consider the patterns formed by regular polygons, namely, regular triangles, squares, and hexagons. Special attention will be given to the hexagons.

Consider initially the equilateral triangle. For this figure it is convenient to choose the set of coordinates as shown in Figure 2.4. The positive portions of the two axes form a 60° angle, and the unit distance along the axes is $R$ (where $R$ is the cell radius). The distance between two points $(u_1, v_1)$ and $(u_2, v_2)$ is given by

$$D^2 = i^2 + ij + j^2 \qquad (2.1)$$

where $i = u_2 - u_1$ and $j = v_2 - v_1$.

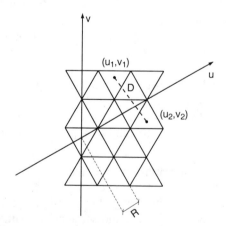

**Figure 2.4.**   Triangular cellular geometry.

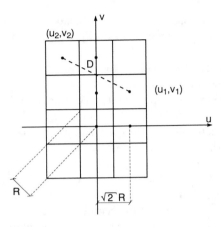

**Figure 2.5.**   Square cellular geometry.

Now consider the square. The set of coordinates lies obviously at the orthogonal axes (Figure 2.5) and the distance $D$ is given by

$$D^2 = i^2 + j^2 \tag{2.2}$$

and the unit distance is $\sqrt{2}\,R$.

For the hexagonal array the $(u, v)$ axes are chosen to have their positive portions crossing at $60°$ angle (Figure 2.6), and the unit distance is $\sqrt{3}\,R$. Then

$$D^2 = i^2 + ij + j^2 \tag{2.3}$$

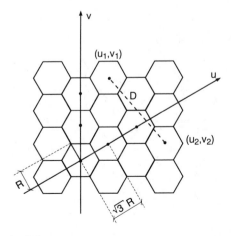

**Figure 2.6.**   Hexagonal cellular geometry.

We shall concentrate our studies on the hexagonal array of cells. Nevertheless, the following discussion can be easily adapted to triangular or square
arrays.

In order to determine the cell pattern, i.e., the cluster, the fundamental
procedure is to place the cocells at equidistant points from a reference cocell.
The next step is to label another cell as a reference and proceed in the same
way as before. It is possible, and in fact more plausible, to simplify this
procedure by just determining the first set of cocells. Then, by translation
without rotation, the first pattern (first set of cocells) can be replicated
around the initial reference.

Now we need to find the loci of points where the equidistant cocells are
located. If we impose that the reuse distance must be isotropic, there are six
hexagons equidistant from the reference hexagons. Using Equation 2.3 it can
be seen that if we position the hexagons at the set of coordinates $(p, q)$,
$(p + q, -p)$, $(-q, p + q)$, $(-p, -q)$, $(-p - q, p)$, and $(q, -p - q)$, their
distances to the reference hexagon (centered at $(0, 0)$) are equally given by
$p^2 + pq + q^2$. Therefore, these are the required cocell positions. An equivalent procedure is to use a system of axes as that shown in Figure 2.6 and then
position the coordinates $(p, q)$ using all six possible pairs of consecutive axes,
i.e., axes $ij, jk, kl, lm, mn, no$. (Note that $i = -l$, $j = -m$, and $k = -n$.) As
an illustration consider $(p, q)$ equal to $(1, 2)$. The corresponding cocells
represented as cell 1 are as shown in Figure 2.7. Now by translating without
rotating the set of cocells, we obtain another set of cocells (named cell
2). This can be done until the whole pattern is filled up. This is shown in
Figure 2.8.

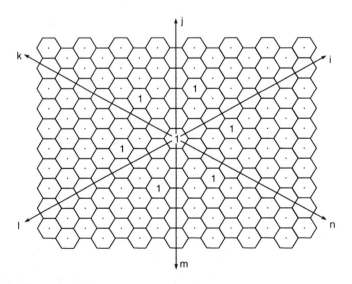

**Figure 2.7.**  New system of axes.

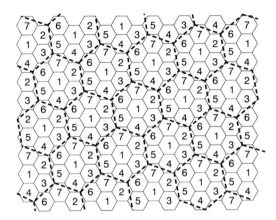

**Figure 2.8.** A seven-cell cluster system.

The theory of cellular pattern is based on the concept that a pattern is set up by replicating the fundamental area known as a cluster. It is assumed that a cluster is a contiguous group of cells. Consequently, all of the clusters must be identical. The shape of the clusters is also of a regular hexagonal form and this is intuitively deduced. The explanation given here is taken from MacDonald,[2] but a more rigorous approach can be found in Heeralall.[7]

The intuitive idea is based on three facts:

1. Each cell has six equidistant nearest cochannel cells.
2. Each adjacent pair of the six axes connecting the center of the reference cell to the centers of the six surrounding cochannel cells has an angle separation of 60°.
3. Observations 1 and 2 are valid for any arbitrary cell.

In short, the main constraints forcing the clusters to have a hexagonal shape are as follows:

1. Reuse distances must be isotropic.
2. A cluster must be a continuous group of cells.

This can be seen in Figure 2.8, where the dotted hexagons represent the clusters.

### 2.6.2  Number of Cells per Cluster

Assuming that a cluster has a hexagonal shape, we want to determine the number $N$ of hexagonal cells per cluster. Let $a$ and $A$ be the areas of the cell

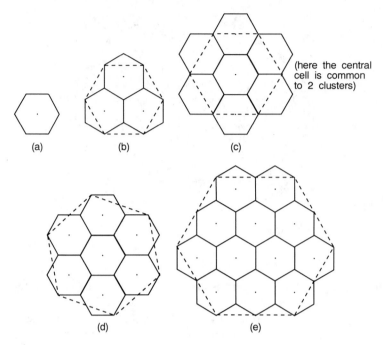

**Figure 2.9.** Common configurations of cellular patterns: (a) 1-cell cluster; (b) 3-cell cluster; (c) 4-cell cluster; (d) 7-cell cluster; (e) 12-cell cluster.

and of the cluster, respectively. The area $a$ is then*

$$a = 3\sqrt{3}\, R^2 / 2 = \sqrt{3} / 2 \qquad (2.4)$$

The distance between the centers of two cocells is $D$. Let us choose these cells to be the centers of the corresponding hexagonal clusters. Hence the area $A$ is

$$A = 3\sqrt{3} \left( \frac{D/2}{\cos 30°} \right)^2 \frac{1}{2} = \sqrt{3}\, \frac{D^2}{2} \qquad (2.5)$$

The number of cells per cluster is obviously

$$N = A/a = D^2 \qquad (2.6)$$

or equivalently, using Equation 2.3

$$N = i^2 + ij + j^2 \qquad (2.7)$$

---

*Note that $\sqrt{3}\,R$ has been assumed to be the distance unit.

Because $i$ and $j$ are integers, the clusters will accommodate only certain numbers of cells, such as $1, 3, 4, 7, 9, 12, 13, 16, 19, 21, \ldots$ .

Patterns 7 and 12 are the most common configurations of the cellular systems. Figure 2.9 shows these patterns and the hypothetical hexagonal shape of the corresponding clusters.

An important parameter of a cellular layout is the $D/R$ ratio, known as the cochannel reuse ratio. This ratio gives an indication of both the transmission quality and also traffic capacity. On the transmission side, the $D/R$ ratio gives an indication of the cochannel interference statistics: the higher the ratio, the lower the interference potential. As far as traffic capacity is concerned the $D/R$ ratio can also give a measure of performance evaluation. This is promptly seen if we express the ratio as a function of the number of cells per clusters. Using the preceding equations,

$$D/R = \sqrt{N} \big/ (1/\sqrt{3}) = \sqrt{3N} \qquad (2.8)$$

The fewer cells per cluster (or, equivalently, the smaller the $D/R$ ratio), the larger the number of channels per cell, i.e., a higher density of channels corresponding to a higher traffic carrying capacity.

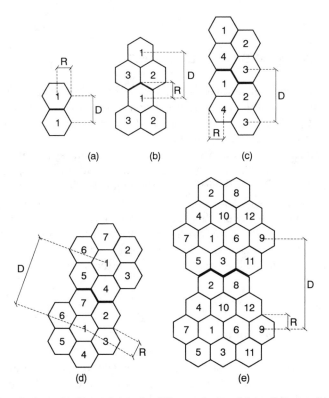

**Figure 2.10.** Cochannel cell separations for different clusters: (a) 1-cell cluster; (b) 3-cell cluster; (c) 4-cell cluster; (d) 7-cell cluster; (e) 12-cell cluster.

**TABLE 2.1. Traffic Capacity and Cochannel Interference**

| Cluster Size | D / R | Channels per Cell | Traffic Capacity | Transmission Quality |
|---|---|---|---|---|
| 1 | 1.73 | 360 | Highest | Lowest |
| 3 | 3.00 | 120 | ↑ | │ |
| 4 | 3.46 | 90 | │ | │ |
| 7 | 4.58 | 51 | │ | ↓ |
| 12 | 6.00 | 30 | Lowest | Highest |

It can be seen that transmission quality and traffic capacity work in opposite directions (see Figure 2.10). Moreover, the smaller the cluster size, the lower the cost of the system. The determination of the $D/R$ ratio is then a trade-off between these factors and, obviously, an intermediate point can be found so that a reasonable grade of service can be accomplished.

Assume, for example, that the total number of channels in the system is 360. Then Table 2.1 depicts a comparison between the traffic capacity and the transmission quality.

## 2.7  SYSTEM EXPANSION TECHNIQUES

One of the many advantages of the cellular architecture is its modular growth capability. A start-up system is usually constituted of hexagonal cells of the largest possible radii. As the demand for service grows, the system will tend to absorb the new users up to a limit where it can still offer a good grade of service. If the quality of the service is initially high, it may be possible to accept an increase of the traffic load and allow a system performance degradation, but still within acceptable levels. This, in fact, constitutes a very convenient way of allowing the system to adapt to a sudden growth of the traffic demand. However, this adaptation is only efficient on a short-term basis, because any additional growth can cause a disastrous degradation. The network will be able to accommodate more subscribers if there is a change in the system itself. We shall examine some of the techniques that may be used.

### 2.7.1  Adding New Channels

This can be done only if there are channels available. Initially, when a system is set up not all the channels need be used, and growth and expansion can be planned in an orderly manner. However, once all the channels have already been used, new expansion methods must be found.

### 2.7.2 Frequency Borrowing

The channels are allocated to the cells according to the geographic distribution of the traffic. If the traffic demand shows a greater concentration in certain regions, i.e., if some cells become more overloaded than others, it may be possible to reallocate channels by transferring frequencies from the less loaded to the more loaded cells. This works on an interim basis because further growth will require that those borrowed frequencies return to their original cell. The advantage of this technique is that it does not require a big change in the hardware of the system.

### 2.7.3 Change of Cell Pattern

It has been previously seen that a smaller number of cells per cluster represents a higher traffic capacity. However, it has also been seen that this corresponds to a degradation of the transmission quality (higher cochannel interference). This technique, nevertheless, is rather costly and seldom applied.

### 2.7.4 Cell Splitting

Cell splitting is a technique aimed at diminishing the size of larger cells by dividing them into a number of smaller cells. By reducing the size of the cells, more cells per area will be available with a consequent increase of channels in such area and an increase in traffic capacity.

Radius reduction by a factor of $K$ reduces the coverage area and increases the number of base stations both by a factor of $K^2$. It is advisable to choose

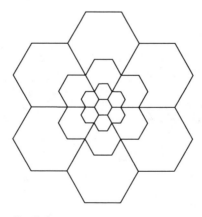

**Figure 2.11.**  Two-stage cell splitting.

$K$ so that mixed cell sizes can coexist, also taking advantage of using the old base stations. Figure 2.11 depicts two stages of cell splitting.

Cell splitting usually takes place at the midpoint of the congested area. Further splitting may occur until the minimum size of the cell radius is reached. This minimum size is determined taking into account (1) the minimum cochannel reuse distance that can be achieved while maintaining voice quality objectives, (2) the problem of siting the base stations, and, obviously, (3) the economic aspects.

### 2.7.5  Overlaid and Underlaid Cells

When cell splitting occurs, there may be cells of different sizes in the system. The start-up system is conceived so that a $D/R$ ratio is maintained throughout the network. With different cell sizes, special attention must be given to cochannel interference problems. Consider the example of Figure 2.12, where the pattern is a seven-cell cluster. The distance between bigger cocells is $4.6R$, where $R$ is the bigger cell radius. The distance between smaller cocells is $4.6r$, where $r = R/2$ is the smaller cell radius. Cochannel interference between the bigger cells must be within tolerable limits as designed at the conception of the system. The same applies to the smaller cells because the cochannel reuse ratio is maintained. The calls within smaller cocells will not cause undue interference on the channels of the bigger cells, because if the cochannel ratio requirement is satisfied for the smaller cells, it is also satisfied for the bigger ones. The opposite situation, however, is not true. Calls being served by the smaller cells will suffer

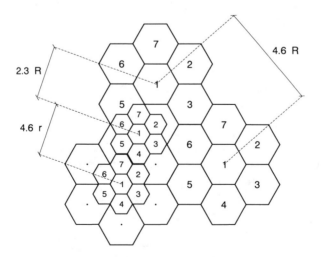

**Figure 2.12.**  Mixed cell sizes in a seven-cell cluster system after splitting.

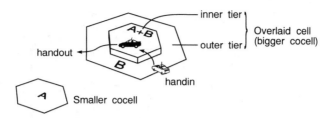

**Figure 2.13.**  Overlaid–underlaid cell concept.

interference from the cochannels of the bigger cells because the distance between them will be $2.3R$ (half of the initial value $4.6R$).

A possible solution for this is to visualize the bigger cell as if it were composed of two tiers (layers) as shown in Figure 2.13. The coverage area of the cell is still the same. The difference now is that the inner layer and the outer layer operate with distinct sets of channels. The main restriction is that the outer tier cannot be served by any of the frequencies used in the smaller cocells. Suppose that the bigger cell has a set of channels $C$ divided into subsets $A$ and $B$ ($C = A + B$). Suppose also that the smaller cocell will use the subset $A$. Consequently the outer layer can only be served by the subset $B$.

The inner cells' coverage area is controlled by transmission power reduction and hand-off control parameters. If a mobile moves out from its layer, a hand-off procedure must take place. Hand-off occurring when the mobile moves from the inner to the outer layer is called handout. The converse is named hand-in. Note that the cell is still served by the same number of frequencies and that the outer voice channels are also available for the mobiles within the inner radius.

## 2.7.6  Sectorization

In theory, the cell splitting process may be carried out indefinitely. In practice, however, there are some very obvious constraints:

1. As the distance between cells reduces, the cochannel interference increases, although the same repeat pattern is kept.
2. Finding a suitable location for the base station may become a difficult task, because the siting tolerance is contracted as well.
3. The total cost of the system is increased because the number of required base stations is increased.

An alternative to cell splitting is *sectorization*. This technique consists of dividing the cell into a number of sectors, each of which is served by a different set of channels and illuminated by a directional antenna. The

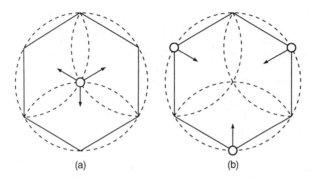

(a)                              (b)

**Figure 2.14.**  Locating the directional antennas (the dotted lines represent the approximate contours of the main lobe of a real directional antenna): (a) center-excited cell; (b) corner-excited cell.

sector, therefore, can be considered as a new cell. The most common arrangements are three and six sectors per cell.

The base stations can be located either at the center or at the corner of the cell. The first case is referred to as center-excited cells and the second as corner-excited cells. In fact, there is no difference between the system conceived in one or the other way. These two approaches are illustrated in Figure 2.14, where the dotted lines sketch the main lobe of a hypothetical directional antenna.

It will be shown later that the use of directional antennas substantially cuts down the cochannel interference, thus allowing the cocells to be more closely spaced. Closer cell spacing implies smaller $D/R$ ratio, corresponding to smaller number of cells per cluster (see Equation 2.8). In other words, more channels per cell can be used, corresponding to a higher capacity, because the total number of available channels in the system is constant. Sectorization is then very attractive because it allows the system to grow with a reduced investment when compared with the cell-splitting technique.

The question that may arise is "Why not start up the system with directional antennas already?" At the inception of a cellular mobile system, the aim is to have as few base stations as possible to cover the required geographical region, because base stations constitute a high investment. Because the use of directional antennas reduces the covered area, not only more antennas but also more transmitters are needed. When the system expands, the aim is to maintain the same cell sites, while still reducing the cell area to increase capacity. This is achieved by means of sectorization with the use of directional antennas.

## 2.8  PERFORMANCE MEASURES AND EFFICIENCY

The systems are usually designed to meet certain specifications. These specifications include some performance parameters directly influencing the network efficiency. The ultimate requirement of any communication network is the *capacity*. Particularly, in a mobile radio system this is related to traffic capacity. Hence two main parameters are of interest:

1. Carrier-to-cochannel interference ratio ($C/I_c$)
2. Blocking probability ($E$) or grade of service (GOS)

The $C/I_c$ ratio, as previously mentioned, varies according to the cellular pattern. It can be easily calculated if the mobiles are assumed to be fixed and positioned for the worst-case performance. In fact, $C/I_c$ is a random variable affected by random phenomena such as location of mobiles, fading (Rayleigh and Gaussian), antenna characteristics, and base station locations. It is indeed a complex parameter to be estimated if all the variables are to be taken into account. The connection between the performance parameter $C/I_c$ and the traffic capacity is that variations in $C/I_c$ can be directly translated into reuse distance ($D/R$ ratio), giving a measure of the compactness of the cellular layout. The larger the $C/I_c$ ratio (less interference), the larger the $D/R$ ratio, leading to a larger cell pattern. On the other hand, the larger the cell pattern the smaller number of channels per cell, leading to a smaller traffic capacity. Design specifications require that a minimum $C/I_c$ (usually 18 dB, 17 dB, or sometimes even less) must be achieved over a large percentage of the coverage area (usually 90%).

Blocking probability or grade of service is a performance parameter easily associated with traffic capacity. Several formulas for blocking probability calculations are available, and they are used according to specific applications. Among them, the most widely used is the Erlang-B, given by

$$\text{GOS} = E(A, N) = \frac{A^N/N!}{\sum_{i=0}^{N} A^i/i!} \tag{2.9}$$

where $A$ is the traffic offered in erlangs (erl) and $N$ is the number of channels. Note, however, that this is a general-purpose formula used only for a quick traffic performance assessment. It does not take into account the various phenomena of the mobile radio system, mainly characterized by the mobility of the subscriber (hand-off, etc.). For the purposes of this section the use of this formula will be sufficient.*

The start-up system usually begins with a grade of service of P.02 (2% of blocking probability), rising up to P.05 as this system grows. If more subscribers are allowed in the system the blocking probability may reach unacceptable values. Accordingly, some system expansion techniques must be used.

It can be seen that both the $C/I_c$ and the GOS constitute isolated parameters not clearly expressing the overall system performance. In particu-

---

*Refer to Chapter 12 for studies of traffic in mobile radio systems.

lar, the GOS can be associated with one specific cell whereas the $C/I_c$ ratio can be determined for a specific region or situation. For instance, $C/I_c$ is high in a large cluster where the blocking probability (GOS) can be very small. At first sight, high $C/I_c$ and low GOS is the ideal situation, but note that in this case a cluster may not be using the channels efficiently. Therefore, an overall measure of efficiency, taking into account all the relevant parameters in a mobile radio system, is required.

There are three main efficiency measures: (1) spectrum efficiency, (2) trunking efficiency, and (3) economic efficiency.

The *spectrum efficiency* $\eta_s$ is a measure of how efficiently space, frequency, and time are used. These three variables are combined into a product form as

$$\eta_s = \frac{\text{number of reuses}}{\text{coverage area}} \times \frac{\text{number of channels}}{\text{bandwidth available}} \times \frac{\text{time the channel is busy}}{\text{total time of the channel}}$$

Hence $\eta_s$ is expressed in units of erlangs per square meter per hertz (erl m$^{-2}$ Hz$^{-1}$). Several techniques can be implemented to improve the spectrum efficiency. We shall examine some of them.

1. Given that the available bandwidth is fixed, the only way to increase the number of channels is to reduce the channel bandwidth. This can be accomplished by means of improving the modulation techniques, filter response, etc.
2. The cell area can be reduced by the cell-splitting technique, or sectorization.
3. The channel usage can be optimized if the messages are compressed or interleaved, taking advantage of the fact that the carried information (speech or data) is not continuously present. Additionally, channel usage can be increased with the application of some dynamic channel assignment or alternative routing strategies.

The *trunking efficiency* measures the number of subscribers that each channel in each cell can accommodate. This, of course, intrinsically depends on the grade of service that the system is prepared to offer. Given a grade of service GOS and number of channels, the total traffic can be calculated using the Erlang-B formula. If, on average, each subscriber generates a certain amount of traffic, then the total number of subscribers is estimated. The graphs of Figure 2.15 give curves for three different grades of service (2, 5, and 10%), for which a traffic of 0.02 erlang per subscriber is assumed.

It can be seen that the trunking efficiency decreases rapidly when the number of channels per cell falls below 20.

The *economic efficiency* depends on many factors and a simple assessment is not available. Its aim is to measure how affordable the mobile service is to people and to the cellular operator. In other words, this is directly related to the costs of service per customer.

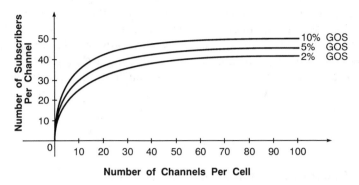

**Figure 2.15.**  Trunking efficiency.

## 2.9   TRAFFIC ENGINEERING

The starting point for engineering the traffic is knowledge of the required grade of service. This is usually specified to be around 2% during the busy hour. The definition of the busy hour may vary according to the license administrator's point of view. There are usually three options:

- Busy hour at the busiest cell
- System busy hour
- System average over all hours

The estimate of the subscriber usage rates is usually made on a demographic basis because this can vary according to the region. The traffic distribution is then worked out and the cell areas identified. The calculations are generally simplified if the traffic is assumed to be evenly distributed. However, in practice, even traffic distribution is quite rare. In urban areas there is a high concentration of traffic at the town center during the rush hour, decreasing smoothly toward its outskirts. This distribution is known as "bell shaped". In the start-up system, however, the traffic distribution is not well known. The calculations are carried out based on the best available traffic estimates. The final system capacity is obtained by grossly exaggerating the calculated figures.

## 2.10   DATA AND CONTROL SIGNALLING

Similar to any other communication network, the mobile radio channels need to handle data transmission, at least for signalling purposes. There are two ways of transmitting data: (1) using the speech channels or (2) using dedicated channels. In the first method, when the channel is not busy with voice transmission, then an idle tone is inserted in order to indicate its availability for data transmission. In the second method, a percentage of the total amount of radio channels is exclusively used for signalling transmission. The first method was initially used by the early mobile radio systems. However, for cellular purposes, the idle tone speech channel method is not appropriate. This is due to the amount of signalling required to handle the enormous quantity of voice channels, cells, and subscribers present in the cellular systems.

### 2.10.1  Control over the Signalling Channels

The proportion of signalling channels may vary with the system capacity. A start-up system requires a small number of these dedicated channels, with this number increasing as the system expands. The allocation of the channels to their functions also varies with the network. In the AMPS system[22] these signalling channels are divided into three groups: (1) *dedicated control channels* (CC), (2) *paging channels* (PC), and (3) *access channels* (AC).

The dedicated control channels are common to all the mobiles throughout the system. Every mobile is programmed to tune to one of the CCs, in fact, to the strongest one. The CCs carry overhead messages giving specific information about the system. The messages include the service area identification, the number of the paging channels available, etc. The paging channels are mainly used for seeking a called mobile. The PCs transmit the identification number of the called subscriber, over the whole service area (the mobile is "paged"). The PCs also carry the number of access channels in that region. These channels are used when a mobile wants to initiate a call. Each AC has a busy/idle bit indicating its availability for use. Now we shall briefly describe a successful simplified call sequence.[22] We first present the initialization process.

Once the mobile unit is switched on, it scans over all the dedicated control channels and it tunes to the strongest one. The mobile is then informed about the available paging channels in that area. It scans these channels and it tunes to the strongest one.

1. Call from mobile to fixed subscriber:
   a. The mobile receives on the PC the number of the AC available in its area.
   b. It scans the ACs and tunes to the strongest one. The selected AC should, most likely, belong to the closest base station.
   c. The mobile then sends the dialled digits and its own identification. The base station receives and routes these digits to the mobile switching center (MSC).
   d. The MSC determines and transmits the voice channel number to be allocated to the mobile, and the mobile tunes to it.

    e. The MSC sends the dialled digits to the public switched telephone network (PSTN).

    f. The PSTN completes the voice path.

2. Call from fixed to mobile subscriber:

    a. The PSTN routes the call to the MSC within which the called mobile is registered.

    b. The MSC impels (causes) all its base stations to transmit, on their PCs, the identification number of the called mobile.

    c. The mobile identifies its number, tunes to the strongest AC, and acknowledges the paging. The base station relays this acknowledgment to the MSC.

    d. The MSC selects an idle voice channel of that base station and the mobile tunes to it.

    e. An alerting tone is sent through the voice channel to the mobile and a ring tone is sent to the calling subscriber.

    f. The voice path will be completed as soon as the called user answers (the alerting and ringing tone are, obviously, removed).

These two sequences are depicted in the Figures 2.16 and 2.17, respectively. In these figures the arrowed lines indicate the direction of the action. The dotted lines indicate signalling and the solid lines represent voice.

### 2.10.2   Control over Speech Channels

The speech channels must also carry some sort of signalling information, so that the established conversation be under control. Suppose, for instance, a mobile moving from one cell into another. How does the hand-off take place? How is the signal strength monitoring carried out? How is the on-hook–off-hook status detected? These control tasks are provided by means of supervision tones, inserted somewhere above the voice band in the speech channels. The AMPS system[2] uses the (1) signalling tone and (2) supervisory audio tone, having their functions briefly described next.

The *signalling tone* (ST) is a one-way (from mobile to base station) 10-kHz tone used in bursts for disconnection, alerting, hand-off, and flashing.

The *supervisory audio tone* (SAT) comprises a set of three continuous tones (6 kHz and 6 kHz ± 30 Hz). Only one frequency, however, is used by the base station of a given cluster. Neighboring clusters employ one of the remaining two audio tones. This allocation scheme assures a more reliable supervision control because interference is greatly reduced. Cochannels using the same supervisory tone are spaced farther apart from each other. In effective terms, as far as supervision is concerned, the size of the cluster is multiplied by 3, and, consequently, the reuse $D/R$ ratio is increased by $\sqrt{3}$. The supervisory audio tone is continuously sent by the base station to the mobile, which loops it back to the base station. If the received tone differs

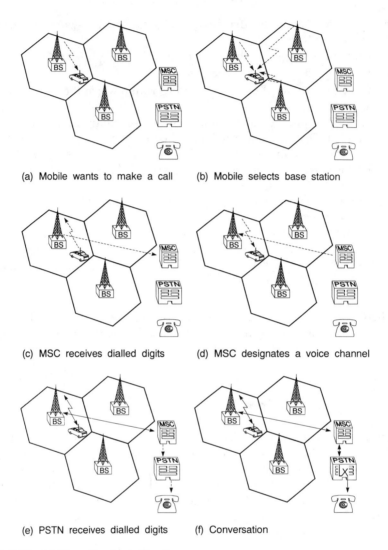

(a) Mobile wants to make a call    (b) Mobile selects base station

(c) MSC receives dialled digits    (d) MSC designates a voice channel

(e) PSTN receives dialled digits    (f) Conversation

**Figure 2.16.**   Mobile calling fixed subscriber.

from the tone that has been sent, some sort of interference may have occurred. If no tone is returned, either the mobile is in fading or its transmitter is off.

The speech channels can also transmit data messages while the mobile is in conversation, this occurring, for instance, in the hand-off process. We shall briefly describe the hand-off procedure:

1. The serving and the surrounding base stations monitor the transmission quality of the channel in use. The mobile switching center receives this

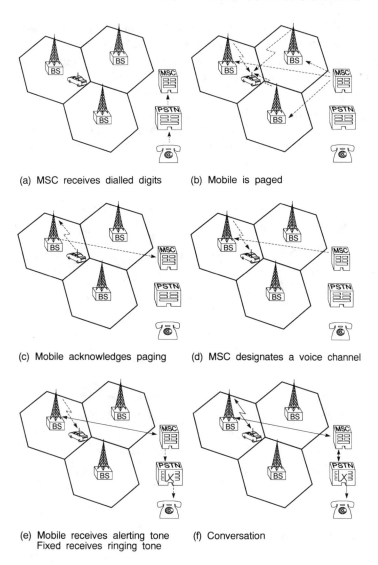

(a) MSC receives dialled digits  (b) Mobile is paged

(c) Mobile acknowledges paging  (d) MSC designates a voice channel

(e) Mobile receives alerting tone  (f) Conversation
Fixed receives ringing tone

**Figure 2.17.** Fixed subscriber calling mobile.

information over the data link. An analysis is then carried out and the MSC decides to which base station the call must be handed off.

2. The MSC informs the new base station that a call is to be handed off to a specified channel. This base station switches on its transmitter and the supervisory audio tone is sent over the channel.

3. The MSC informs the serving base station that such call is to be handed off to a new channel having a given supervisory audio tone. This information is passed over to the mobile in the speech channel in use.

4. The mobile sends a burst of the signalling tone, turns its transmitter on, tunes to the new channel, and loops back the audio tone.
5. The former base station clears the call; the new base station, by recognizing the looped-back audio tone, activates the call.

These steps are depicted in Figure 2.18.

The algorithms used for hand-off can vary according to the system. The simplest criterion is based only on the signal strength measurement. The

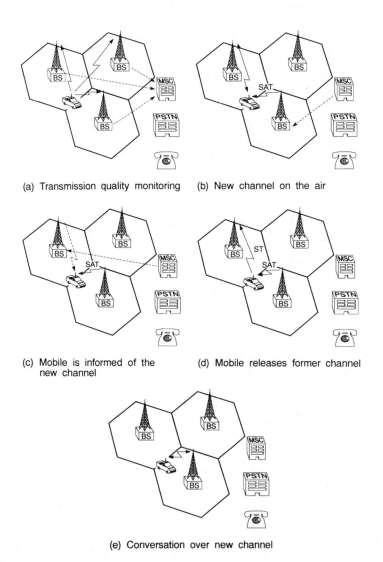

(a) Transmission quality monitoring    (b) New channel on the air

(c) Mobile is informed of the new channel    (d) Mobile releases former channel

(e) Conversation over new channel

**Figure 2.18.**  Hand-off sequence.

supervisory audio tone may also be involved, to determine the distance from the mobile to the base stations.

## 2.11 CELLULAR SYSTEM REQUIREMENTS AND ENGINEERING

The basic objective of the cellular mobile radio is to provide flexibility to the subscriber, who is initially familiar with the fixed telephone network. Therefore, the basic specifications require the services to be offered with telephone quality. As far as traffic is concerned, the blocking probability during the busy hour should be kept below 2%. In fact, the 5% figure is quite acceptable, but there are systems already working with 10% or more. On the

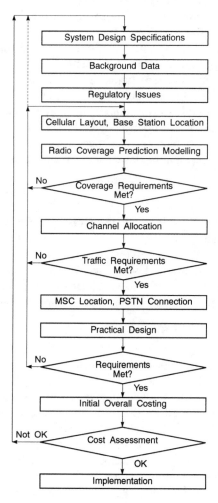

**Figure 2.19.** Basic steps in a cellular system design.

transmission aspect the aim is to provide most of the customers with a good transmission quality for at least 90% of the time. This "good quality" is something rather subjective, and many tests are carried out so that different opinions are gathered among people who have volunteered to be submitted to those tests.[10, 12]

The transmission quality in cellular systems depends mainly on (1) the signal-to-cochannel interference ($S/I_c$) ratio and (2) the adjacent-channel selectivity (ACS). The $S/I_c$ is a subjective measure, usually taken to be around 17 dB.[2] The ACS is a specification of the CCIR having a minimum of 70 dB. The cochannel-to-interference ratio $C/I_c$ depends on the modulation scheme. For a 25-kHz FM the $C/I_c$ is around 8 dB and for a 12.5-kHz FM this is 12 dB.[10] The minimum signal-to-interference ($S/I = S/(\text{noise} + I_c)$) requirement is 18 dB. An improvement of this figure does not result in a significant voice quality enhancement. Below this figure there would probably be "cross-talk" and further below, the call is cut off.[13]

Engineering a cellular system to meet the aforementioned goals is not a straightforward task. It requires a great deal of information, such as (1) market demographics, (2) area to be served, (3) traffic offered, etc., not usually available in the earlier stages of system design. If some data are available, they are usually not enough to achieve an optimized planning. Anyhow, the data available at the time of the system inception are always the best (only) input and must be used. As the network evolves, additional statistics will help the system performance assessment and replanning. The main steps in a cellular system design are shown in the flow chart form of Figure 2.19.

## 2.12    ALTERNATIVE TRAFFIC PERFORMANCE ENHANCEMENT TECHNIQUES

In Section 2.6 we mentioned some system expansion techniques. In this section we describe some traffic-oriented procedures that can be explored. The aim is just to highlight the main points, leaving a deeper discussion to be given in Chapter 12.

### 2.12.1    Channel Assignment Algorithms

The efficient use of channels determines the good performance of the system and can be obtained by different channel assignment techniques. A great deal of studies and computations have shown that there is a noticeable increase in channel occupancy as some dynamic assignment algorithms are applied. The main techniques for allocating channels are briefly described next. A full performance analysis of the following techniques and their variations is carried out by Sanchez,[18] where realistic models and conditions are applied.

### 2.12.1.1 *Fixed Channel Allocation*

In this technique a subset of the channels available in the system is assigned to each cell. The same subset is reassigned to the cell with the required reuse distance (to avoid cochannel interference). If all the channels are busy in one cell the call is blocked. When the traffic profile is well known, the allocation of the channels can be optimized to give the best performance. However, any sudden traffic variation can cause disturbance.

### 2.12.1.2 *Dynamic Channel Allocation*

This algorithm, in fact, comprises a number of strategies with the common characteristics that all the channels in the system are available to all the cells. The assignment is carried out according to the dynamic traffic demand of the subscribers. This algorithm can cope with the varying nonuniform spatial traffic distribution but gives poor results at high load.[15, 16]

### 2.12.1.3 *Hybrid Channel Allocation*

This technique is a combination of the two previous techniques. Each cell has a fixed percentage of pre-allocated channels, whereas the rest of the channels are assigned dynamically. The performance then depends on both the traffic distribution as well as on the proportion of fixed-to-dynamic channels.

### 2.12.1.4 *Borrowed Channel Allocation*

In this approach a cell having all of its channels busy looks for a free channel in the neighboring cell. If there are no channels available, then the call is blocked. It is possible to improve this by "forcing" the borrowing from the adjacent cell even if that cell is blocked. This blocked cell will, in its turn, force a borrowing from another cell and so on.[17]

### 2.12.2   Fuzzy Cell Boundaries

The implementation of any of the preceding channel assignment techniques (except for the fixed one) implies a complete involvement of the central processing control unit with any call having to be handled through it. It may be possible to consider a more local approach, with the decision needing to be taken not so far away (i.e., centrally) perhaps within the local MSC itself. If there is traffic available for alternative routing between adjacent cells, then its management can lie within the concerned traffic area. It is just a question of determining the cell and the amount of traffic that this cell must be offered so that, by sharing the load among neighboring cells, the system can be led to a traffic balance condition.

Many aspects of cellular radio design and performance are studied on the assumption of fixed cell boundaries. Practically, the cell boundaries are uncertain and shifting because radio propagation is variable both in space and time. Hand-off procedures take some account of this, normally being based on the relative signal levels of two paths from a mobile to two base stations. However, in general, planning and design have not been optimized for the practical situation in which boundaries are fuzzy and base station service areas overlap.

If a mobile is near a cell boundary it may well have adequate communication with more than one base station. It should then be possible to use this property as the basis for alternative routing techniques, in the light of information such as the current channel occupancy in each cell, and the mean or forecast traffic in each cell, so as to maximize the joint traffic capacity of a set of contiguous cells.

The first step in evaluating alternative routing strategies is to estimate the proportion of traffic that might reasonably be treated as available for alternative routing. This will be done in Chapter 3, but it is still possible to have a rough idea, just by using a very simple geometrical approach as follows.

### 2.12.2.1 *Mobiles with More Than One Path — A Rough Estimate*

The hexagonal shape of the cells in a mobile radio system is just considered as an ideal model. It cannot be obtained in the real world. With the use of omnidirectional antennas, the coverage area can ideally be approximated by a circle. In a hexagonal cell array, if each cell is superseded by circles (Figure 2.20) there will be overlapped areas representing regions being served by two base stations. Note that this is just a simplistic model because in a real system the mobile can be served by several base stations. Usually, the best relative signal strength is chosen for communication. What we will show here is the potential benefits of choosing the channel not only based on the relative signal strength criteria, but also on the traffic load of the system. Chapters 3 and 12 treat this subject in more detail.

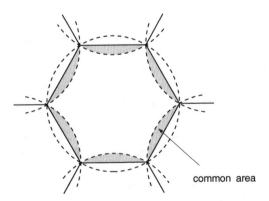

common area

**Figure 2.20.** Overlapped areas between cells.

Define $\gamma$ as the proportion of overlapped area with seven base stations taken two at a time, i.e.,

$$\gamma = \frac{\text{overlapped area}}{\text{total hexagonal area}}$$

By simple area calculations, from Figure 2.20, $\gamma$ is obtained as approximately 21%. Assuming that the mobiles are uniformly distributed within the cell, it can be concluded that 21% of the subscribers are within the fuzzy area, i.e., can be served by two base stations.

Even if it were physically possible, there is no interest or advantage, for the sake of the system performance, in making the boundaries between cells completely rigid. That is, there is no interest or advantage in having the coverage area of one base station start exactly at the point where the area served by an adjacent base station ends. An overlapping region has to be allowed so that hand-off can take place safely. There should be a certain *flexibility* for the continuity of the calls from mobiles crossing the border between cells. Such flexible area assures the system a practical *adaptability* to conditions when calls cannot be handed off immediately (e.g., when there are no free channels in the other cell or when any other higher-priority task is being processed at the moment of the request). Accordingly, if the cell coverage area is expanded to allow for more flexibility, there will be overlapped areas common to three cells. Therefore, the traffic at this region may have access to three base stations. Let $\delta$ be the proportion of the overlapped area common to the three neighboring cells. It is obvious that $\delta$ is a function of $\gamma$. Moreover, a linear variation in $\gamma$ implies a quadratic variation in $\delta$, because this is reflected into area variation. Hence $\delta \simeq \gamma^2$. For the purposes of this section we can use

$$\delta \simeq \gamma^2 \tag{2.10}$$

It will be shown in Chapter 3 that Equation 2.10 is in fact an underestimate of $\delta$.

### 2.12.2.2 *Potential Benefits from Alternative Routing for Three Contiguous Cells*

In order to have an insight into the improvement in traffic performance that can be achieved with the use of the fuzzy traffic, consider a three-cell system with a traffic distribution as follows: cell 1 with 0.75 erl, cell 2 with 4.25 erl, and cell 3 with 2.5 erl. If $\gamma$ is the proportion of mobiles with two or more paths and $\delta$ is the proportion of mobiles with three or more paths, then the proportion of mobiles with only two paths is $\gamma - \delta$. Assuming the approximation given by Equation 2.10, with $\gamma = 45\%$, then the distribution of the fixed and flexible traffic is as shown in Figure 2.21. The streams of

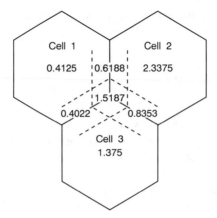

**Figure 2.21.** Two- and three-paths adaptable traffic for a flexibility of 45%.

traffic have been calculated as follows:

$$\text{Flexible traffic between cell 1 and cell 2} = (0.75 + 4.25)\left(\frac{\delta - \gamma}{2}\right)$$

$$\text{Flexible traffic between cell 2 and cell 3} = (4.25 + 2.5)\left(\frac{\delta - \gamma}{2}\right)$$

$$\text{Flexible traffic between cell 3 and cell 1} = (2.5 + 0.75)\left(\frac{\delta - \gamma}{2}\right)$$

$$\text{Flexible traffic between cell 1 and cell 2 and cell 3} = (0.75 + 4.25 + 2.5)\delta$$

We have assumed the flexible traffic of each cell to be equally distributed between its boundaries. This is the reason for the factor $\frac{1}{2}$ in the preceding calculations. Note that if the traffic in Figure 2.21 is conveniently redistributed between the cells, the system can restore the balance with each cell being offered 2.5 erl. If each cell has five channels, the initial mean blocking probability of the system (when no fuzzy traffic is used) is 16% whereas this blocking falls to 7% after the traffic redistribution.

## 2.13  SUMMARY AND CONCLUSIONS

Mobile communications have been in use since the early days of the invention of the radio, and the cellular concept is rather old. However, cellular systems have only appeared more recently as a consequence of the high demand for mobile services, radio frequency spectrum limitation, and availability of technological resources.

Radio is a rather complex communication medium, and there is no universally accepted propagation model applicable to all the situations. A considerable number of parameters must be taken into account and usually not all the necessary input data are easily available. The cellular architecture imposes additional complexity when interferences are to be considered. Hence special attention in assigning the frequencies has to be paid so that interference may be kept within acceptable limits.

There are many ways of assessing the system performance. The main efficiency measure, however, combines the amount of traffic per area and bandwidth. The aim, then, is to include as many channels as possible in cell as small as possible, and to make efficient use of the channels. This is not a straightforward task because system capacity and interference work in opposite directions. The question then is to achieve a more efficient use of the available channels, keeping the interference within acceptable tolerances. The tolerance limits are subjective measures, requiring tests involving people. This has already been done in some pioneer systems and those figures are generally accepted as standards. In terms of blocking probability the objective is to maintain the same requirements as for the fixed network. Nevertheless, due to the increasing demand, this is seldom achieved and the subscriber may experience blocking up to 10 times as high as that of the initial requirement.

In order to increase the channel occupancy in the cellular system, traffic enhancement techniques have been proposed. They all aim at improving the performance and may be a very good option for a near-future application.

## REFERENCES

1. Steele, R. and Prabru, V. K., High-user-density digital cellular mobile radio systems, *IEEE Proc.*, Part F, 132(8), 396, 1985.
2. MacDonald, V. H., Advanced mobile phone service: the cellular concept, *Bell Syst. Tech. J.*, 58(1), 1979.
3. Coxeter, W. B., *Introduction to Geometry*, 2nd ed., John Wiley & Sons, New York, 1969.
4. Lockwood, E. H., and Macmillan, R. H., *Geometric Symmetry*, Cambridge University Press, Cambridge, U.K., 1978.
5. Shubnikov, A. V., et al., *Colored Symmetry*, Pergamon Press, Oxford, 1964.
6. Cox, D. C., Cochannel interference considerations in frequency reuse small-coverage area radio systems, *IEEE Trans. Vehicular Tech.*, VT-32(3), 217–224, 1983.
7. Heeralall, S., The Applications of Directional Antennas in Cellular Mobile Radio Systems, Ph.D. thesis, University of Essex, England, 1988.
8. Paschke, D., RF Engineering Considerations for Cellular Systems (a presentation for TELEBRAS), Novatel, Canada, October 1988.
9. Cown, D. R. and Maciezko, R., User density for 800 MHz microwave mobile radiotelephone communications, 1980, in *International Zurich Seminar on Digital Communications*, C2.1–3, March 1980.

10. French, R. C., The effect of fading and shadowing on channel reuse in mobile radio, *IEEE Trans. Vehicular Tech.*, VT-28, 171–181, August 1979.

11. Hoff, J., Mobile telephony in the next decade, 2nd Nordic Seminar on Digital Land Mobile Radio Communication, Stockholm, October 1986.

12. Garner, P. J., Co-channel and quasi-synchronous characteristics of SSB relative to FM in mobile radio communication systems, *IEE Commun. 80*, 16–18, April 1980; also in *Communications Equipment and Systems*, Conference Publication No. 184, 174–177.

13. Sowell, K. and Kolano, J., Factor in the cellular system capacity formula, U.S. West presentation, in *2nd Int. Seminar on Cellular Mobile Telephony* (in Portuguese), 28–29, November 1988, RNT, São Paulo, S.P., Brazil.

14. Johansson, V., The best is yet to come, *Communications Systems Worldwide*, October 1987.

15. Cox, D. C. and Reudinik, D. O., A comparison of channel assignment strategies in large-scale mobile communications systems, *IEEE Trans. Commun.*, COM-20(2), April 1972.

16. Cox, D. C. and Reudinik, D. O., Increasing channel occupancy in large-scale mobile radio systems: dynamic channel reassignment, *IEEE Trans. Commun.*, COM-21(11), November 1973.

17. Antennas Specialists, Dynamic Frequency Allocations Increase Cellular Efficiency, *Communications Engineering International*, November 1986.

18. Sanchez, V., J. H., Traffic Performance of Cellular Mobile Radio Systems, Ph.D. Thesis, University of Essex, 1988.

19. Gross, D. and Harris, C. M., *Fundamental of Queueing Theory*, John Wiley & Sons, New York, 1974.

20. Yacoub, M. D., Mobile Radio with Fuzzy Cell Boundaries, Ph.D. Thesis, University of Essex, 1988.

21. Young, W. R., Advanced mobile phone service: introduction, background and objectives, *Bell Syst. Tech. J.*, 58, January 1979.

22. Fluhr, Z. C. and Porter, P., Advanced mobile phone service: control architecture, *Bell Syst. Tech. J.*, 58, January 1979.

# PART II
# Mobile Radio Channel

# CHAPTER 3

# Mobile Radio Propagation Model

This chapter is concerned with the description of the models used to characterize propagation phenomena in a mobile radio environment. It starts by reviewing some fundamentals of antenna theory, with the aim of obtaining the Friis free-space transmission formula. This formula constitutes the basis for the propagation path loss models presented next.

The propagation models can be divided into two groups, namely, theoretical models and empirical models. The theoretical models are usually described by means of closed-form expressions, whereas the empirical ones are derived from field measurements, taken at different conditions. In the first case, many approximations are carried out, so that the models are not directly applicable to real situations. In the second case many parameters are taken into account, so that the models are usually very complex. A combination of these groups of models gives rise to a simplified prediction model with excellent results, if high accuracy is not required. The various parameters affecting the mobile radio propagation are also discussed and analyzed.

We then describe the stochastic behavior of the mobile radio signal by means of its statistical distributions. By combining these statistics with some of the main results of the propagation path loss models, we determine the signal coverage area (base station service area). In the same way we determine the fuzzy boundaries between cells. It is shown that the proportion of overlapped areas between pairs of cells or groups of three cells is rather substantial.

## 3.1  INTRODUCTION

Communication with mobile points is a rather difficult task and, with today's technology, can only be achieved by means of radio waves. This transmission medium is greatly dependent on the environment. Accordingly, it can be affected by an "infinitude" of parameters describing the environment.

Mobile radio systems have been driven to use high radio frequencies due to the congestion at the low part of the frequency spectrum. Nevertheless, dealing with high frequencies usually leads to intricate problems. The theoretical analysis of the involved phenomena are, in general, very complex due to the number of variables that must be taken into consideration. Many of these variables are simply neglected for low-frequency applications. As an example, consider a radio transmission at 60 MHz, corresponding to a wavelength of 5 m. If the sizes of the obstructions encountered by this radio frequency are equivalent over many wavelengths, these obstructions may work as scatterers. If we imagine a radio transmission at 900 MHz, obstructions with a size of only tens of centimeters can work as scatterers.

Due to multipath propagation the radio signal fades rapidly, with minima reaching more than 40 dB below the mean signal level. Fading also occurs due to shadowing provoked by hills, tunnels, and other obstructions. Path loss, shadowing, and rapid fading may deteriorate the propagated signal in such a way that, if the system is not carefully engineered, loss of communication may occur quite often.

In addition to this, the mobile communication may also experience cochannel and adjacent-channel interferences together with the various types of noise generated within the radio equipment and by the environment. The motion of the vehicle also imposes random variations in the signal, deteriorating the communication even more.

It can be seen that a deterministic solution for the variability and prediction of mobile radio signal strength is rather impracticable. Instead, a statistical treatment of the signal can be used to describe the various phenomena.

## 3.2   ANTENNA FUNDAMENTALS

In this section we briefly recall some of the basic concepts and formulas of antenna theory. Our aim is to use this to arrive at the Friis free-space transmission formula. For a thorough and rigorous approach refer to Kraus.[7]

### 3.2.1   Basic Concepts

(1) An *isotropic source* is a source radiating energy in all directions.

(2) The *Poynting vector* (**P**) or *power density* is the electromagnetic power flow per unit area (watts per square meter).

(3) The *radiated power crossing a surface S* (*W*) is defined by

$$W = \iint \mathbf{P} \cdot d\mathbf{s} = \iint P \, ds \tag{3.1}$$

where $P$ is the radial component of **P**, and $d\mathbf{s}$ is an infinitesimal area in $S$. An isotropic source radiates uniformly through a spherical surface. Consequently, $W = P 4\pi d^2$, where $d$ is the sphere's radius.

(4) The *free space transmission formula* is obtained from the preceding definition: In a lossless medium, the received power density $P$, at a distance $d$, is

$$P = \frac{W}{4\pi d^2} \tag{3.2}$$

(5) The *radiation intensity* $(U)$ is the power per unit solid angle. At a distance $d$

$$U = d^2 P \tag{3.3}$$

For an isotropic source $U_0 = d^2 P = W/4\pi$.

(6) The *radiation pattern* is the geographical distribution of the power radiated by the source. It is usually shown on both azimuthal (horizontal) and elevation (vertical) planes. An isotropic source's radiation pattern is a circle in both planes. A real antenna, however, will illuminate one region (the target) more intensely than another. In the corresponding power pattern this is shown as the main lobe (for the target) and side lobes (otherwise). Antennas having a nearly circular azimuthal radiation pattern are known as omnidirectional. Antennas with directional properties are referred to as directional.

(7) The *directivity* $(D)$ is the ratio between the maximum radiation intensity $(U_M)$ from the source under consideration and the radiation intensity from an isotropic source $(U_0)$ radiating the same power

$$D = \frac{U_M}{U_0} \tag{3.4}$$

Accordingly, the directivity of an isotropic source is equal to 1. In the same way, a source with a hemispherical power pattern has directivity 2. The directivity can also be expressed as

$$D = \frac{\text{maximum radiation intensity}}{\text{average radiation intensity}} \tag{3.5}$$

The average radiation intensity is the total power averaged over the total solid angle $4\pi$. Then

$$D = \frac{U_M}{W/4\pi} = \frac{4\pi}{B} \tag{3.6}$$

where

$$B \triangleq \frac{W}{U_M} \tag{3.7}$$

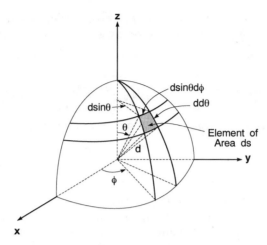

**Figure 3.1.** Incremental area *ds* of a surface *S* in spherical coordinates.

is the beam area. Rigorously speaking, $B$ is dimensionless, but it corresponds to the solid angle of the considered region. Consequently, it may be expressed either without any unit or as square radians. Using Equation 3.3 in Equation 3.1 and then Equation 3.1 in Equation 3.7, we obtain

$$B = \frac{\iint (U/d^2)\, ds}{U_M} \tag{3.8}$$

In spherical coordinates the infinitesimal area $ds$ is found to be (see Figure 3.1)

$$ds = d^2 \sin \theta\, d\theta\, d\phi \tag{3.9}$$

Consequently, the beam area $B$ is given by

$$B = \frac{\iint U\, d\Omega}{U_M} \tag{3.10}$$

where

$$d\Omega \triangleq \sin \theta\, d\theta\, d\phi \tag{3.11}$$

is the infinitesimal element of the solid angle $\Omega$.

Suppose, as an example, that we want to calculate the directivity of a source having a conical radiation pattern as sketched in the Figure 3.2.

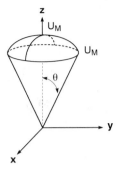

**Figure 3.2.**  Conical radiation pattern.

The total power $W$ within the conical region is given by

$$W = \int_0^{2\pi}\int_0^\theta U_M \sin\theta' \, d\theta' \, d\phi = 2\pi(1 - \cos\theta)U_M$$

Consequently

$$D = \frac{4\pi}{2\pi(1 - \cos\theta)}$$

For a hemisphere $\theta = \pi/2$, so $D = 2$. For a complete sphere (isotropic source) $\theta = \pi$, so $D = 1$. For more complex geometries an approximate formula for the beam area can be used as follows[7]:

$$B \simeq \Delta\theta\,\Delta\phi \qquad\qquad (3.12)$$

where $\Delta\theta$ and $\Delta\phi$ are half-power beamwidths, in radians.

As an example, consider a unidirectional cosine power pattern source $U$, such that $U = U_M \cos\theta$ (see Figure 3.3).

The total power $W$ is given by

$$W = \int_0^{2\pi}\int_0^{\pi/2} U_M \cos\theta \sin\theta \, d\theta \, d\phi = \pi U_M$$

The directivity is then $D = 4$. Using the approximate method, the half-power is obtained when $\theta = \phi = \pi/3$. Consequently, $\Delta\theta = \Delta\phi = \pi/3 - (-\pi/3) = 2\pi/3$, and the approximate directivity is $D \simeq 4\pi/(2\pi/3)^2 = 2.87$.

(8) The *gain* ($G$) gives a measure of the antenna's efficiency. It is expressed in relation to a reference source. Usually the gain is referred to an

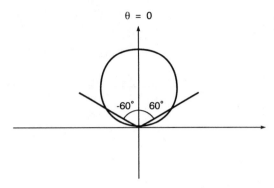

**Figure 3.3.** Unidirectional cosine radiation pattern shown in cross section. (The complete pattern is obtained by revolving the circle around the vertical axis.)

isotropic source, such that

$$G \triangleq \frac{\text{maximum radiation intensity}}{\text{radiation intensity from an isotropic source}} \quad (3.13)$$

Let $U'_M$ be the maximum radiation intensity of a real (lossy) antenna and let $U_M$ be this intensity for an ideal (lossless) 100%-efficient antenna. Then $U'_M = \eta U_M$, where $0 \leq \eta \leq 1$ is the efficiency. Accordingly, using Equations 3.13 and 3.4, we have

$$G = \frac{U'_M}{U_0} = \eta \frac{U_M}{U_0} = \eta D \quad (3.14)$$

For a lossless isotropic source $G = D = 1$. In general, gain and directivity are given in decibels, i.e., $10 \log G$ and $10 \log D$, respectively.

(9) To find the *antenna aperture*, a receiving antenna can be modelled as having a terminating and an intrinsic impedance. The terminating impedance corresponds to the load receiving the power delivered to the antenna. The intrinsic impedance is responsible for the losses, corresponding to the power dissipated by heat or reradiated by the antenna. The antenna aperture (given in square meters) is defined as the ratio between the power loss (in watts) and the power density (in watts per square meter) of the wave. The aperture may be interpreted as the virtual area of the antenna immersed in an electromagnetic field. Depending on the loss to be considered, we may have different types of antenna aperture as follows:

- *Effective aperture*—the ratio between the power delivered to the terminating impedance and the power density; the maximum effective aperture is known as the *effective area*
- *Loss aperture*—the ratio between the power dissipated as heat and the power density

- *Scattering aperture*—the ratio between the power reflected back by the antenna and the power density
- *Collecting aperture*—the sum of the effective, loss, and scattering apertures
- *Physical aperture*—the physical size of the antenna

(10) The *gain, directivity,* and *aperture* are intrinsically related to each other. The relation between gain and directivity is expressed by Equation 3.14. The common parameter between directivity and aperture is the power. In a receiving antenna the power extracted from the electromagnetic field increases with the increase of both directivity and effective area. Therefore,

$$D = KA \qquad (3.15)$$

where $K$ is a constant and $A$ is the effective area of the antenna. Accordingly, for two antennas having directivities $D_i$ and $D_j$ and effective areas $A_i$ and $A_j$,

$$\frac{D_i}{D_j} = \frac{A_i}{A_j} \qquad (3.16)$$

Using Equations 3.14 and 3.16, we have

$$\frac{\eta_j G_i}{\eta_i G_j} = \frac{D_i}{D_j} = \frac{A_i}{A_j} \qquad (3.17)$$

where $\eta_i, \eta_j$ and $G_i, G_j$ are the efficiencies and gains of antennas $i$ and $j$, respectively.

These parameters have been determined for some antennas of interest[7]:

- *Isotropic antenna*—$A = \lambda^2/4\pi$, $D = 1$
- *Short dipole antenna*—$A = 3\lambda^2/8\pi$, $D = 3/2$
- *Dipole* (linear half-wavelength)—$A = 30/73\pi$, $D = 1.64$

Using Equation 3.16 and any pair of $A$ and $D$ of these antennas it is straightforward to show that, for any antenna $i$,

$$A_i = \frac{\lambda^2}{4\pi} D_i \qquad (3.18)$$

(11) The *Friis free-space transmission formula* is obtained as follows: Let $W_t$ be the input power of a transmitting antenna having a gain $G_t$. The radiated power is $G_t W_t$ and the power density $P$ at a distance $d$ is

$$P = \frac{G_t W_t}{4\pi d^2} \qquad (3.19)$$

The total power $W_r$ received by the load at $d$ is

$$W_r = A_r P = \frac{A_r G_t W_t}{4\pi d^2} \tag{3.20}$$

where $A_r$ is the antenna aperture. With Equation 3.18 in Equation 3.20 we obtain

$$\frac{W_r}{W_t} = G_t D_r \left( \frac{\lambda}{4\pi d} \right)^2$$

If the gain is equal to the directivity ($\eta = 1$) at the receiver, then

$$\frac{W_r}{W_t} = G_t G_r \left( \frac{\lambda}{4\pi d} \right)^2 \tag{3.21}$$

Equation 3.21 is known as the Friis free-space transmission formula.

## 3.3  PROPAGATION PATH LOSS

A measure of great interest in radio propagation studies is the *path loss* ($l$). This is defined as the ratio between the received power ($W_r$) and the transmitted power,

$$l \triangleq W_r / W_t \tag{3.22}$$

In decibels, path loss is given as

$$L = -10 \log l = -10 \log W_r + 10 \log W_t$$

An exact estimate of the path loss in a mobile radio environment is not available. In the following subsections we consider some theoretical models applicable to very special cases. Then we modify these models to give an approximate measure of the path loss in the mobile environment. Finally, we describe the various empirical methods and their applicability.

### 3.3.1  Free-Space Path Loss

The ratio between the received and transmitted powers in a free-space propagation condition is given by the Friis free-space transmission formula. Hence

$$\frac{W_r}{W_t} = G_t G_r \left( \frac{\lambda}{4\pi d} \right)^2 \tag{3.23}$$

Accordingly, the path loss (in decibels) is

$$L = -10 \log G_t - 10 \log G_r - 20 \log \lambda + 20 \log d + 21.98$$

Using lossless and isotropic antennas ($G_t = G_r = 1$), with the frequency in megahertz and the distance in kilometers, we have

$$L = 20 \log f + 20 \log d + 32.44 \text{ dB} \tag{3.24}$$

## 3.3.2  Plane Earth Path Loss

Consider the propagation of a radio wave in a flat-terrain environment. The transmitted signal may reach the receiving antenna by several ways:

- Through a direct path
- Through an indirect path, consisting of the radio wave reflected by the ground
- Through an indirect path, consisting of surface wave
- Through other secondary means.

The received signal is a combination of all of these waves, with a resultant power equal to the sum of their individual powers. The signal power due to the direct path is given by the Friis free-space formula. The reflected wave will have a power given by the Friis formula but attenuated by a factor equal to the ground reflection coefficient $\rho$. Moreover, the reflected signal will be shifted by a phase $\Delta\varphi$ due to the indirect path. The ground wave is provoked by the signal that has been absorbed by the ground. The proportion of the absorbed signal is given by $(1 - \rho)$, corresponding to the nonreflected signal times an attenuation factor $A$. Therefore, the ratio between the received and transmitted powers in a flat-terrain environment is

$$\frac{W_r}{W_t} = G_t G_r \left( \frac{\lambda}{4\pi d} \right)^2 \left| 1 + \rho e^{j\Delta\varphi} + (1 - \rho) A e^{j\Delta\varphi} + \cdots \right|^2 \tag{3.25}$$

Both $\rho$ and $A$ depend on various factors such as incidence angle, polarization, earth constants, and frequency. In particular,

$$\rho = \frac{\sin\theta - K}{\sin\theta + K} \tag{3.26}$$

where $\theta$ is the incidence angle and $K$ varies with all of the parameters previously mentioned. For a grazing angle $\theta \simeq 0°$ (corresponding to the case where the distance between base station and mobile is much greater than the antenna height) then $\rho \simeq -1$ (see Equation 3.26). Moreover, $\rho$ tends to $-1$ for frequencies above 100 MHz and incidence angles less than $10°$.[2] The

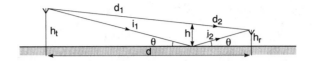

**Figure 3.4.**    Direct and indirect paths on a flat-terrain environment.

effects of the ground waves are sensed only a few wavelengths above the ground. Therefore, they can be neglected for the case of mobile radio using microwave frequencies. Consequently, Equation 3.25 is reduced to

$$\frac{W_r}{W_t} \simeq G_t G_r \left( \frac{\lambda}{4\pi d} \right)^2 |1 - e^{j\Delta\varphi}|^2 \tag{3.27}$$

Consider transmitting and receiving antennas of heights $h_t$ and $h_r$, respectively, separated by a distance $d$, as shown in Figure 3.4.

If the time delay between the direct and indirect waves is $\Delta t$, then the phase shift $\Delta\varphi$ is given by $\Delta\varphi = 2\pi f \Delta t$, where $f$ is the signal frequency. Therefore,

$$\Delta\varphi = 2\pi \frac{\Delta d}{\lambda} \tag{3.28}$$

where

$$\Delta d = (i_1 + i_2) - (d_1 + d_2) \tag{3.29}$$

is the difference between the indirect and direct paths. The distance difference $\Delta d$ can be easily written as a function of $h_t$, $h_r$, and $d$. Then

$$\Delta\varphi = \frac{2\pi d}{\lambda} \left\{ \left[ \left( \frac{h_t + h_r}{d} \right)^2 + 1 \right]^{1/2} - \left[ \left( \frac{h_t - h_r}{d} \right)^2 + 1 \right]^{1/2} \right\} \tag{3.30}$$

Using the approximation $(1 + x)^{1/2} \simeq 1 + x/2$, for small $x$, from Equation 3.30 we obtain

$$\Delta\varphi \simeq 4\pi \frac{h_t h_r}{\lambda d} \tag{3.31}$$

The squared modulus in Equation 3.27 can be expanded as

$$|1 - e^{j\Delta\varphi}|^2 = 2(1 - \cos \Delta\varphi) = 4\sin^2 \frac{\Delta\varphi}{2} \tag{3.32}$$

For small $\Delta\varphi$, $\sin(\Delta\varphi/2) \simeq \Delta\varphi/2$. Then

$$\sin^2 \frac{\Delta\varphi}{2} \simeq \left(\frac{\Delta\varphi}{2}\right)^2 \tag{3.33}$$

With the appropriate use of Equations 3.33, 3.32, 3.31, and 3.27 we finally obtain

$$\frac{W_r}{W_t} = G_t G_r \left(\frac{h_t h_r}{d^2}\right)^2 \tag{3.34}$$

This is the inverse fourth-power loss formula. The corresponding path loss in decibels is

$$L = -10 \log G_t - 10 \log G_r - 20 \log(h_t h_r) + 40 \log d \tag{3.35}$$

Note that there is a loss of 12 dB when the distance is doubled. On the other hand, there is a gain of 6 dB when the antenna height is doubled.*

The flat-terrain model can be applied to an environment considered to be smooth. The measure of the terrain smoothness, $S$, depends on various parameters, including the frequency. One such measure is given by the Rayleigh criterion as[2]

$$S = \frac{4\pi\sigma\theta}{\lambda} \tag{3.36}$$

where $\sigma$ is the standard deviation of the mean heights of the irregularities, $\theta$ is the incidence angle, and $\lambda$ is the wavelength. A surface with $S < 0.1$ is considered to be smooth, whereas $S > 10$ is considered as rough.

### 3.3.3  Knife-Edge Diffraction Loss

A radio wave radiated from a transmitting antenna can be intercepted by obstructions such as hills, buildings, trees, and others. In this case the signal reaching the receiving antenna may arrive as a diffracted ray as shown in Figure 3.5. If the radiated electric field strength is $E_0$, the diffracted field $E$ is[2]

$$E = E_0 F e^{j\Delta\varphi} \tag{3.37}$$

where $F$ is the diffraction coefficient and $\Delta\varphi$ is the phase difference between

---

*Field measurements have shown that this is only true for the base station's antenna. For the mobile's antenna the gain is only 3 dB if $h_r < 3$ m or 6 dB if $3 \text{ m} \le h_r \le 10$ m (refer to Section 3.3.6 for the Okumura model).

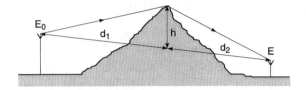

**Figure 3.5.**   Knife-edge diffraction geometry.

the indirect and direct paths. These parameters are given by[2,3]

$$F = \frac{S(x) + 0.5}{\sqrt{2}\,\sin(\Delta\varphi + \pi/4)}$$  (3.38)

and

$$\Delta\varphi = \tan^{-1}\left[\frac{S(x) + 0.5}{C(x) + 0.5}\right] - \frac{\pi}{4}$$  (3.39)

where $C(x)$ and $S(x)$ are the Fresnel cosine and sine integrals, respectively,

$$C(x) = \int_0^x \cos\left(\frac{\pi}{2}u^2\right) du$$  (3.40)

$$S(x) = \int_0^x \sin\left(\frac{\pi}{2}u^2\right) du$$  (3.41)

and

$$x = -h\sqrt{\frac{2}{\lambda}\left(\frac{d_1 + d_2}{d_1 d_2}\right)}$$  (3.42)

Note that the height $h$ can also be negative as shown in Figure 3.6. In such a case there will be a direct path between the two antennas. From Equation

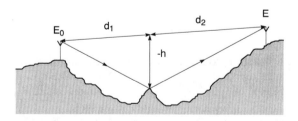

**Figure 3.6.**   Knife-edge diffraction with negative height.

3.37 the loss due to the diffraction is

$$L = 10 \log \left| \frac{E}{E_0} \right|^2 = 20 \log F \qquad (3.43)$$

The Fresnel integrals can be evaluated by means of series expansion as

$$S(x) = \sqrt{\frac{2}{\pi}} \sum_{n=0}^{\infty} (-1)^{n+2} \frac{\left(\sqrt{\pi/2}\,x\right)^{3+4n}}{(1+2n)!(3+4n)} \qquad (3.44)$$

$$C(x) = \sqrt{\frac{2}{\pi}} \sum_{n=0}^{\infty} (-1)^{n+2} \frac{\left(\sqrt{\pi/2}\,x\right)^{1+4n}}{(2n)!(1+4n)} \qquad (3.45)$$

It is interesting to note that $S(-x) = -S(x)$ and $C(-x) = -C(x)$. Moreover, $S(\infty) = C(\infty) = 0.5$ and $S(0) = C(0) = 0$. Consequently,

$$\lim_{x \to \infty} F = 1, \qquad \lim_{x \to -\infty} F = 0 \quad \text{and} \quad F = 0.5 \quad \text{when } x = 0$$

A sketch of diffraction loss is shown in Figure 3.7.

Equation 3.42 can be written as $x = -hK$, where $K$ is a nonnegative constant. A positive height $(+h)$ implies a negative $x$ corresponding to the situation shown in Figure 3.5. In this case the receiving antenna is in a shadowed region and the loss curve is that of the left semiplane of Figure 3.7. The greater the height the larger the loss. On the other hand, a negative height implies a positive $x$ corresponding to the situation shown in Figure

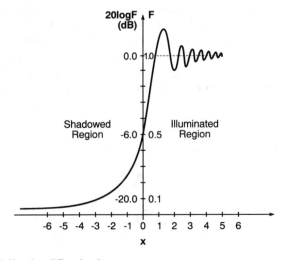

**Figure 3.7.**   Knife-edge diffraction loss.

3.6. In this case the receiving antenna is in the illuminated region. If the obstruction is just "touching" the direct path ($x = h = 0$), the received electric field will have half of the magnitude of the transmitted field. In other words, the received power is one-fourth ($0.5^2$) of the transmitted power.

### 3.3.4 Multiple Knife-Edge Diffraction Loss

In a real situation a radiated wave may be diffracted not only by one obstruction but by several. A mathematical approach to deal with this situation is rather complicated and no simple solution is available. There are some approximate methods as we shall examine next.

#### 3.3.4.1 Bullington's Model

This model was primarily developed to tackle the two-obstruction path loss case. Consequently, the results are better for this situation. As the number of obstructions increases the model gives a poorer performance due to the oversimplification. The aim of the model is to reduce an $n$-dimensional problem to one dimension by replacing the $n$ obstructions by one equivalent knife edge. This corresponds to find an equivalent height producing the same effects as those produced by all of the other knife edges together. This method is illustrated in Figure 3.8. Note that some obstructions are ignored.

#### 3.3.4.2 Epstein–Peterson Model

This model considers each knife edge individually and approximates the total loss as the sum of the individual losses, as shown in Figure 3.9.

The first loss is calculated by considering the path "transmitter, first obstruction, second obstruction". The other loss considers the subsequent path constituted by "first obstruction, second obstruction, receiver". If the obstructions are very close to each other, some of the heights (at least one) will not be correctly determined. In this case this method gives poor results.

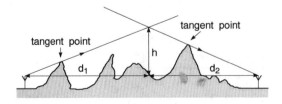

**Figure 3.8.**   Equivalent knife edge of height $h$ in Bullington's model.

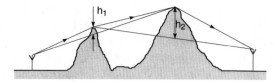

**Figure 3.9.** Epstein–Peterson model.

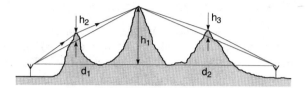

**Figure 3.10.** Deygout's model.

### 3.3.4.3 *Deygout's Model*

This model initially estimates a path loss by considering the dominant obstruction in the environment. The other losses due to the remaining knife edges are determined with respect to this dominant edge, as shown in Figure 3.10.

### 3.3.5  Path Clearance Conditions

As we have seen, the path loss depends on several parameters, including antenna heights, distances between the antennas, frequency, and others. For a given set of these parameters we may estimate the proportion of the path loss caused by the diffracted waves as well as by the reflected waves. Consider the situation depicted in Figure 3.11.

If the height $h$ is sufficiently large, the received wave will not be affected by the diffracted waves. However, the indirect path may be long enough to

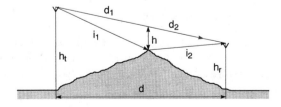

**Figure 3.11.** Influence of the diffracted and reflected waves.

provoke a phase difference between the direct and indirect signals so that the resultant signal can be substantially deteriorated. In this case the path loss is dominated by the reflected signal. On the other hand, if the height $h$ is sufficiently small, there will not be much difference between the direct and indirect paths. Accordingly, the phase shift in the indirect wave will be small. However, the effects of the knife edge can be significant. In this case the path loss is dominated by the diffracted waves. We shall investigate these two situations.

### 3.3.5.1 *Loss Due to Reflection and Refraction*

From Figure 3.7 we understand that the condition $x > 0$ corresponds to have the receiving antenna in the illuminated region. Moreover, a careful look at this figure (or equivalently, at Equations 3.38 and 3.39) shows that, for $x > \sqrt{2}$, the diffraction coefficient $F$ (diffraction loss) tends to unity. In other words, for $x > \sqrt{2}$ the diffraction loss tends to be negligible. Therefore, using Equation 3.42 and the constraint $x > \sqrt{2}$, we have*

$$h\sqrt{\frac{2}{\lambda}\left(\frac{d_1 + d_2}{d_1 d_2}\right)} > \sqrt{2}$$

Therefore,

$$h > \sqrt{\lambda\left(\frac{d_1 d_2}{d_1 + d_2}\right)} \tag{3.46}$$

The condition $x > \sqrt{2}$ corresponds to $\Delta\varphi > \pi/2$. Accordingly, we can say that when $\Delta\varphi > \pi/2$ the indirect wave is not substantially affected by diffraction. We may express Equation 3.46 in terms of $d$, $h_t$, and $h_r$ by using the simple geometric relations obtained from Figure 3.11. Then,

$$d < 4h_t h_r/\lambda \tag{3.47}$$

Hence, if Equation 3.46 (or equivalently Equation 3.47) is satisfied, then the loss is mainly due to the reflected wave. Otherwise, the loss is due to the diffracted wave.

### 3.3.5.2 *Avoiding Nulls at Reception*

At the receiver a null will occur if the direct and indirect waves arrive with equal amplitudes and opposite phases. Both the power of the resultant signal as well as the phase shift of the reflected wave have been determined in

---

*Note that in Figure 3.11 the height $h$ corresponds to $-h$ in Figure 3.6.

Section 3.3.2. From Equation 3.27 we see that a null will occur if $|1 - e^{j\Delta\varphi}|^2 = 0$. Equivalently, using Equation 3.32 the null will occur if $4\sin^2(\Delta\varphi/2) = 0$, resulting in $\Delta\varphi = 2n\pi$, where $n$ is an integer. Accordingly, with $\Delta\varphi = 2n\pi$ in Equation 3.31 we obtain

$$d = \frac{2h_t h_r}{n\lambda}$$

Note that in the preceding equation the distance $d$ decreases with increasing $n$ (maximum obtained when $n = 1$). Therefore, we conclude that for distances

$$d > \frac{2h_t h_r}{\lambda} \tag{3.48}$$

nulls will not occur any longer. Expressing Equation 3.48 in terms of $h$, $d_1$, and $d_2$ (refer to Figure 3.4) we obtain

$$h < \sqrt{2\lambda\left(\frac{d_1 d_2}{d_1 + d_2}\right)} \tag{3.49}$$

### 3.3.6 Prediction Models

Field-strength prediction models are usually based on some of the previously described path loss models, modified by parameters obtained in field measurements. The models take into account information about the topography of the terrain, including the orography (description of the hills) and the category of land usage (built-up area, forest, open area, or water). The degree of terrain undulation is given by the parameter known as interdecile range ($\Delta h$), measured, for instance, at a distance of 10 km, as illustrated in Figure 3.12.

Some estimates of $\Delta h$ according to the terrain topography are shown in the diagram of Figure 3.13.[2]

The prediction algorithms usually deal with a significant amount of data, requiring a computer to process these data. The choice of the model will

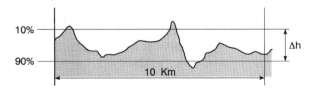

**Figure 3.12.** Interdecile range ($\Delta h$).

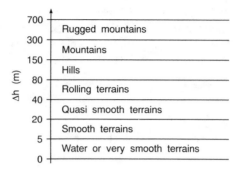

**Figure 3.13.**   Some examples of interdecile range. (*Source:* W. C. Jakes, *Microwave Mobile Communications*, John Wiley & Sons, New York, 1974.)

mainly depend on what is required: a rough estimate or a precise prediction. Moreover, the availability of data plays an important role in this. Accordingly, the algorithms can be chosen within the range from a trivial equation to a very sophisticated (and expensive) software package. After the prediction estimates, field measurements must be carried out in order to validate the model. This step will probably require the parameters to be readjusted. In this section we chose to describe some of the main prediction models. A thorough survey of these models would probably require a separate book.

### 3.3.6.1 *Egli Method*

This method is mainly based on the fourth-power loss (plane earth path loss). It uses some perturbation factors such as frequency, antenna height, and polarization to improve the results. It predicts the field strength at distances up to 60 km with a frequency range of 40–900 MHz. The median transmission loss for a 1.5-m-high mobile antenna is estimated as

$$L = 139.1 - 20 \log h_t + 40 \log d \tag{3.50}$$

In Equation 3.50, $d$ is the distance between transmitting and receiving antennas and $h_t$ is the base station antenna height, given in kilometers and meters, respectively.

### 3.3.6.2 *Blomquist–Ladell Method*

This method considers the free-space loss ($L_0$), plane-earth loss ($L_p$), and knife-edge loss ($L_k$) combined to give the total loss ($L$) as follows:

$$L = L_0 + \max(L_p, L_k) \tag{3.51a}$$

or

$$L = L_0 + \sqrt{L_p^2 + L_k^2} \tag{3.51b}$$

Equation 3.51b seems to give better results, but lacks theoretical justification. The model applies to frequencies within the range 30–900 MHz and distances between 5–22 km.

### 3.3.6.3 *Longley–Rice Method*

This method predicts the attenuation relative to the free-space loss. It requires parameters such as frequency, heights of the transmitting and receiving antennas, distance between antennas, mean surface refractivity, earth's conductivity, earth's dielectric constant, polarization, and description of the terrain. This is a computer-based algorithm working in the following ranges:

$$20 \text{ MHz} \leq \text{frequency} \leq 40 \text{ GHz}$$

$$0.5 \text{ m} \leq \text{antenna height} \leq 3 \text{ km}$$

$$1 \text{ km} \leq \text{distance} \leq 2000 \text{ km}$$

One of the interesting outcomes of the work of Longley and Rice is that the degree of terrain undulation was found as a function of the distance as

$$\Delta h' = \Delta h[1 - \exp(-0.02d)] \tag{3.52}$$

where $\Delta h'$ and $\Delta h$ are given in meters and $d$ in kilometers. Note that $\Delta h' = \Delta h$ for $d = 0$. Hence, if the profile of the terrain is known, the interdecile range can be more accurately estimated.

### 3.3.6.4 *Okumura Method*

This model is based on field measurements taken in the Tokyo area. It provides an initial path loss estimate for a quasismooth terrain ($\Delta h \simeq 20$ m). Then some correction factors must be used to adapt these results to other conditions. This initial estimate comprises a set of curves of attenuation $A(f,d)$ versus frequency $f$, for a base station antenna height $h_t$ of 200 m and a mobile station antenna height of 3 m having the distance $d$ as a parameter, as shown in Figure 3.14. These prediction curves are relative to the free-space loss. The correction factor $G_{area}$ (gain) for the type of terrain is also given as a function of the frequency, as depicted in Figure 3.15. For different antenna heights the correction factors (gain) are as follows:

• Gain $G(h_t)$ of 6 dB per octave for base station antenna height, i.e., of

$$G(h_t) = 20 \log(h_t/200) \qquad h_t > 10 \text{ m}$$

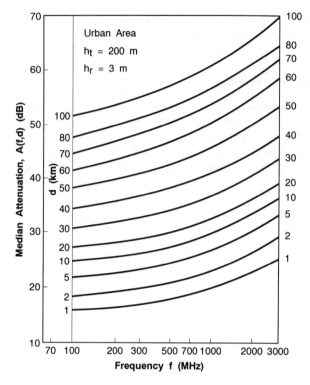

**Figure 3.14.** Median attenuation relative to free-space loss over a quasismooth terrain (using Okumura method). (*Source:* After Y. Okumura, E. Ohmori, T. Kawano, and K. Fukuda, Field strength and its variability in VHF and UHF land mobile service, *Rev. Elec. Comm. Lab.*, pp. 825–873, Sept.–Oct. 1968.)

• Gain $G(h_r)$ of 3 dB or 6 dB per octave for mobile station antenna height depending on the range, i.e.,

$$G(h_r) = 10 \log(h_r/3) \qquad h_r < 3 \text{ m}$$

$$G(h_r) = 20 \log(h_r/3) \qquad 3 \text{ m} \le h_r \le 10 \text{ m}$$

The procedure for estimating the resultant path loss is the following:

1. Given the distance $d$ and frequency $f$, find the attenuation $A(f, d)$ using the curves of Figure 3.14.
2. According to the type of terrain, for the same frequency, find the correction factor $G_{area}$, subtracting it from $A(f, d)$ obtained in step 1.
3. Determine the correction factors $G(h_t)$ and $G(h_r)$ according to the antenna heights $h_t$ and $h_r$, subtracting them from the result in step 2.
4. The loss obtained in step 3 is added to the free-space path loss $(L_0)$ to obtain the overall loss.

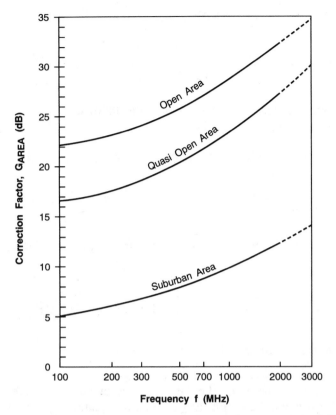

**Figure 3.15.**  Correction factor for different types of terrain (Okumura method). (*Source:* After Y. Okumura, E. Ohmori, T. Kawano, and K. Fukuda, Field strength and its variability in VHF and UHF land mobile service, *Rev. Elec. Comm. Lab.*, pp. 825–873, Sept.–Oct. 1968.)

Therefore, the overall loss is given by

$$L = L_0 + A(f, d) - G_{\text{area}} - G(h_t) - G(h_r) \qquad (3.53)$$

### 3.3.6.5 *Hata's Formula*

The Okumura method cannot be easily automated, because it involves various curves. An empirical formula based on Okumura's results has been developed by Hata.[32] It gives predictions almost indistinguishable from those given by Okumura's method over the limited range for which it applies. The loss $L$ is

$$L = 69.55 + 26.16 \log f - 13.82 \log h_t - A(h_r)$$

$$+ (44.9 - 6.55 \log h_t) \log d \text{ dB} \qquad (3.54a)$$

where $\qquad$ $150\text{ MHz} \leq f \leq 1500\text{ MHz}$

$$30\text{ m} \leq h_t \leq 300\text{ m}$$

$$1\text{ km} \leq d \leq 20\text{ km}$$

The correction factor $A(h_r)$ is computed as follows. For a small- or medium-sized city,

$$A(h_r) = (1.1\log f - 0.7)h_r - (1.56\log f - 0.8)\text{ dB} \qquad (3.54b)$$

where $1\text{ m} \leq h_r \leq 10\text{ m}$.

For a large city,

$$A(h_r) = 8.29\log^2(1.54h_r) - 1.1\text{ dB} \qquad (f \leq 200\text{ MHz}) \quad (3.54c)$$

and

$$A(h_r) = 3.2\log^2(11.75h_r) - 4.97\text{ dB }(f \leq 200\text{ MHz}) \qquad (3.54d)$$

### 3.3.6.6 *Ibrahim–Parsons Method*

This method was designed for path loss in the city of London. It proposes an empirical model based on the data collected by measurements and also a semiempirical model based on the plane-earth equation. In the first case an equation is developed to fit the collected data. In the second case the fourth-power path loss equation is multiplied by a clutter factor varying with the land usage.[8, 15, 16]

### 3.3.7  A Simplified Path Loss Model

Field measurements have shown that, with appropriate correction factors, the plane earth path loss formula can be used to predict the field strength. In this section we use some of the results obtained from the prediction models to modify the plane earth loss formula in order to introduce a simplified path loss model. For convenience we rewrite Equation 3.34 as

$$\frac{W_r}{W_t} = G_t G_r \left(\frac{h_t h_r}{d^2}\right)^2 \qquad (3.55)$$

The Okumura method showed that there is a gain of 6 dB per octave for the base station antenna height. In order words the factor $h_t^2$ in Equation 3.55 is correct. As far as the mobile station antenna is concerned this gain varies from 3 to 6 dB. Accordingly, the loss is proportional to $h_r^x$, where

$x = 1$ or $x = 2$, as follows:

$$\frac{W_r}{W_t} \approx h_r^x \qquad x = \begin{cases} 1 & \text{for } h_r < 3 \text{ m} \\ 2 & \text{for } 3 \text{ m} \le h_r \le 10 \text{ m} \end{cases}$$

The loss with the distance $d$ varies with $d^{-\alpha}$, where $2 \le \alpha \le 4$. The parameter $\alpha$ depends on the topography of the terrain. Then

$$\frac{W_r}{W_t} \approx \frac{1}{d^\alpha} \qquad 2 \le \alpha \le 4$$

The main limitation of Equation 3.55 is that it suggests an independence between path loss and frequency. From the Okumura method it was shown that

$$\frac{W_r}{W_t} \approx \frac{1}{f^y} \qquad 2 \le y \le 3$$

The parameter $y$ depends on both the environment and the frequency itself. It was found to be equal to 2 by Young[17] and equal to 3 by Okumura, but under different conditions.

With all of these considerations we may rewrite Equation 3.55 as

$$\frac{W_r}{W_t} = K \frac{G_t G_r h_t^2 h_r^x}{d^\alpha f^y} \qquad (3.56)$$

where $K$ is a constant for a given environment, and

$$2 \le \alpha \le 4$$

$$2 \le y \le 3$$

$$x = \begin{cases} 1 & \text{for } h_r < 3 \text{ m} \\ 2 & \text{for } 3 \text{ m} \le h_r \le 10 \text{ m} \end{cases}$$

In particular $y \simeq 2$ both in a suburban area and in an open area with $f < 450$ MHz, and $y \simeq 3$ in an urban area with $f > 450$ MHz. For any given environment Equation 3.56 can be used to predict the path loss if a high accuracy is not required. However, it is mainly used for comparative purposes, as we shall see. Consider that, for a given situation, the parameters in Equation 3.56 assume the values $G_{ti}$, $G_{ri}$, $h_{ti}$, $h_{ri}$, $d_i$, and $f_i$. Therefore, the received power $W_{ri}$, given that the power $W_{ti}$ has been transmitted, is

$$W_{ri} = K \frac{W_{ti} G_{ti} G_{ri} h_{ti}^2 h_{ri}^x}{d_i^\alpha f_i^y} \qquad (3.57)$$

If, for the same environment, these parameters are $G_{tj}$, $G_{rj}$, $h_{tj}$, $h_{rj}$, $d_{rj}$, and $f_j$, then the ratio between the received powers $W_{rj}$ and $W_{ri}$ is

$$\frac{W_{rj}}{W_{ri}} = \left(\frac{d_j}{d_i}\right)^{-\alpha} \beta_j \qquad (3.58)$$

where

$$\beta_j = \frac{W_{tj}G_{tj}G_{rj}}{W_{ti}G_{ti}G_{ri}} \left(\frac{h_{tj}}{h_{ti}}\right)^2 \left(\frac{h_{rj}}{h_{ri}}\right)^x \left(\frac{f_j}{f_i}\right)^{-y} \qquad (3.59)$$

Suppose that the transmitted power, antenna gains, antenna heights, and frequencies are kept in both situations. In this case $\beta_j = 1$ and the ratio between the received powers is given by

$$\frac{W_{rj}}{W_{ri}} = \left(\frac{d_j}{d_i}\right)^{-\alpha} \qquad (3.60a)$$

Expressed in decibels,

$$10 \log\left(\frac{W_{rj}}{W_{ri}}\right) = -\alpha 10 \log\left(\frac{d_j}{d_i}\right) \qquad (3.60b)$$

Note that the curve $10 \log(W_{rj}/W_{ri})$ versus $10 \log(d_j/d_i)$ is a straight line with slope $-\alpha$. Accordingly, the parameter $\alpha$ is known as the path loss slope. For the free-space condition the path loss slope is $\alpha = 2$, and for conductive terrain this is $\alpha = 4$. A graph sketching the path loss curve is shown in Figure 3.16.

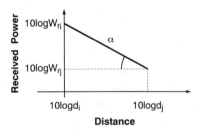

**Figure 3.16.**   A typical path loss curve.

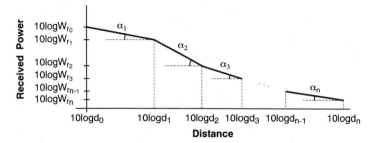

**Figure 3.17.**   Areas with different path loss slopes.

Consider that the mobile travels through regions with different path loss slopes as shown in Figure 3.17. It is easily seen that

$$\frac{W_{r1}}{W_{r0}} = \left(\frac{d_1}{d_0}\right)^{-\alpha_1}$$

$$\frac{W_{r2}}{W_{r1}} = \left(\frac{d_2}{d_1}\right)^{-\alpha_2}$$

$$\vdots$$

$$\frac{W_{rn}}{W_{rn-1}} = \left(\frac{d_n}{d_{n-1}}\right)^{-\alpha_n}$$

Then

$$\frac{W_{rn}}{W_{r0}} = \prod_{i=1}^{n}\left(\frac{d_i}{d_{i-1}}\right)^{-\alpha_i} \tag{3.61a}$$

Equation 3.61a can be rewritten as

$$\frac{W_{rn}}{W_{r0}} = \left(\frac{d_n}{d_0}\right)^{-\alpha_1} C_1 \tag{3.61b}$$

where

$$C_1 = d_1^{-\alpha_1} d_n^{\alpha_1} \prod_{i=2}^{n}\left(\frac{d_i}{d_{i-1}}\right)^{-\alpha_i}$$

The constant $C_1$ may be seen as a correction factor due to the various parameters affecting the radio propagation (foliage, street orientation, sloping terrain, and others). If the mobile remains within the same environment where the path loss is equal to $\alpha_1$, then $C_1 = 1$.

### 3.3.8 Considerations on Other Effects

Several other factors affect the radio propagation conditions. We shall comment on some of these factors.

#### 3.3.8.1 Atmospheric Conditions

The moisture in the atmosphere attenuates the radio signal, depending on the radio frequency. Above 10 GHz the loss due to rain is already considerable. There is a peak of absorption due to water vapor at 24 GHz and due to oxygen at 60 GHz.[2]

#### 3.3.8.2 Foliage

Trees work as obstructions diffracting, reflecting, and absorbing the radio signal. In urban areas, where there is a small concentration of trees, the effects are negligible. Estimates of the attenuation due to trees are rather complex because there is a considerable number of variables involved. Attenuation may vary with height, shape, density, season, humidity, etc. The loss is negligible at low frequencies (low VHF) but is substantial at higher frequencies (UHF). Moreover, the loss varies with the field polarization as shown by Equation 3.62a, for vertical polarization, and by Equation 3.62b, for horizontal polarization,[33]

$$L = 1637\sigma + \frac{\exp(-90/f)\log(1 + f/10)}{2.99} \quad \frac{\mathrm{dB}}{\mathrm{m}} \qquad (3.62a)$$

and

$$L = 1637\sigma + \frac{\exp(-210/f)\log(1 + f/200)}{2.34} \quad \frac{\mathrm{dB}}{\mathrm{m}} \qquad (3.62b)$$

where $\sigma$ is the conductivity (Siemens per meter) and $f$ is the frequency (in megahertz). Some values of $\sigma$ are given in Table 3.1.[33]

The tree tops may be considered as a knife edge at distances bigger than the mean tree height above the transmitting antenna. As far as radio coverage planning is concerned, a 10-dB tolerance is usually allowed when the service area contains trees varying from bare in the winter to full leaf in the summer.

TABLE 3.1. Conductivity of Various Types of Foliage (after Camwell[33]).

| Foliage | $\sigma(\times 10^{-5})$ | |
| --- | --- | --- |
| | Dry | Wet |
| Bare tree, branches | 0.5–1 | 2–10 |
| Deciduous, full leaf | 1 | 5–20 |
| Evergreen forest | 2–5 | 5–20 |
| Thin jungle scrub | 1–10 | 3–20 |
| Dense rain forest | 10–50 | 50 |

### 3.3.8.3 *Street Orientation*

The buildings lining streets work as "wave guides" affecting the propagation direction of the radio wave. Mobiles on streets running radially from the base station, or on the streets parallel to these, may receive a signal 10–20 dB higher than that received when they run on the perpendicular streets. This effect is more significant in the vicinity of the base station (up to 2 km away), becoming negligible at distances above 10 km.[2]

### 3.3.8.4 *Tunnel*

Microwave frequencies are substantially attenuated by the structure of tunnels. This attenuation can reach 20 dB or more, greatly affecting radio communication. On the other hand, tunnels may work as wave guides, channelling the radio signal. Reudink[18] carried out an investigation where he placed a transmitter at approximately 300 m inside a tunnel, taking measurements at a distance of 600 m inside the tunnel in a line-of-sight path. Some of his results are shown in Figure 3.18.

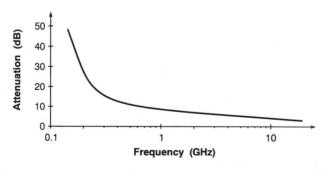

**Figure 3.18.** Attenuation in a tunnel. (*Source:* W. C. Jakes, *Microwave Mobile Communications*, John Wiley & Sons, New York, 1974.)

## 3.4  STATISTICAL DISTRIBUTIONS OF THE MOBILE RADIO SIGNAL

In the previous sections we described some of the parameters affecting mobile radio signals. From our limited survey of the subject it is not difficult to infer that, in fact, a countless number of factors may influence the radio signal. Accordingly, a single deterministic treatment of this signal will certainly be reducing the problem to an oversimplified model. Therefore, we may treat the signal on a statistical basis and interpret the results as random events occurring with a given probability.

Three distributions are closely related to the mobile radio statistics: lognormal, Rayleigh, and Ricean. The log normal distribution describes the envelope of the received signal shadowed by obstructions such as hills, buildings, and others. The Rayleigh distribution describes the envelope of the received signal resulting from multipath propagation. The Ricean distribution considers the envelope of the received signal with multipath propagation plus a line-of-sight component. We may also combine the lognormal and Rayleigh distributions to obtain an overall distribution, known as a Suzuki distribution.

In this section we obtain, describe, and analyze these probability distributions. The way of obtaining these statistics, by means of field measurements, is described in Chapter 4.

### 3.4.1  Lognormal Distribution

#### 3.4.1.1  *Wave Equations*

The wave equations for electromagnetic fields in a dispersive medium with electric permittivity $\varepsilon$, magnetic permeability $\mu$, and conductivity $\sigma$, are obtained from Maxwell's equations as

$$\nabla^2 \mathbf{e} - \mu\varepsilon \frac{\partial^2 \mathbf{e}}{\partial t^2} - \mu\sigma \frac{\partial \mathbf{e}}{\partial t} = 0 \tag{3.63a}$$

$$\nabla^2 \mathbf{h} - \mu\varepsilon \frac{\partial^2 \mathbf{h}}{\partial t^2} - \mu\sigma \frac{\partial \mathbf{h}}{\partial t} = 0 \tag{3.63b}$$

If the field has a harmonic time dependence, then the derivative $\partial/\partial t$ is replaced by* $j\omega$. Hence, for the electric field,

$$\nabla^2 \mathbf{E} + \omega^2 \mu\varepsilon \left(1 + \frac{\sigma}{j\omega\varepsilon}\right)\mathbf{E} = 0 \tag{3.64}$$

(for the magnetic field the results are analogous).

---

*For a harmonic time-dependent electric field, $\mathbf{e} \triangleq \mathbf{E} = \mathbf{E}_0 e^{j(\omega t + \theta)}$, and $\partial \mathbf{e}/\partial t = j\omega \mathbf{E}$.

It is not difficult to prove that the solution

$$\mathbf{E} = \mathbf{E}_0 e^{\pm \boldsymbol{\gamma} \cdot \mathbf{r}} \tag{3.65}$$

satisfies Equation 3.64. In Equation 3.65 $\mathbf{E}_0$ is the amplitude of the electric field in the propagation medium (free space), $\mathbf{r}$ is the direction of the propagation, and $\boldsymbol{\gamma}$ is the vector propagation constant. Substituting Equation 3.65 into Equation 3.64, we conclude that

$$\gamma = \sqrt{(\sigma + j\omega\varepsilon)j\omega\mu} = \alpha + j\beta \tag{3.66}$$

where

$$\alpha^2 = \omega^2 \frac{\mu\varepsilon}{2}\left(\sqrt{1 + \left(\frac{\sigma}{\omega\varepsilon}\right)^2} - 1\right)$$

$$\beta^2 = \omega^2 \frac{\mu\varepsilon}{2}\left(\sqrt{1 + \left(\frac{\sigma}{\omega\varepsilon}\right)^2} + 1\right)$$

The parameters $\alpha$ and $\beta$ are the attenuation constant (neper per meter or decibel per meter) and the phase constant (radian per meter), respectively.

If we consider that the field propagates at the direction of $\mathbf{r}$, then, from Equation 3.65, the magnitude of $\mathbf{E}$ is

$$E_M = E_0 \exp\left[-(\alpha + j\beta)r\right] \tag{3.67}$$

From Equation 3.67 the modulus of $E_M$ is

$$E = E_0 \exp(-\alpha r) \tag{3.68}$$

### 3.4.1.2 *Lognormal Distribution*

Consider a radio wave propagating in a mobile radio environment. When reaching the mobile station the radio wave will have travelled through different obstructions such as buildings, tunnels, hills, trees, etc. Each obstruction presents its own attenuation constant as well as thickness. Suppose that the $i$th obstruction has an attenuation constant $\alpha_i$ and thickness $\Delta r_i$. If the amplitude of the wave entering this obstruction is $E_{i-1}$, and $E_i$ is this amplitude after the obstruction, then

$$E_i = E_{i-1} \exp(-\alpha_i \Delta r_i) \tag{3.69}$$

Using recursivity, it follows that the signal leaving the $n$th obstruction is given by

$$E_n = E_0 \exp\left(-\sum_{i=1}^{n} \alpha_i \Delta r_i\right) \tag{3.70}$$

It is reasonable to admit that $\alpha_i$ and $\Delta r_i$ vary randomly from obstruction to obstruction. Define $x$ as

$$x \triangleq - \sum_{i=1}^{n} \alpha_i \Delta r_i \qquad (3.71)$$

Then

$$E_n = E_0 \exp(x) \qquad (3.72)$$

If the number of obstructions is large enough ($n \to \infty$ in Equation 3.71) we can use the central limit theorem* to state that the random variable $x$ has a normal distribution $p(x)$ such that

$$p(x) = \frac{1}{\sqrt{2\pi}\,\sigma_x} \exp\left[ -\frac{1}{2}\left(\frac{x - m_x}{\sigma_x}\right)^2 \right] \qquad (3.73)$$

where $m_x$ is the mean value of $x$ and $\sigma_x^2$ is its variance. Now let us find the distribution of the ratio $y$ of the fields $E_n/E_0$ as expressed in Equation 3.72

$$y = \frac{E_n}{E_0} = \exp(x)$$

Then, define $Y$ as

$$Y \triangleq \log y = x \log e \qquad (3.74a)$$

The mean and the variance of $Y$ are $M_Y$ and $\sigma_Y^2$, respectively, and are defined as

$$M_Y \triangleq \log m_Y = m_x \log e \qquad (3.74b)$$

$$\sigma_Y \triangleq \log \sigma_Y = \sigma_x \log e \qquad (3.74c)$$

In order to find the probability density $p(Y)$ of $Y$ we equate the areas under the densities $p(Y)$ and $p(x)$, such that

$$p(Y)|dY| = p(x)|dx| \qquad (3.75)$$

From Equation 3.74a

$$|dY| = \log e |dx| \qquad (3.76)$$

---

*"The probability distribution function of a sum of independent random variables approaches that of a Gaussian random variable as the number of independent random variables increases without limit".[20] In our case we are also assuming the obstructions to be independent.

Then, using Equation 3.76 in Equation 3.75 and taking $p(x)$ from Equation 3.73, we obtain

$$p(Y) = \frac{1}{\log e} p(x) = \frac{1}{\sqrt{2\pi} \, \sigma_x \log e} \exp\left[ -\frac{1}{2}\left( \frac{x - m_x}{\sigma_x} \right)^2 \right]$$

Therefore,

$$p(Y) = \frac{1}{\sqrt{2\pi} \, \sigma_Y} \exp\left[ -\frac{1}{2}\left( \frac{Y - M_Y}{\sigma_Y} \right)^2 \right] \qquad (3.77a)$$

It can be seen that $Y$ also has a normal distribution but, because $Y$, $M_Y$, and $\sigma_Y$ are given in a logarithmic form, this function is known as a *lognormal probability density function*. We then conclude that the excess path loss, i.e., the difference in decibels between the received signal and the free-space signal, $20\log(E_n/E_0)$, has a lognormal distribution. It is obvious that the received signal $R$, when measured in decibels, also has a lognormal density function (the proof follows the same steps) given by

$$p(R) = \frac{1}{\sqrt{2\pi} \, \sigma_R} \exp\left[ -\frac{1}{2}\left( \frac{R - M_R}{\sigma_R} \right)^2 \right] \qquad (3.77b)$$

where $M_R$ and $\sigma_R^2$ are, respectively, the mean and variance of $R$ given in decibels. Measurements[2] have shown that the standard deviation $\sigma_R$ is in the range of 4–10 dB. Note that the same reasoning may be used to obtain

$$p(y) = \frac{1}{\sqrt{2\pi} \, y\sigma_x} \exp\left[ -\frac{1}{2}\left( \frac{\ln y - m_x}{\sigma_x} \right)^2 \right] \qquad (3.78a)$$

In the same way, if $R = \log r$, then

$$p(r) = \frac{1}{\sqrt{2\pi} \, r\sigma_x} \exp\left[ -\frac{\ln^2(r/m_r)}{2\sigma_x^2} \right] \qquad (3.78b)$$

The cumulative distribution is

$$P(Y_0) = \text{prob}(Y \le Y_0) = \int_{-\infty}^{Y_0} p(Y) \, dY \qquad (3.79)$$

Both functions are sketched in Figure 3.19.

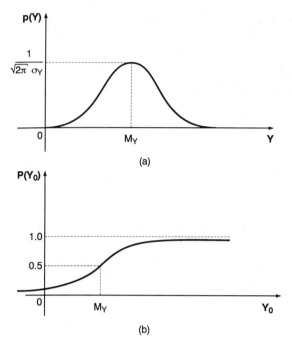

**Figure 3.19.** Long-term fading statistics: (a) the lognormal density function; (b) the lognormal cumulative distribution.

### 3.4.2 Rayleigh Distribution

The received signal at a mobile will rarely have a direct line-of-sight to a transmitter. It is the sum of the signals formed by the transmitted signal scattered by randomly placed obstructions imposing different attenuations and phases on the resultant signals. This is known as multipath propagation. It is plausible to suppose that the phases of the scattered waves are uniformly distributed from 0 to $2\pi$ rad and that amplitudes and phases are statistically independent from each other. Consequently, we may expect that at a certain instant the waves will be in phase, producing a large amplitude (constructive interference), whereas at another instant they will be out of phase, producing a small amplitude (destructive interference). In this section we shall determine the statistics of this fading signal.

Consider a carrier signal $s$ at a frequency $\omega_0$ and with an amplitude $a$ written in its exponential form

$$s = a \exp(j\omega_0 t) \tag{3.80}$$

(The actual signal is given by either the real part or the imaginary part of $s$.)

Let $a_i$ and $\theta_i$ be the amplitude and the phase of the $i$th scattered wave, respectively. The resultant signal $s_r$ at the mobile is the sum of the $n$ scattered waves:

$$s_r = \sum_{i=1}^{n} a_i \exp\left[ j(\omega_0 t + \theta_i) \right] \tag{3.81a}$$

Equivalently,

$$s_r = r \exp\left[ j(\omega_0 t + \theta) \right] \tag{3.81b}$$

where

$$r \exp(j\theta) = \sum_{i=1}^{n} a_i \exp(j\theta_i)$$

However,

$$r \exp(j\theta) = \sum_{i=1}^{n} a_i \cos\theta_i + j \sum_{i=1}^{n} a_i \sin\theta_i \triangleq x + jy \tag{3.82a}$$

Then

$$x \triangleq \sum_{i=1}^{n} a_i \cos\theta_i \quad \text{and} \quad y \triangleq \sum_{i=1}^{n} a_i \sin\theta_i$$

where

$$r^2 = x^2 + y^2 \tag{3.82b}$$

$$x = r \cos\theta \tag{3.82c}$$

$$y = r \sin\theta \tag{3.82d}$$

Because (1) $n$ is usually very large, (2) the individual amplitudes $a_i$ are random, and (3) the phases $\theta_i$ have a uniform distribution, it can be assumed that (calling on the central limit theorem again) $x$ and $y$ are both Gaussian variates with means equal to zero and variances $\sigma_x^2 = \sigma_y^2 \triangleq \sigma_r^2$. Consequently, their distributions are

$$p(z) = \frac{1}{\sqrt{2\pi}\,\sigma_z} \exp\left( -\frac{z^2}{2\sigma_z^2} \right) \tag{3.83}$$

where $z = x$ or $z = y$ as required.

It is shown in Section 4.5 that $x$ and $y$ are independent random variables, despite being Gaussian with the same standard deviation. Then the joint distribution $p(x, y)$ is

$$p(x, y) = p(x)p(y) = \frac{1}{2\pi\sigma_r^2} \exp\left(-\frac{x^2 + y^2}{2\sigma_r^2}\right) \tag{3.84}$$

The distribution $p(r, \theta)$ can be written as a function of $p(x, y)$ as follows:

$$p(r, \theta) = |J|p(x, y) \tag{3.85a}$$

where

$$J \triangleq \begin{vmatrix} \partial x/\partial r & \partial x/\partial \theta \\ \partial y/\partial r & \partial y/\partial \theta \end{vmatrix} \tag{3.85b}$$

is the Jacobian of the transformation of the random variables $x, y$ into $r, \theta$. Using Equation 3.82c and 3.82d in Equation 3.85b, we obtain $J = r$. Therefore, with Equation 3.84 in Equation 3.85a we have

$$p(r, \theta) = \frac{r}{2\pi\sigma_r} \exp\left(-\frac{r^2}{2\sigma_r^2}\right) \tag{3.86}$$

The density $p(r)$ is obtained by averaging $p(r, \theta)$ over the range of variation of $\theta$. Hence

$$p(r) = \int_0^{2\pi} p(r, \theta) \, d\theta$$

$$= \begin{cases} \dfrac{r}{\sigma_r^2} \exp\left(-\dfrac{r^2}{2\sigma_r^2}\right) & r \geq 0 \\ 0 & \text{otherwise} \end{cases} \tag{3.87}$$

Equation 3.87 is the Rayleigh probability density function. Its distribution $P(r_0)$ is

$$P(r_0) = \text{prob}(r \leq r_0) = \int_0^{r_0} p(r) \, dr = 1 - \exp\left(-\frac{r_0^2}{2\sigma_r^2}\right) \tag{3.88}$$

Both Functions 3.87 and 3.88 are shown in Figure 3.20.
Some important points of this distribution are as follows:

- Mean value $= E[r] = \int_0^\infty rp(r) \, dr = \sqrt{\pi/2}\,\sigma_r$.
- Most likely value $= \max\{p(r)\} = \sigma_r$.
- Second moment (mean squared value) $= E[r^2] = \int_0^\infty r^2 p(r) \, dr = 2\sigma_r^2$; then its rms value is $\sqrt{2}\,\sigma_r$.
- Variance $= E[r^2] - E^2[r] = (2 - \pi/2)\sigma_r^2$.
- Median, defined as the value $r_0$ obtained when $\int_{r_0}^\infty p(r) \, dr = 0.5$. Then, $r_0 = 1.18\sigma_r$.

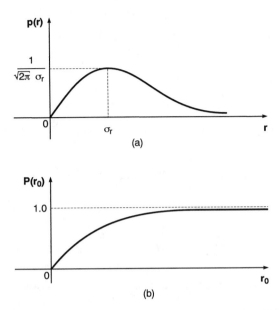

**Figure 3.20.** Short-term fading statistics: (a) Rayleigh density function; (b) Rayleigh distribution function.

### 3.4.3 Ricean Distribution

The Rayleigh fading model holds only in the case where there is a large number of indirect paths and they greatly predominate over the direct path. However, in some circumstances, when there is a line-of-sight propagation, the direct path predominates over the indirect ones. This may happen, for instance, within a building: Most buildings have a reinforced central core, curtain walls and ceilings and floors containing a large amount of metal. It is plausible to expect that, if a transmitting antenna is placed on a floor, some ducting of waves will occur, added to multipath scattering of the waves. Consequently, the received signal is a sum of the scattered and direct signals. We want to investigate how the statistics of the received envelope will vary according to the proportion of direct waves to scattered waves.

Using Equations 3.81b and 3.80, the received signal $s_r$ is

$$s_r = \overbrace{r \exp(j\omega_0 t + \theta)}^{\text{scattered waves}} + \overbrace{a \exp(j\omega_0 t)}^{\text{direct waves}} \qquad (3.89a)$$

or, equivalently,

$$s_r = [(x + a) + jy]\exp(j\omega_0 t) \qquad (3.89b)$$

Note that, in this case,

$$r^2 = (x + a)^2 + y^2 \qquad (3.89c)$$

$$x + a = r \cos \theta \qquad (3.89d)$$

$$y = r \sin \theta \qquad (3.89e)$$

By following the same steps described in Section 3.4.2 we obtain

$$p(r) = \frac{r}{\sigma_r^2} \exp\left(-\frac{r^2 + a^2}{2\sigma_r^2}\right) I_0\left(\frac{ar}{\sigma_r^2}\right) \qquad (3.90a)$$

where

$$I_0\left(\frac{ar}{\sigma_r^2}\right) = \frac{1}{2\pi} \int_0^{2\pi} \exp\left(\frac{ar \cos \theta}{\sigma_r^2}\right) d\theta \qquad (3.90b)$$

is the modified zeroth-order Bessel function. This function can be found in tabulated form (e.g., see Abramowitz and Stegun[31]) or can be evaluated numerically by[31]

$$I_0(x) = \sum_{i=0}^{\infty} \left(\frac{x^i}{i!2^i}\right) \qquad (3.90c)$$

Equation 3.90a corresponds to the Ricean distribution. Note that if $a = 0$, we obtain the Rayleigh distribution. If the ratio $a/\sigma$ is large enough, the in-phase component $(x + a)$ will predominate over the quadrature component $y$ of the signal. Therefore, the distribution $p(r)$, of $r = x + a$, is approximately equal to that of $p(x)$, but with a mean value equal to $a$ (see Figure 3.21).

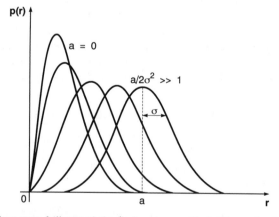

**Figure 3.21.**   Short-term fading statistics (indoor propagation): Ricean density function.

### 3.4.4  Suzuki Distribution

As we have seen, the long-term fading signal has a lognormal distribution, whereas the short-term fading signal has a Rayleigh distribution.* It is reasonable to expect the overall distribution of the received signal to be a mixture of Rayleigh and lognormal. In this section we obtain such a distribution.

Let $R$ be the local mean value of the received signal $r$, and let $M_R$ be its area mean value. The probability density function of $R$ is lognormal as shown in Equation 3.77b. The density of $r$ conditional on the mean value $R$ is the Rayleigh probability density function, i.e.,

$$p(r|R) = \frac{r}{\sigma_r^2} \exp\left(-\frac{r^2}{2\sigma_r^2}\right) \tag{3.91}$$

The mean of this density, as shown in Section 3.4.2, is $E[r|R] = \sqrt{\pi/2}\,\sigma_r$. Such a mean equals $R$ when expressed in logarithmic form. Therefore

$$20\log(E[r|R]) = 20\log\left(\sqrt{\pi/2}\,\sigma_r\right) = R \tag{3.92}$$

or, equivalently,

$$\sigma_r = \sqrt{2/\pi}\,10^{R/20} \tag{3.93}$$

Using Equation 3.93 in Equation 3.91 we obtain

$$p(r|R) = \frac{\pi r}{2 \times 10^{R/10}} \exp\left(-\frac{\pi r^2}{4 \times 10^{R/10}}\right) \tag{3.94}$$

The unconditional probability $p(r)$ is obtained by averaging the conditional probability $p(r|R)$ over all possible values of $R$. Hence

$$p(r) = \int_{-\infty}^{\infty} p(r|R)p(R)\,dR$$

Finally, using Equations 3.94 and 3.77b in the preceding equation, we obtain

$$p(r) = \sqrt{\frac{\pi}{8\sigma_R^2}} \int_{-\infty}^{\infty} \frac{r}{10^{R/10}} \exp\left(-\frac{\pi r^2}{4 \times 10^{R/10}}\right) \exp\left[-\frac{1}{2}\left(\frac{R - M_R}{\sigma_R}\right)^2\right] dR \tag{3.95}$$

Equation 3.95 is known as the Suzuki distribution.[33]

---

* We are assuming outdoor propagation. The short-term fading signal has a Ricean distribution for indoor propagation.

## 3.5   SIGNAL COVERAGE AREA (CELL AREA)

Let $w_l$ be the mean power received by the mobile at a distance $l$ from the base station. Let $w_L$ be the mean power received by the mobile positioned at the cell's border (radius $L$). If at $l$ the environment is not exactly the same as that at $L$, then, from Equation 3.61b, the ratio between these two powers is

$$\frac{w_l}{w_L} = \left(\frac{l}{L}\right)^{-\alpha} c \tag{3.96}$$

where $\alpha$ is the path loss slope at $l$ and $c$ is a correction factor, due to the change of environment. Define $m_w$ as the mean signal power such that

$$m_w \triangleq k(l/L)^{-\alpha} \tag{3.97a}$$

where $k = cw_L$. Expressed in decibels,

$$M_w \triangleq 10 \log m_w = K - 10\alpha \log(l/L) \tag{3.97b}$$

where $K = 10 \log k$.

The problem of estimating the cell area (or, equivalently, the signal coverage area) can be approached in two different ways. In the first approach we may wish to determine the proportion of the locations at $L$ (circumference defined by $L$) where the received signal power $w$ is above a power threshold $w_0$. In the second approach we may determine the proportion of the circular area defined by $L$ where the received signal is above $w_0$. Note that in the first case this proportion is averaged over the perimeter of the circumference, whereas in the second case the average is over the entire circular area.

In both cases it is assumed that the probability density $p(w)$ of $w$ and the mean signal strength $m_w$ at a given distance $l$ are known. The mean signal strength can be determined by one of the prediction models or by field measurements. As for the distribution of $w$, any of the statistics described in Section 3.4 can be used, as appropriate. In the analyses that follow we shall consider separately the fading signal to follow (1) a lognormal or (2) a Rayleigh distribution.

**Proportion of Locations at Distance L.** The proportion $\beta$ of locations at this distance where a mobile would experience a received signal above the threshold $w_0$ (or, equivalently, the probability $\beta$ that the received signal at $L$ is greater than $w_0$) is

$$\beta \triangleq \text{prob}(w \geq w_0) = \int_{w_0}^{\infty} p(w)\, dw \tag{3.98a}$$

or

$$\beta \triangleq \text{prob}(W \geq W_0) = \int_{W_0}^{\infty} p(W) \, dW \qquad (3.98b)$$

where $W = 10 \log w$ and $W_0 = 10 \log w_0$.

*Proportion of Locations Within the Area Defined by L.* If we assume that the mobiles are uniformly distributed within the cell area, then the proportion of locations within the circle defined by $L$, receiving a signal above the threshold $w_0$, also corresponds to the proportion of mobiles receiving such signal. Equivalently, this proportion gives the probability that a mobile within the defined area receives a signal above the threshold $w_0$. Let $\mu$ be such proportion and let $\text{prob}(w \geq w_0)$ be the probability that $w$ exceeds $w_0$ within an infinitesimal area $dA$. Therefore, the wanted proportion $\mu$ is the probability $\text{prob}(w \geq w_0)$ averaged over the entire circular area. Thus

$$\mu = \frac{1}{A} \int_A \text{prob}(w \geq w_0) \, dA \qquad (3.99)$$

where $A = \pi L^2$ and $dA = l \, dl \, d\theta$. Therefore

$$\mu = \frac{1}{\pi L^2} \int_0^L \int_0^{2\pi} \text{prob}(w \geq w_0) l \, dl \, d\theta$$

$$= \frac{2}{L^2} \int_0^L \text{prob}(w \geq w_0) l \, dl \qquad (3.100a)$$

If $W$ and $W_0$ are expressed in decibels,

$$\mu = \frac{2}{L^2} \int_0^L \text{prob}(W \geq W_0) l \, dl \qquad (3.100b)$$

### 3.5.1 Lognormal Fading Case

Consider a pure lognormal fading environment. The envelope $R$ of the received signal follows a normal distribution such as that given by Equation 3.77b. Because $R$ is given in decibels, then $W = R + R_0$, where $R_0$ is a constant, and the density $p(W)$ of $W$ is

$$p(W) = \frac{1}{\sqrt{2\pi}\, \sigma_w} \exp\left[ -\frac{1}{2} \left( \frac{W - M_w}{\sigma_w} \right)^2 \right] \qquad (3.101)$$

With Equation 3.101 in Equation 3.98b we obtain

$$\beta = \text{prob}(W \geq W_0) = \frac{1}{2}\left[1 - \text{erf}\left(\frac{W_0 - M_w}{\sqrt{2}\,\sigma_w}\right)\right] \qquad (3.102)$$

where erf$(\cdot)$ is the error function.

Using Equations 3.97b and 3.102 in Equation 3.100b yields

$$\mu = \frac{2}{L^2}\int_0^L \int_{u_0}^\infty \frac{1}{\sqrt{\pi}}\exp(-u^2)\,du\,l\,dl \qquad (3.103)$$

where

$$u = \frac{W - K + 10\alpha\log(l/L)}{\sqrt{2}\,\sigma_w}$$

and

$$u_0 = \frac{W_0 - K + 10\alpha\log(l/L)}{\sqrt{2}\,\sigma_w} \qquad (3.104)$$

We may express Equation 3.103 in terms of the error function erf$(u_0)$,

$$\text{erf}(u_0) \triangleq 2\int_0^{u_0}\frac{1}{\sqrt{\pi}}\exp(-v^2)\,dv$$

such that

$$\mu = \frac{2}{L^2}\int_0^L\left[\frac{1}{2} - \frac{1}{2}\text{erf}(u_0)\right]l\,dl = \frac{1}{2} - \frac{1}{L^2}\int_0^L \text{erf}(u_0)l\,dl \quad (3.105)$$

From Equation 3.104,

$$l\,dl = \frac{\sqrt{2}\,\sigma_w}{10\alpha\log e}l^2\,du_0 \qquad (3.106)$$

With Equations 3.104 and 3.106 in Equation 3.105 we get

$$\mu = \frac{1}{2} - \frac{\sqrt{2}\,\sigma_w}{10\alpha\log e}\exp\left(-2\frac{W_0 - K}{10\alpha\log e}\right)$$

$$\times \int_{-\infty}^{(W_0-K)/(\sqrt{2}\sigma_w)}\exp\left(\frac{2\sqrt{2}\,u_0}{10\alpha\log e}\right)\text{erf}(u_0)\,du_0 \qquad (3.107)$$

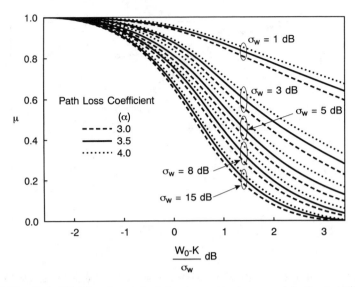

**Figure 3.22.** Probability that the received signal power is above a certain threshold (long-term fading).

This integral is then rearranged to be found in integral tables. Finally,

$$\mu = \frac{1}{2}\left\{1 + \mathrm{erf}\left(\frac{K - W_0}{\sqrt{2}\,\sigma_w}\right) + \exp\left[\frac{2(K - W_0)10\alpha \log e + 2\sigma_w^2}{100\alpha^2 \log^2 e}\right]\right.$$

$$\left. \times \left[1 - \mathrm{erf}\left(\frac{(K - W_0)10\alpha \log e + 2\sigma_w^2}{(10\alpha \log e)\sqrt{2}\,\sigma_w}\right)\right]\right\} \quad (3.108)$$

The proportion $\mu$ given by Equation 3.108 is plotted versus $(W_0 - K)/\sigma_w$ in Figure 3.22 for some values of $\alpha$ and $\sigma_w$.

### 3.5.2 Rayleigh Fading Case

In a pure Rayleigh fading environment the envelope $r$ of the received signal has a probability density $p(r)$ as given by Equation 3.87. Accordingly, the density $p(w)$ of $w = r^2/2$ is such that

$$p(w)|dw| = p(r)|dr|$$

Because $dw = r\,dr$, then

$$p(w) = \frac{1}{m_w}\exp\left(-\frac{w}{m_w}\right) \quad (3.109)$$

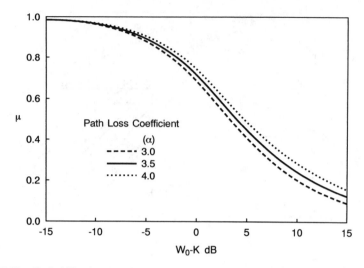

**Figure 3.23.** Probability that the received signal power is above certain threshold (short-term fading).

Note that both the mean and the standard deviation of $w$ are equal to $m_w$. With Equation 3.109 in Equation 3.98a we obtain

$$\beta = \text{prob}(w \geq w_0) = \exp\left(-\frac{w_0}{m_w}\right) \tag{3.110}$$

Now, using Equations 3.97a and 3.110 in Equation 3.100a, we obtain

$$\mu = \frac{2}{\alpha}\left(\frac{k}{w_0}\right)^{2/\alpha}\Gamma\left(\frac{2}{\alpha}, \frac{w_0}{k}\right) \tag{3.111}$$

where

$$\Gamma(x, y) = \int_0^y t^{x-1}\exp(-t)\,dt \tag{3.112}$$

is the incomplete gamma function. The proportion $\mu$ given by Equation 3.111 is plotted versus $W_0 - K$ in Figure 3.23 for some values of $\alpha$.

### 3.5.3  Some Examples

(1) Let the mean signal strength at $l = L$ be $M_w = K = -100$ dBm* in an environment where $\alpha = 3.5$. We want to estimate the probability that the

---

*If $P$ is a power given in watts and $p$ is the same power expressed in milliwatts, then $10\log p = 10\log(P/10^3)$ is given in dBm.

received signal exceeds a threshold $W_0 = -105$ dBm (a) within the circular area delimited by $L$ and (b) at the perimeter of the corresponding circumference.

- Lognormal fading case—Assume that $\sigma_w = 5$ dB. Hence $(W_0 - K)/\sigma_w = (W_0 - M_w)/\sigma_w = -1$. (a) From Figure 3.22, $\mu = 96\%$. (b) From Equation 3.102, $\beta = 84.13\%$.
- Rayleigh fading case—Because $W_0 - K = -5$ dBm, (a) from Figure 3.23, $\mu = 90\%$; (b) from Equation 3.110, $\beta = 73\%$.

(2) Let the mean signal strength at $l = 10$ km be $M_w = -100$ dBm in an environment where $\alpha = 3.5$. We want to estimate the cell radius $L$ such that the mobile stations receive a signal power above $W_0 = -110$ dBm 90% of the time (a) within the circular area delimited by $L$ and (b) at the perimeter of the corresponding circumference.

- Lognormal fading case—Assume that $\sigma_w = 5$ dB. (a) From Figure 3.22 with $\mu = 90\%$ we find $(W_0 - K)/\sigma_w = -0.47$, yielding $K = -107.65$ dBm. Because $M_w - K = -10\alpha \log(l/L)$, then $L = 16.5$ km. (b) From Equation 3.102 with $\beta = 90\%$ we find $(W_0 - K)/\sigma_w = -1.28$, yielding $K = -103.6$ dBm. Then $L = 12.7$ km.
- Rayleigh fading case—(a) From Figure 3.23 with $\mu = 90\%$ we find $W_0 - K = -5.2$ dBm, yielding $K = -104.8$ dBm. Then $L = 13.7$ km. (b) From Equation 3.110 with $\beta = 90\%$ we find $W_0 - K \approx -100$ dBm. Then $L = 10$ km.

## 3.6   BOUNDARIES BETWEEN CELLS

The cells of the mobile radio systems are not clearly defined but have fuzzy boundaries because of the statistical fluctuation in radio path losses. If the mobiles are near the cell border, they may well have adequate communication with more than one base station. The proportion of the cell area within which mobiles are considered to have more than one radio path depends on the fading distribution, on the permissible tolerance in path loss for satisfactory communication, and on the geographical distribution of cells and mobiles. This section examines the joint statistics of paths to two and three base stations and estimates such proportion. We impose that the path losses shall differ by not more than $A$ dB, while neither is more than $C$ dB below the long-term mean of these paths. That is, by choice of parameters it may be stipulated that the paths are sufficiently similar and also neither of them is in deep fade.

### 3.6.1   Joint Rayleigh Fading

Consider two Rayleigh-fading signals with instantaneous amplitude $r_i$, mean amplitude $\sqrt{\pi/2}\, m_i$, and probability density

$$p(r_i) = \frac{r_i}{m_i^2} \exp\left[\frac{-r_i^2}{2m_i^2}\right] \qquad i = 1, 2 \qquad (3.113)$$

If the fadings are independent, the joint density is simply

$$p(r_1, r_2) = p(r_1)p(r_2) \qquad (3.114)$$

An important parameter of the joint distribution is the difference in mean levels, denoted by

$$b = \frac{m_2}{m_1}$$

$$B = 20 \log b$$

It is intended to estimate the probability $P(a, b)$ that the two signals differ in instantaneous level by not more than $A$ dB, where

$$A = 20 \log a$$

It is perhaps desirable to require that neither of the two signals be below a threshold level $c$. The required probability can be determined by integrating Equation 3.114 over the region shown in Figure 3.24. The parameter $c$ is

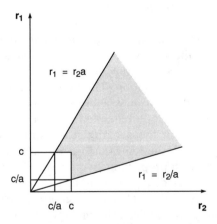

**Figure 3.24.**   Regions of integrations for the joint Rayleigh distribution.

defined by

$$C = -10 \log\left(\frac{c^2}{m_1 m_2}\right)$$

By successive applications of the integral

$$\int r \exp(kr^2)\, dr = -\frac{\exp(-kr^2)}{2k}$$

it can be shown that

$$P(a,b,c) = \frac{\exp\left[-\dfrac{c^2}{2m_2^2}\left(1+\dfrac{b^2}{a^2}\right)\right]}{1+\dfrac{b^2}{a^2}} + \frac{\exp\left[-\dfrac{c^2}{2m_1^2}\left(1+\dfrac{1}{a^2 b^2}\right)\right]}{1+\dfrac{1}{a^2 b^2}}$$

$$- \exp\left[-\frac{c^2}{2}\left(\frac{1}{m_1^2}+\frac{1}{m_2^2}\right)\right] \qquad (3.115a)$$

By expansion of the exponential function in $P(a,b,c)$ it may be shown that

$$P(a,b,c) - P(a,b,0) = \frac{c^4}{4m_1^2 m_2^2}\left(1-\frac{1}{a^2}\right) + O(c^6)$$

which is small for likely values of $c$. Moreover, exact calculation shows that the term in $c^4$ exceeds the true probability. For example, with $A = C = 6$ dB the approximation gives 0.0118, while the exact expression gives 0.0103 for $B = 0$, falling to 0.0089 for $B = 12$ dB. The overall effect on the results is small, and the simpler expression $P(a,b,0)$ gives a good guide. Then

$$P(a,b,0) = \frac{a^2 - 1/a^2}{(a^2 + 1/a^2) + (b^2 + 1/b^2)} \qquad (3.115b)$$

## 3.6.2 Joint Lognormal Fading

With this distribution it is simpler to use the logarithmic variables, each of which having a Gaussian (normal) distribution. Consider two signals with instantaneous levels $R_i$ and mean levels $M_i$, expressed in decibels. Their probability densities are

$$p(R_i) = \varphi\left(\frac{R_i - M_i}{\sigma_i}\right) \tag{3.116}$$

where the notation $\varphi(\cdot)$ signifies the Gaussian distribution

$$\varphi(u) = \frac{1}{(2\pi)^{1/2}} \exp\left(-\frac{u^2}{2}\right) \tag{3.117}$$

$$\phi(u) = \int_{-\infty}^{u} \varphi(x)\, dx \tag{3.118}$$

The main concern is the ratio of the instantaneous amplitudes, which, when expressed in logarithmic measure, is the difference

$$R = R_2 - R_1$$

Similarly, the parameter $B$ is the difference in mean levels

$$B = M_2 - M_1$$

Now, the difference between two Gaussian random variables is another Gaussian random variable. In particular, $R$ has the density $\varphi[(R - M)/\sigma]$, where

$$M = M_2 - M_1 = B$$

and

$$\sigma^2 = \sigma_1^2 + \sigma_2^2 - 2\rho_R\sigma_1\sigma_2$$

In this equation $\rho_R$ is the correlation coefficient of the two variables. If these variables are independent, $\rho_R = 0$. This is the case explored in this section.

In estimating the probability that the two signals have a level difference not exceeding $A$ dB, it is not necessary to distinguish between the cases $R_1 > R_2$ and $R_2 > R_1$. Thus the distribution of $A$ is effectively that of the modulus $|R|$. Its density is

$$p(A) = \varphi\left(\frac{A - B}{\sigma}\right) + \varphi\left(\frac{A + B}{\sigma}\right) \tag{3.119}$$

and the cumulative distribution is

$$P(A) = \phi\left(\frac{B + A}{\sigma}\right) - \phi\left(\frac{B - A}{\sigma}\right) \tag{3.120}$$

where always $A \geq 0$.

### 3.6.3 The Geographical Distribution of the Mean Power Ratio

Suppose that the mean signal power diminishes with distance $x$ as $x^{-\alpha}$, where $\alpha$ is the path loss slope. Consider a mobile at distances $x_1$ and $x_2$ from two base stations of equal power. Recalling from Section 3.3.7 (Equation 3.60a the ratio $m_2^2/m_1^2$ of the received powers is

$$b^2 = \left(\frac{m_2}{m_1}\right)^2 = \left(\frac{x_1}{x_2}\right)^\alpha \qquad (3.121a)$$

The locus of points with a given $b$ is, therefore, that of points with a given distance ratio

$$h = \frac{x_1}{x_2} = b^{2/\alpha} \qquad (3.121b)$$

The locus may be found by simple algebra. Expressing Equation 3.121b in Cartesian coordinates with origin midway between the base stations (Figure 3.25) and rearranging as a quadratic equation we obtain

$$y^2 + \left(x - \frac{h^2 + 1}{h^2 - 1}\right)^2 = \left(\frac{2h}{h^2 - 1}\right)^2 \qquad (3.122)$$

Equation 3.122 defines a circle, as sketched in Figure 3.25.

Now it is intended to estimate the proportion of cell area where the difference in mean path loss to seven base stations, taken two at a time, is within some assigned tolerance $B$ dB. Consider two adjacent base stations as in Figure 3.26a. By symmetry, it is necessary to consider only a triangular sector (triangle OAB as shown in Figure 3.26b) comprising $\frac{1}{12}$ of a cell. For the given tolerance, the power ratio $b$ can be calculated and thence the distance ratio $h$. This defines a circular locus with center C, of which DE is an arc. Within the area OBED, the ratio of the mean powers will be within

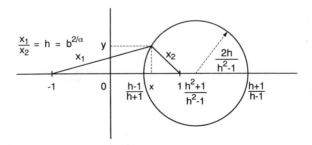

**Figure 3.25.** Geometry of loci of constant $b$.

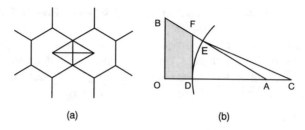

**Figure 3.26.** Hexagonal array geometry: (a) cell array; (b) sector of a cell.

the tolerance. Averaged over the whole array, the proportion of area with such power ratio will be

$$\gamma = \frac{\text{area OBED}}{\text{area OBA}} \tag{3.123}$$

Clearly Equation 3.123 is a function of the tolerance $B$. With the scale used here OBA = $1/2\sqrt{3}$. By simple geometry we have

$$\gamma = 1 + \frac{2\sqrt{3}\,Y_E}{h^2 - 1} - 4\sqrt{3}\left(\frac{h}{h^2 - 1}\right)^2 \sin^{-1}\left(\frac{Y_E(h^2 - 1)}{2h}\right) \tag{3.124a}$$

where

$$2\sqrt{3}\,Y_E = -3\left(\frac{1}{h^2 - 1}\right) + \sqrt{\left(\frac{2h^2 + 1}{h^2 + 1}\right)^2 - 4} \tag{3.124b}$$

As a check and convenient approximation, it can be seen that the area OBED is approximated from below by OBFD, where FD is a tangent to the circle. Consequently,

$$\gamma > \gamma_u = 1 - \left(\frac{2}{h + 1}\right)^2$$

In the same way, the area OBED can be approximated from above if the arc DE is approximated by a straight line. Then

$$\gamma < \gamma_0 = 1 - \frac{2\sqrt{3}\,Y_E}{h + 1}$$

The parameters $\gamma$, $\gamma_0$, and $\gamma_u$ are plotted in Section 3.6.4 (Figure 3.28, where $\alpha$ is assumed to be equal to 3.5). If mobiles are considered to be evenly distributed over the cell, on the average, then these areas give directly the proportion of mobiles with access to two paths whose mean losses differ by only a small amount. It is not difficult to make similar calculations for uneven distribution of mobiles, for example, a gradient in density causing adjacent cells to have traffic imbalance. Unless very extreme, this leads to minor changes in the proportions.

### 3.6.4 The Geographical Distribution of Instantaneous Power Ratio

The occurrence of fading will modify the distribution of the signal strength ratio, as compared with the simple distribution of Section 3.6.3. Not all mobiles within the border zone will have paths within the assigned tolerance, because one or both path losses may depart from the mean value. On the other hand, some mobiles outside the border zone will have such pairs of paths. It is possible to estimate the overall proportion of mobiles with satisfactory path-pairs on an instantaneous rather than a mean criterion, by combining some of the methods of Section 3.6.3 with those of Sections 3.6.2 and 3.6.1.

Refer to Figure 3.25, where $x$ is a position variable. Let the density of mobiles in the vicinity of $x$ be $d(x)$. In a hexagonal cell array $d(x) = 2(1 - x)$ gives a good approximation to the distribution of an uniform cell. This density would be obtained by using OBFD as an approximation to the border zone OBED (as has been seen, it gives a small underestimate of the proportion of mobiles in the border zone). The mean signal strength ratio $b$ is given in Equation 3.121b, and, conditional on the value of $b$, the probability $P(a, b)$ that the path losses differ by not more than $A$ dB can be calculated using Equation 3.115b or 3.120, as desired. It is noteworthy that both joint probability functions (Equations 3.115b and 3.120) give approximately the same results if we use $\sigma = 5$ dB in Equation 3.120. Then the probability $P(a, x)^*$ can be obtained conditional on the variable $x$. This is the probability that a mobile, having a location described by the position variable $x$, has a path-pair within the tolerance $A$ dB. The unconditional probability

$$\gamma = \int_0^1 d(x) P(a, x) \, dx \tag{3.125}$$

is the mean proportion of mobiles within a cell having a path-pair within

---

*Note that $b^2 = (x_1/x_2)^\alpha \simeq [(1 + x)/(1 - x)]^\alpha$, if we consider that geometry for underestimated approximation of $\gamma$.

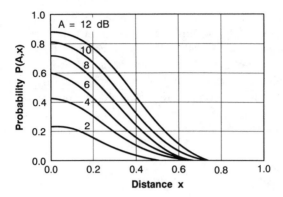

**Figure 3.27.**  Probability of instantaneous path loss difference being $\leq A$ dB, conditional on position.

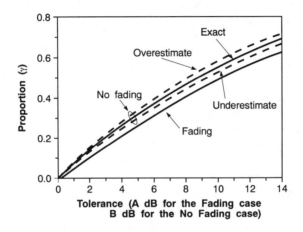

**Figure 3.28.**  Proportion of mobiles with path loss difference $\leq A$ dB.

tolerance, allowing for fading. The conditional probability $P(a, x)$ is shown as a function of $x$ in Figure 3.27.

The integral of Equation 3.125 does not appear to lend itself to a closed-form evaluation, but it has been evaluated numerically and is shown in Figure 3.28.

### 3.6.5  Proportion of the Cell Area With Three Alternative Paths

It is intended in the following sections to extend this model in order to consider the joint statistics to three base stations and then to estimate the proportion of the cell area with three alternative radio paths. It may be

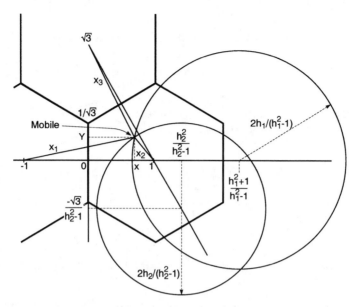

**Figure 3.29.**  Distances of a mobile to three base stations.

possible to find an approximate function relating the proportion of the cell area with three paths to that with two paths. The theory has been worked out for (1) no fading and (2) lognormal fading with standard deviation $\sigma = 5$ dB.

### 3.6.6   The Geographical Distribution of Mean Power Ratio

Consider a mobile at distances $x_1$, $x_2$, and $x_3$ from three base stations of equal power, as shown in Figure 3.29. The ratios of received powers are

$$b_1^2 = \left(\frac{m_2}{m_1}\right)^2 = \left(\frac{x_1}{x_2}\right)^{\alpha} \tag{3.126a}$$

$$b_2^2 = \left(\frac{m_2}{m_3}\right)^2 = \left(\frac{x_3}{x_2}\right)^{\alpha} \tag{3.126b}$$

The locus of points with a given $b_1$ is, therefore, that of points with the distance ratio

$$h_1 = \frac{x_1}{x_2} = b_1^{2/\alpha} \tag{3.127a}$$

Similarly,

$$h_2 = \frac{x_3}{x_2} = b_2^{2/\alpha} \tag{3.127b}$$

Expressing the distance in Cartesian coordinates and rearranging the expressions in a quadratic form we obtain

$$y^2 + \left( x - \frac{h_1^2 + 1}{h_1^2 - 1} \right)^2 = \left( \frac{2h_1^2}{h_1^2 - 1} \right)^2 \tag{3.128a}$$

$$\left( y + \frac{\sqrt{3}}{h_2^2 - 1} \right)^2 + \left( x - \frac{h_2^2}{h_2^2 - 1} \right)^2 = \left( \frac{2h_2^2}{h_2^2 - 1} \right)^2 \tag{3.128b}$$

which define the circles as plotted in Figure 3.29.

Consider three adjacent base stations in a hexagonal array of cells. It is desired to estimate the proportion of cell area where the difference in mean path losses to three adjacent stations is within some assigned tolerance $B = 20 \log b$. By symmetry it is necessary to consider only a triangular sector OBA as shown in Figure 3.30. For the given tolerance, the power ratio $b$ and hence the distance ratio $h$ can be calculated. This defines two circular loci with centers at C and C', respectively. Within the area OBEG the ratios of mean powers will be within tolerance. Averaged over the whole cell, the

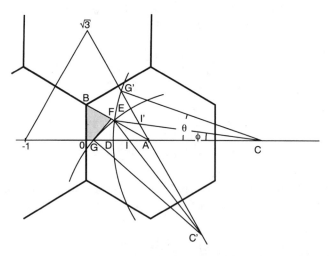

**Figure 3.30.** Area where the mean path losses are $\leq B$ dB (shaded area) ($h \geq \sqrt{3}$).

proportion of area with such a power ratio is

$$\delta = \frac{\text{area OBEG}}{\text{area OBA}}$$

With simple geometry (but not straightforward expressions)

$$\delta = 1 + \frac{2\sqrt{3}\,Y_G'}{h^2 - 1} - \sqrt{3}\,CD^2 \left[ \sin^{-1}\left(\frac{Y_G'}{CD}\right) - \sin^{-1}\left(\frac{Y_E}{CD}\right) \right]$$

$$- \frac{4\tan(\phi)\left[\sqrt{3}\,Y_E(h^2 - 1) + 3\right]}{(h^2 - 1)^2(\sqrt{3} - \tan(\phi))} \tag{3.129a}$$

where $Y_G'$ and $Y_E$ are the ordinates of the points G′ and E, respectively, and

$$Y_E = \frac{-\sqrt{3} + \sqrt{4h^2 - 1}}{2(h^2 - 1)} \tag{3.129b}$$

$$Y_G' = \frac{\sqrt{3}\left(-1 + \sqrt{4h^2 - 3}\right)}{2(h^2 - 1)} \tag{3.129c}$$

$$CD = \frac{2h}{h^2 - 1} \tag{3.129d}$$

$$\tan(\phi) = \frac{Y_E}{\sqrt{CD^2 - Y_E^2}} \tag{3.129e}$$

Equation 3.129a applies within the range AG ≤ 1, that is,

$$AG = AG' = \frac{Y_G'}{\cos(30°)} \leq 1$$

Thus this geometry is valid for $h \geq \sqrt{3}$.

Now consider the case where $1 \leq h \leq \sqrt{3}$. The overlapped area is shown in Figure 3.31.

In a similar way,

$$\delta = \frac{\text{area BEG}}{\text{area OBA}} \tag{3.130}$$

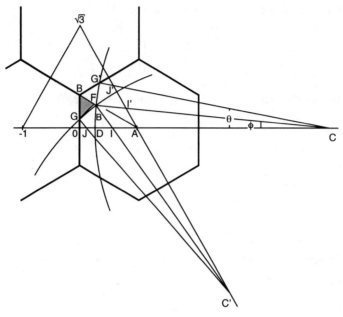

**Figure 3.31.** Area where the mean path losses are $\leq B$ dB (shaded area) $(1 \leq h \leq \sqrt{3})$.

With simple geometry (but again not straightforward expressions)

$$\delta = 1 + \frac{12\tan(\theta)}{(\sqrt{3} - \tan(\theta))(h^2 - 1)^2} - \sqrt{3}\,CD^2\left[\sin^{-1}\left(\frac{Y'_G}{CD}\right) - \sin^{-1}\left(\frac{Y_E}{CD}\right)\right]$$

$$- \frac{4\tan(\phi)\left[\sqrt{3}\,Y_E(h^2 - 1) - 3\right]}{(h^2 - 1)^2(\sqrt{3} - \tan(\phi))}$$

$$- (3 - 2\sqrt{3}\,Y'_G)\left[1 - \frac{4\tan(\theta)}{(h^2 - 1)(\sqrt{3} - \tan(\theta))}\right] \qquad (3.131a)$$

where $Y_E$, $CD$, and $\tan(\phi)$ are given by Equations 3.129b, 3.129d, and 3.129e, respectively, and

$$Y'_G = \frac{h(\sqrt{3}\,h - \sqrt{4 - h^2})}{2(h^2 - 1)} \qquad (3.131b)$$

$$\tan(\theta) = \frac{Y'_G}{\sqrt{CD^2 - Y'^2_G}} \qquad (3.131c)$$

The expressions for $\delta$ given by Equations 3.131a ($1 \leq h \leq \sqrt{3}$) and 3.129a ($\sqrt{3} \leq h$) are rather complex. It is convenient to determine simpler equations that, although approximated, may yield quicker means of evaluation and checking. This is carried out as follows.

*Underestimated $\delta$.* Refer to Figure 3.30 for $h \geq \sqrt{3}$. The area OBEG is approximated from below by OBFG, where FG is tangent to the circle of which GE is an arc. Hence

$$\delta > \delta_u = 1 - \frac{2\left(-1 + \sqrt{4h^2 - 3}\right)}{(h + 1)(h^2 - 1)} \tag{3.132a}$$

Now refer to Figure 3.31 for $1 \leq h \leq \sqrt{3}$. In the same way the area BEG is approximated from below by BFG. Then

$$\delta > \delta_u = 2\sqrt{3}\left(\frac{h - 1}{h + 1}\right)\left(Y_G' - \frac{1}{\sqrt{3}}\right) \tag{3.132b}$$

where $Y_G'$ is given by Equation 3.131b.

*Overestimated $\delta$.* Refer to Figure 3.30 ($h \geq \sqrt{3}$). If the area OBEG is delimited by straight lines, then the approximation is from above. Thus

$$\delta < \delta_0 = 1 - 2Y_E Y_G' \tag{3.133a}$$

where $Y_E$ and $Y_G'$ are given by Equations 3.129b and 3.129c, respectively.

The same approach can be used for $1 \leq h \leq \sqrt{3}$ (Figure 3.30), i.e., with BEG approximated by straight lines. Then

$$\delta < \delta_0 = 6\left(Y_G' - \frac{1}{\sqrt{3}}\right)\left(Y_E - \frac{1}{\sqrt{3}}\right) \tag{3.133b}$$

where $Y_E$ and $Y_G'$ are given by Equations 3.129b and 3.131b, respectively.

Equations 3.131a and 3.129a for the exact $\delta$, Equations 3.132b and 3.132a for the underestimated $\delta$, and Equations 3.133b and 3.133a for the overestimated $\delta$ are plotted as functions of $B$ in Section 3.6.8 (Figure 3.33).

### 3.6.7 Joint Lognormal Fading

This section aims at determining the joint probability density of three lognormal fading signals and calculating its cumulative distribution. As we shall see, such a distribution can be put in the form of a bivariate normal distribution. The bivariate normal distribution function is well established in the form of tables. The use of these tables requires specific procedures, which vary according to the application. The application used here is detailed in Yacoub.[23]

Consider three lognormal fading signals with instantaneous level $R_i$ and mean levels $M_i$ expressed in decibels. Their probability densities are as given by Equation 3.116, for $i = 1, 2, 3$. Assuming independent fading signals with the same standard deviation $\sigma$, the joint density is given by

$$p(R_1, R_2, R_3) = \frac{1}{\left(\sqrt{2\pi}\,\sigma\right)^3}$$

$$\times \exp\left\{\frac{-1}{2\sigma^2}\left[(R_1 - M_1)^2 + (R_2 - M_2)^2 + (R_3 - M_3)^2\right]\right\}$$

$$(3.134)$$

The next step is to determine the probability that each pair of signals has an instantaneous level difference not exceeding $A$ dB, that is,

$$P(R_1, R_2, R_3) = \iiint_S p(R_1, R_2, R_3)\, dR_1\, dR_2\, dR_3 \qquad (3.135)$$

where $S$ is the solid determined by

$$|R_1 - R_2| \le A$$

$$|R_2 - R_3| \le A \qquad (3.136)$$

$$|R_3 - R_1| \le A$$

Such solid is a hexagonal cylinder centered along the axis $R_1 = R_2 = R_3$.

By means of linear transformation (rotation of axes) it is possible to make one of the Gaussians coincide with one of the axes of the new system of coordinates, so that Equation 3.135 is reduced to an integration of a two-dimensional Gaussian over a hexagon. Let $(x, y, z)$ be the new system of coordinates with the axis of the hexagonal cylinder centered along $z$. Then

$$P(R_1, R_2, R_3) = \int_{-\infty}^{\infty} \iint_H p'(x, y, z)\,|J|\, dx\, dy\, dz \qquad (3.137)$$

where

(1) $p'(x, y, z) = \dfrac{1}{\left(\sqrt{2\pi}\,\sigma\right)^3} \exp\left\{\dfrac{-1}{2\sigma^2}\left[(x - \mu_x)^2 + (y - \mu_y)^2 + (z - \mu_z)^2\right]\right\}$

$$(3.138)$$

(2) $\mu_x, \mu_y, \mu_z$ are the means in the new system; (3) $J$ is the Jacobian of the transformation; and (4) $H$ is the hexagon of integration.

From Equation 3.137

$$P(R_1, R_2, R_3) = \frac{1}{2\pi\sigma^2} \iint_{\overset{\cdot}{H}} \exp\left\{\frac{-1}{2\sigma^2}\left[(x - \mu_x)^2 + (y - \mu_y)^2\right]\right\} |J| \, dx \, dy$$

(3.139)

With the proposed rotation of axes

$$\begin{bmatrix} R_1 \\ R_2 \\ R_3 \end{bmatrix} = \begin{bmatrix} -2/\sqrt{6} & 0 & 1/\sqrt{3} \\ 1/\sqrt{6} & 1/\sqrt{2} & 1/\sqrt{3} \\ 1/\sqrt{6} & -1/\sqrt{2} & 1/\sqrt{3} \end{bmatrix} \begin{bmatrix} x \\ y \\ z \end{bmatrix}$$

(3.140)

Consequently, the Jacobian of the transformation is given by determinant of the matrix shown in Equation 3.140. Thus $|J| = 1$. The means $M_1, M_2, M_3$ relate to $\mu_x, \mu_y, \mu_z$ as in Equation 3.140. The region delimited by Equation 3.136 in the new coordinates defines a hexagonal area as follows:

$$|-3/\sqrt{6}\,x - 1/\sqrt{2}\,y| \le A$$

$$|3/\sqrt{6}\,x - 1/\sqrt{2}\,y| \le A$$

(3.141)

$$|\sqrt{2}\,y| \le A$$

Due to symmetry we shall explore only one triangular sector of the integration hexagon. Let $x = 0$ and $y = \sqrt{3}x$ define this sector. Equation 3.139 can be evaluated by taking the means $\mu_x$ and $\mu_y$ from the lines $x = 0$ and $y = \sqrt{3}x$. The probabilities $P(R_1, R_2, R_3)$ are shown as functions of $\mu_{xy}$ in Figure 3.32,[23] where

$$\mu_{xy}^2 \triangleq \mu_x^2 + \mu_y^2$$

(3.142)

Note that the probabilities with the means taken from the line $x = 0$ and $y = \sqrt{3}x$ do not significantly differ from each other. Accordingly, we may use $\mu_{xy}$ as an overall mean to estimate $P(R_1, R_2, R_3)$.

### 3.6.8 The Geographical Distribution of the Instantaneous Power Ratio

In this section we shall estimate the overall proportion of the cell area with satisfactory paths to three base stations on an instantaneous rather than on a mean criterion, by combining some of the methods described in Sections 3.6.6 and 3.6.7. The mean signal strength is a function of the position of the

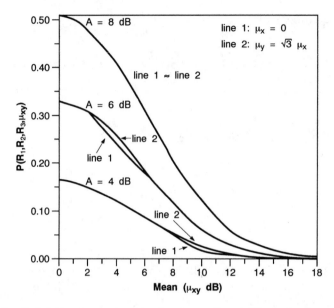

**Figure 3.32.**   Probability of any pair of signals not exceeding $A$ dB, as a function of the mean.

mobile in the cell

$$\mu_{xy} = f(x, y) \tag{3.143}$$

Conditional on the value of $\mu_{xy}$, the probability $P(R_1, R_2, R_3, \mu_{xy})$ that each pair of signals has an instantaneous level difference not exceeding $A$ dB can be estimated by using the curves of Figure 3.32. Hence $P(R_1, R_2, R_3, x, y)$ conditional on the position $(x, y)$ can be estimated. The unconditional probability

$$\delta = \frac{1}{T} \iint_T P(R_1, R_2, R_3, x, y) \, dx \, dy \tag{3.144}$$

(where $T$ is a triangular sector) is the mean proportion of the cell area with a signal strength within a tolerance allowing for fading. Again by symmetry only $\frac{1}{12}$ of the hexagon needs to be considered. Rewriting Equation 3.142 as a function of the means $M_1$, $M_2$ and $M_3$ yields

$$\mu_{xy}^2 = \tfrac{1}{6}(2M_1 - M_2 - M_3)^2 + \tfrac{1}{2}(M_2 - M_3)^2 \tag{3.145}$$

Define differences of means as

$$D_1 = M_1 - M_2$$

$$D_2 = M_2 - M_3 \qquad (3.146)$$

$$D_3 = M_3 - M_1$$

Thus

$$\mu_{xy}^2 = \tfrac{2}{3}\left(D_1^2 + D_2^2 + D_1 D_2\right) \qquad (3.147)$$

Consider a mobile in the position $(-x, y)$ (this is the mirror image of the situation of Figure 3.29). The mobiles in this triangular sector of the hexagon will have $M_1 \geq M_2 \geq M_3$ (the other five possible combinations correspond to the respective five other triangular sectors around the common corner of the three hexagons). The ratios of the received powers are

$$h_1^\alpha = \left(\frac{x_2}{x_1}\right)^\alpha = \left(\frac{m_1}{m_2}\right)^2 = d_1 = \left(\frac{y^2 + (x-1)^2}{y^2 + (x+1)^2}\right)^{\alpha/2}$$

$$h_2^\alpha = \left(\frac{x_3}{x_2}\right)^\alpha = \left(\frac{m_2}{m_3}\right)^2 = d_2 = \left(\frac{\left(y - \sqrt{3}\right)^2 + x^2}{y^2 + (x-1)^2}\right)^{\alpha/2} \qquad (3.148)$$

$$h_3^\alpha = \left(\frac{x_3}{x_1}\right)^\alpha = \left(\frac{m_1}{m_3}\right)^2 = d_3 = \left(\frac{\left(y - \sqrt{3}\right)^2 + x^2}{y^2 + (x+1)^2}\right)^{\alpha/2}$$

with $D_i = 20 \log d_i = 10\alpha \log h_i$. From Equations 3.148 and 3.147

$$\mu_{xy}^2 = \frac{200\alpha^2\left[\log^2(h_3) - \log(h_2)\log(h_1)\right]}{3} \qquad (3.149)$$

In order to calculate the mean $\mu_{xy}$ in that sector, each one of the two sides composing the right angle of the triangle was divided into 20 equal distances, and for each point $(-x, y)$ Equation 3.149 was applied (a total of 231 points). For each calculated $\mu_{xy}$, the probability distribution of the signals within tolerance in that sector is determined. Equation 3.144 is evaluated by

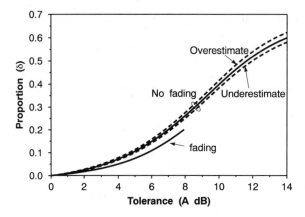

**Figure 3.33.**  Proportion of mobiles with radio paths to three base stations.

numerical methods from

$$\delta \simeq \frac{\Delta x\,\Delta y}{T} \sum_{i=0}^{1/\Delta x} \sum_{j=0}^{(1/\Delta x)-i} W\left(-i\,\Delta x, j\frac{\Delta x}{\sqrt{3}}\right) P\left(-i\,\Delta x, j\frac{\Delta x}{\sqrt{3}}\right) \quad (3.150)$$

where $W(i, j)$ are the weights of $P(i, j)$ and $\Delta x$ and $\Delta y$ are the incremental distances into which the sector has been split, in $x$ and $y$ directions, respectively. The result is shown in Figure 3.33.

### 3.6.9  δ as Function of γ

There are two distinct situations to be considered, namely, $\delta = f(\gamma)$ without fading and $\delta = f(\gamma)$ with fading.

$\delta = f(\gamma)$ *without fading.*  Refer to Sections 3.6.3 and 3.6.6, where the functions $\gamma = f(h)$ and $\delta = f(h)$, respectively, are determined for the case of no fading. It is obvious that $\delta = f(\gamma)$ can be found, but not necessarily in a closed form. If the exact functions are considered, then neither $h = f^{-1}(\gamma)$ nor $h = f^{-1}(\delta)$ can be put in closed form. Nevertheless this is straightforward for the approximate formulas (underestimate and overestimate approximations). The exact curve is plotted in Figure 3.34 (the underestimate and overestimate curves are almost coincident with the exact curve).

$\delta = f(\gamma)$ *with lognormal fading.*  Refer to Sections 3.6.4 and 3.6.8, where $\gamma = f(A)$ and $\delta = f(A)$ are determined. Here definitely $\delta = f(\gamma)$ cannot be expressed in a closed form, but it can be plotted with the help of the respective graphs having the tolerance $A$ dB as a common variable. Figure

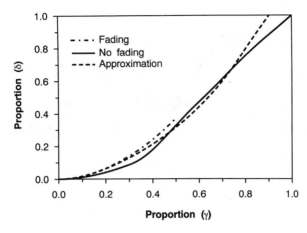

**Figure 3.34.** Proportion $\gamma$ and $\delta$ with and without fading, and the approximation.

3.34 shows $\delta = f(\gamma)$ with and without fading, and suggests that $\delta = 1.25\gamma^2$ is a good approximation in the range of $0 \leq \gamma \leq 0.8$.

## 3.7   SUMMARY AND CONCLUSIONS

The mobile radio propagation phenomena are very complex and cannot be entirely described by a single model. Many signal strength prediction algorithms are available but a good prediction is only accomplished with a reasonable amount of input data. Some of these algorithms are applicable to very specific situations whereas others can be used in a wider range of conditions when correction factors must be employed. A simplified model is also available and can be used mainly for comparative purposes.

The prediction models yield the mean signal strength at a given distance from the base station. However, due to the statistical fluctuations of the various phenomena involving mobile radio propagation, the mobile radio signal cannot be treated only by deterministic methods. Accordingly, we may establish a three-part model to describe the various phenomena better. This model has proved to be very efficient and is in very good agreement with the field measurements, as follows:

1. Mean signal decreases with distance $d$ as $d^{-\alpha}$, where $\alpha$ is a parameter typically in the range 2–4, depending on the environment.
2. Slow fading due to shadowing has an approximately lognormal distribution with standard deviation in the range 4–10 dB.
3. Fast fading due to multipath propagation has a Rayleigh distribution. Within buildings, where both multipath and line-of-sight waves can be found, the fast fading follows a Ricean distribution.

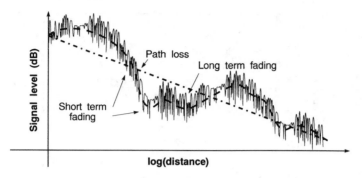

**Figure 3.35.**  The total fading signal.

The resultant (received) signal is a combination of all of these phenomena, as sketched in Figure 3.35.

As far as boundaries between cells are concerned we have calculated two statistics:

1. The probability that a mobile user in some definite set of locations has two or more paths whose mean losses differ only within a given tolerance; the averaging is implicitly over both time (long-term) and space (within the set of locations).
2. The probability that a mobile user, at a randomly chosen point within some definite set of locations, has at a randomly chosen time two or more paths whose losses differ only within a given tolerance. At first sight it might appear that this differs from statistic 1 in that it is averaged over space but not over time. However, the matter is more complex, because the loss fluctuations occur both in time and in space, and those are coupled by the movement of the user.

These two statistics are not very dissimilar from one another in quantitative terms. Hence we may argue that a true evaluation of the proportion of the cell area where mobiles may have two or more alternative paths lies somewhere between these statistics.

It can be seen from the graphs of Figure 3.28 that, if it is accepted that paths within 6 dB of each other are valid alternatives, the proportion of the cell area where mobiles may have access to at least two base stations is somewhere in the range 30–40%. For the same tolerance the proportion with possible access to three base stations is 12–20%.

In fact, it is usually accepted[27] that, in order to avoid unnecessary hand-off, the signal strength has to be set as high as 10–15 dB, implying that the proportion of overlapped area is even larger.

# REFERENCES

1. Parsons, J. D., Editorial—land mobile radio, *IEE Proc.* 132(5), Part F, 1985.
2. Jakes, W. C., *Microwave Mobile Communications*, John Wiley & Sons, New York, 1974.
3. Lee, W. C. Y., *Mobile Communications Engineering*, McGraw-Hill, New York, 1982.
4. Lee, W. C. Y., *Mobile Communications Design Fundamentals*, Howard W. Sams & Co., Indianapolis, IN, 1986.
5. Camwell, P. L. and McRory, J. G., Experimental Results of In-Building Anisotropic Propagation at 835 MHz Using Leaky Feeders and Dipole Antennas, Montech, 1987.
6. Collins, R. E., *Antennas and Radiowave Propagation*, McGraw-Hill, New York, 1985.
7. Kraus, J. D., *Antennas*, McGraw-Hill, New York, 1950.
8. Holbeche, R. J., Ed., *Land Mobile Radio Systems*, Peter Peregrinus, London, 1985.
9. Ramo, S., Whinnery, J. R., and Duzer, V., *Fields and Waves in Communication Electronics*, John Wiley & Sons, New York, 1965.
10. Okumura, Y., Ohmori, E., Kawano, T., and Fukuda, K., Field strength and its variability in VHF and UHF land mobile service, *Rev. Elec. Comm. Lab.*, 16, 825–873, September–October 1968.
11. Rice, P. L., Longley, A. G., Norton, K. A., and Barsis, A. P., Transmission loss prediction for tropospheric communication circuits, *NBS Tech. Note No. 101*, vols. 1 and 2, 1967.
12. Bullington, K., Radio propagation fundamentals, *Bell Syst. Tech. J.*, 36, May 1957.
13. Longley, A. G. and Rice, P. E., Prediction of tropospheric radio transmission loss over irregular terrain, in *A Computer Method—1968, ESSA Research Laboratories*, ERL79-ITS67, U.S. Government Printing Office, Washington, D.C., 1968.
14. Lorenz, R. W., Frequency planning of cellular radio by the use of topographical data base, in *35th IEEE Vehicular Technology Conf.*, 1985.
15. Ibrahim, M. F. and Parsons, J. D., Urban mobile radio propagation at 900 MHz, *Electron. Lett.* 18(3), 113–115, 1982.
16. Ibrahim, M. F., Edwards, J. A., and Parsons, J. D., Automated logging and analysis system for VHF and UHF signal strength measurements, in *Conf. on Radio Receivers and Associated Systems*, Leeds, *IERE Conf. Proc. No. 50*, July 1981.
17. Young, W. R., Mobile radio transmission compared at 150 to 3700 MC, *Bell Syst. Tech. J.*, 31, 1068–1085, November 1952.
18. Reudink, D. O., Mobile radio propagation in tunnels, IEEE Vehicular Technology Group Conference, San Francisco, December 2–4, 1968.
19. Aurand, J. F. and Post, R. E., A comparison of prediction methods for 800 MHz mobile radio propagation, *IEEE Trans. Vehicular Tech.*, VT-34(4), November 1985.
20. Davenport, W. B., Jr., *Probability and Random Processes*, International Student Edition, McGraw-Hill KogaKusha, Ltd., Tokyo, 1970.

21. Schwartz, M., *Information Transmission, Modulation and Noise*, McGraw-Hill, New York, 1959.

22. Ng, E. W., and Geller, M., A table of integrals of the error functions, *J. Res. NBS—B*, 73B, January–March 1969.

23. Yacoub, M. D., Mobile Radio with Fuzzy Cell Boundaries, Ph.D. thesis, University of Essex, England, April 1988.

24. Yacoub, M. D., Rodriguez, D. M. and Cattermole, K. W., Alternative routing in cellular mobile radio, 3rd Teletraffic Symposium, Colchester, England, June 1986.

25. Yacoub, M. D. and Cattermole, K. W., Cellular mobile radio with fuzzy cell boundaries, 4th Teletraffic Symposium, Bristol, England, May 1987.

26. Tables of Bivariate Normal Distribution Function, U.S. National Bureau of Standards, Washington, D.C., 1959.

27. Garcia-Hernandez, C. F. and Hughes, C. J., Simulation of handovers conditions for cellular radio systems, *Electron. Lett.*, 22(17), 904–905, 1986.

28. Loew, K., Boundaries between radio cells—influence of buildings and vegetation, *IEE Proc.*, 132(5), Part F, 1985.

29. Davis, B. R. and Bogner, R. E., Propagation at 500 MHz for mobile radio, *IEE Proc.*, 132(5), Part F, 1985.

30. Gradshteyn, I. S. and Ryahik, I. W., *Table of Integrals, Series and Products*, Academic, New York, 1965.

31. Abramowitz, M. and Stegun, I. A., Eds., *Handbook of Mathematical Functions with Formulas, Graphs and Mathematical Tables*, National Bureau of Standards, Applied Mathematical Series 55, U.S. Government Printing Office, Washington, D.C., June 1964.

32. Hata, M., Empirical formula for propagation loss in land-mobile radio services, *IEEE Trans. Vehicular Tech.*, VT-29, 317–325, 1980.

CHAPTER **4**

# Multipath Propagation Effects

In this chapter we select for analysis some of the most relevant aspects to be considered in a multipath propagation environment. We start with the basic concept of wave velocity to derive, thereafter, a relationship describing the frequency deviation of the propagated signal due to vehicle motion. The simple formula is, in fact, an approximation of a more complex one given by the theory of relativity. We then study the statistical behavior of the time delay and the delay spread of the propagated signal. The effect of such delay is analyzed in order to determine how strongly correlated two received signals are, given a time delay or frequency separation between them. The frequency separation for a given correlation factor is named coherence bandwidth. Because in a multipath medium we deal with fast fading, it is interesting to determine how often the fades occur given a threshold voltage level; this is known as level crossing rate. The duration of the fades is also of interest. The next relevant topic is concerned with the study of the randomness of the signal's frequency (phase) variation and its effect on the quality of the received signal. It is shown that this phenomenon, known as random frequency modulation, introduces noise having a power varying as a quadratic function of vehicle speed. We also determine the power spectra of the signal and their dependence of the antenna gain. All of these phenomena can then be observed by means of field measurements. We outline the main points to be considered in the acquisition and analysis of data. We finally show the various manners of implementing fading simulators, including the analog and digital approaches.

## 4.1  INTRODUCTION

A radio signal transmitted from the base station reaches the mobile station as a large number of scattered waves. The scattering can be provoked by multiple reflections over irregular terrains, presence of a great number of obstructions, variations of the medium's dielectric constant, etc. Due to the randomness of these phenomena, the mobile radio signal is usually treated in a statistical basis. The envelope, phase, and frequency of the received signal vary randomly according to some well-known probability distributions. In particular, the envelope follows the Rayleigh distribution, whereas the phase is uniformly distributed in the range $0-2\pi$ rad.

Other statistics must be determined for better characterization of the Rayleigh fading signal. One such statistic is related to the time variation of the received envelope. In the same way, a joint distribution of two Rayleigh fading signals is also of interest. With these two statistics it is possible to describe and characterize some of the various multipath propagation effects, such as coherence bandwidth, level crossing rate, average duration of fades, random frequency modulation, and others. Accordingly, we dedicate some sections of this chapter to obtaining these distributions.

Other related topics treated in this chapter include the Doppler effect, delay spread, signal power spectra, field measurement, and fading simulation.

## 4.2    VELOCITIES OF WAVE PROPAGATION

Let $e$ be the instantaneous value of the electric field propagating in a dispersive medium. From Section 3.4.1 (Equation 3.67),

$$e = E_M \exp(j\omega t) = E_0 \exp(-\alpha r)\exp[j(\omega t - \beta r)] \qquad (4.1)$$

where $E_0$ is the electric field in free space, $r$ is the distance, $\omega$ is the frequency, $\alpha$ is the attenuation constant, and $\beta$ is the phase constant.

### 4.2.1    Phase Velocity

Phase velocity $(v_p)$ corresponds to the velocity the wave must travel in order to keep its instantaneous phase constant. Hence, for a single-frequency electric field we must have

$$\omega t - \beta r = \text{constant}$$

Therefore,

$$v_p = \frac{dr}{dt} = \frac{\omega}{\beta} \qquad (4.2)$$

In a lossless medium the conductivity $\sigma$ equals 0. Hence from Equation 3.66, $\alpha = 0$ and $\beta = \omega\sqrt{\mu\varepsilon}$, where $\mu$ is the magnetic permeability and $\varepsilon$ is the electric permittivity. Thus

$$v_p = 1/\sqrt{\mu\varepsilon} \qquad (4.3)$$

Because the wavelength $\lambda$ is defined as the distance the wave propagates within one period $(T = 1/f)$, then

$$\lambda = v_p \frac{1}{f} = \frac{\omega}{\beta} \frac{1}{f} = \frac{2\pi}{\beta} \qquad (4.4)$$

## 4.2.2 Group Velocity

In a dispersive medium the spectral components of an arbitrary wave may travel with different speeds, characterizing the phenomenon known as dispersion. In such a case the use of the group velocity $v_g$, as defined in Equation 4.5, is more appropriate:

$$v_g = d\omega/d\beta \qquad (4.5)$$

The group velocity approximately represents the velocity of the wave and is often referred to as the velocity of energy travel. From Equations 4.5 and 4.4 it can be shown that

$$v_g = \frac{v_p}{1 - (\omega/v_p)(dv_p/d\omega)} \qquad (4.6)$$

## 4.3 DOPPLER FREQUENCY

The Doppler frequency refers to the apparent shift in frequency of the carrier as experienced by a vehicle moving under free-space conditions[1].

### 4.3.1 Insight into the Problem

Consider a vehicle moving at speed $v$ and a carrier with phase velocity $v_p$ arriving at an angle $\theta$ as shown in Figure 4.1. The speed of the vehicle will impose an apparent phase velocity $v_p'$ on the wave such that

$$v_p' = v_p - v \cos \theta \qquad (4.7)$$

**Figure 4.1.** Doppler effect.

From Equation 4.4 $v'_p = \lambda f'$ and $v_p = \lambda f$, where $f'$ is the apparent frequency and $f$ is the propagated frequency. Using this in Equation 4.7 we have

$$f' = f - f_D \qquad (4.8)$$

where

$$f_D = \frac{v}{\lambda} \cos \theta \qquad (4.9a)$$

is the Doppler shift. Writing Equation 4.9a in terms of the angular frequency, we obtain

$$\omega_D = \omega_m \cos \theta \qquad (4.9b)$$

where $\omega_m \triangleq \beta v$ is the maximum Doppler shift.

### 4.3.2 Relativity Theory

The approach just described is very simplistic and does not give the exact solution to the problem. This question is much more intricate and the exact solution is given by relativity theory. The predicted Doppler frequency $f'$ for $\theta = 0°$ is[2]

$$f' = f \frac{1 \pm v/c}{\sqrt{1 - (v/c)^2}} \qquad (4.10)$$

where $c$ is the speed of light, $v$ is the speed of the observer, and $f$ is the propagated frequency. The plus sign in the numerator applies to the case when observer and source are moving in opposite directions. The minus sign refers to the opposite situation. Using series expansion of Equation 4.10, for the minus sign case we have

$$f' = f \left[ 1 - \frac{v}{c} + \frac{1}{2}\left(\frac{v}{c}\right)^2 - \cdots \right] \qquad (4.11)$$

Because the ratio $v/c$ is usually very small, we may neglect the higher-order terms. Therefore,

$$f' \approx f - fv/c$$

Given that $c = \lambda f$, then

$$f' \approx f - v/\lambda \qquad (4.12)$$

as would be obtained by the "simplistic" solution given by Equations 4.8 and 4.9a when $\theta = 0$.

As an example consider a vehicle moving at 72 km/h (20 m/s) and frequency of 1000 MHz ($\lambda = 0.3$ m). Then the maximum Doppler shift is $20/0.3 \simeq 66.7$ Hz.

## 4.4   DELAY SPREAD

Due to multipath propagation, a transmitted signal arrives at the receiver at different time instants and with different amplitudes. If an impulse $a\delta(t)$, having "amplitude" $a$ is propagated at time instant $t = 0$, the received signal will be $h(t)$, such that

$$h(t) = \sum_{i=1}^{n} a_i \delta(t - T_i) \tag{4.13}$$

where $n$ is the number of scatterers, $a_i$ is the "amplitude" of the received impulse due to the $i$th path, and $T_i$ is the time delay of the $i$th arrived impulse. The longer the path, the smaller the received amplitude, the longer the time delay. Accordingly, we may expect the received signals to have a profile as sketched in Figure 4.2.

The impulse arrival time $T$ is usually characterized by a probability density function. Accordingly, the *mean time delay* $\overline{T}$ is the mean of this density function, and the *delay spread* $\sigma_T$ corresponds to its standard deviation. Let $p(T)$ be the probability density function of $T$. Its $k$th moment is

$$E[T^k] \triangleq \int_0^{\infty} T^k p(T)\, dT$$

Therefore, the mean time delay is

$$\overline{T} = E[T] \tag{4.14}$$

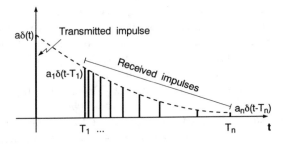

**Figure 4.2.**   Delay profile.

and the delay spread is

$$\sigma_T = \{E[T^2] - E^2[T]\}^{1/2} \qquad (4.15)$$

An exact characterization of the impulse arrival time is not available. In general, a negative exponential distribution is used to describe its density function. Accordingly,

$$p(T) = \frac{1}{\overline{T}} \exp\left(-\frac{T}{\overline{T}}\right) \qquad (4.16)$$

With such distribution both the mean time delay and the delay spread are equal to $\overline{T}$.

In practice, the delay spreads vary from fractions of microseconds to many microseconds. In urban areas the delays are usually longer (greater than 3 $\mu$s) whereas in suburban areas and in open areas they are shorter (0.5 $\mu$s and less than 0.2 $\mu$s, respectively). The characterization of the delay spread is very important for digital mobile radio applications where intersymbol interference may occur (refer to Chapters 6 and 10).

## 4.5   COHERENCE BANDWIDTH

Consider an arbitrary radio signal with a given bandwidth, propagating in a mobile radio environment. Due to the multipath effect, each frequency component of this signal may reach the destination (the mobile station) with different time delays. It is, therefore, essential to determine the maximum frequency separation for which the signals are still considered to be correlated. This frequency separation is named *coherence bandwidth*.

Systems operating with channels substantially narrower than the coherence bandwidth are known as *narrowband systems*. *Wideband systems* are those operating with channels wider than the coherence bandwidth. In the narrowband systems all of the components of the signal are equally influenced by multipath propagation. Accordingly, although with different amplitudes and affected by noise and interference, the received signal is essentially "the same" as the transmitted signal. In other words, the radio signal experiences nonselective fading. In the wideband systems the various frequency components of the signal may be differently affected by the fading, characterizing the phenomenon known as selective fading.

In order to estimate the coherence bandwidth, the first step is to determine the envelope and phase correlations between two signals arriving at two different time instants. This can only be carried out if the joint probability density function of these signals is known. Accordingly, in the next sections we shall (1) model the received signal, (2) determine their joint density function, (3) calculate the envelope correlation, (4) calculate phase correlation, and, finally, (5) calculate the coherence bandwidth.

## 4.5.1 Received Signal

Let $e_0 = E_0 \exp(j\omega t)$ be a transmitted wave, propagating in the mobile radio environment. If we consider only one radio path, the received wave is that given by Equation 4.1. However, due to the multipath propagation, each path of length $r_i$ will impose a phase shift equal to $\beta r_i = (\omega/v_p)r_i = \omega T_i$, where $T_i$ is the time delay of the $i$th path. Moreover, each delayed wave $i$, arriving at an angle $\theta_i$, will contribute to a Doppler shift equal to $\omega_i = \beta v \cos \theta_i$. Therefore, the $i$th received wave $e_i$ can be written as

$$e_i = E_0 \exp[ j(\omega t + \omega_i t - \omega T_i)] \qquad (4.17)$$

where the effects of shadowing and path loss have not been taken into account.

If we consider that the propagated wave has been reflected by $n$ scatterers, then within an infinitesimal time delay $dT$ and arrival angle $d\theta$, the resultant signal received by an omnidirectional antenna, is*

$$e = E_0 \sum_{i=1}^{n} a_i \exp[ j(\omega t + \omega_i t - \omega T_i)] \qquad (4.18a)$$

where $a_i^2 = p(\theta_i) \, d\theta \, p(T_i) \, dT$; the term $p(\theta_i) \, d\theta \, p(T_i) \, dT$ represents the fraction of the incoming power within $d\theta$ of the angle $\theta$ and within $dT$ of the time $T$ in the limit with $i$ very large.

Assuming a uniform distribution in angle of the incident power, $p(\theta) = 1/2\pi$ for $0 \leq \theta \leq 2\pi$ rad, and $p(\theta) = 0$, otherwise. Using Equation 4.16 for the density of $T$, it follows that

$$\lim_{i \to \infty} a_i^2 = \frac{1}{2\pi \overline{T}} \exp\left( -\frac{T}{\overline{T}} \right) d\theta \, dT \qquad (4.18b)$$

---

*Rigorously speaking, for each scatterer $i$ causing the reflected wave to arrive at an angle $\theta_i$ and with a time delay $T_i$, there are $m$ other scatterers causing $m$ other waves to arrive at $m$ different angles but with the same time delay. Consequently, the received wave is

$$e = E_0 \sum_{i=1}^{n} \sum_{j=1}^{m} a_{ij} \exp[ j(\omega t + \omega_{ij} t - \omega T_i)]$$

where $a_{ij}^2 = p(\theta_{ij}) \, d\theta \, p(T_i) \, dT$ and $\omega_{ij} = \beta v \cos \theta_{ij}$.

The only implication of having a double sum in the expression for the received wave is that in our calculations with $n \to \infty$ and $m \to \infty$, the double sum becomes double integrals. We obviously take this into account, as shall be seen in the following paragraphs. We decided, however, to keep Equation 4.18a with a single sum just for simplicity.

Taking the real part of the signal $e$ given by Equation 4.18a, we obtain

$$s(t) \triangleq \mathrm{Re}(e) = X \cos \omega t - Y \sin \omega t \qquad (4.19a)$$

where

$$X = E_0 \sum_{i=1}^{n} a_i \cos(\omega_i t - \omega T_i) \triangleq r \cos \varphi$$

$$Y = E_0 \sum_{i=1}^{n} a_i \sin(\omega_i t - \omega T_i) \triangleq r \sin \varphi \qquad (4.19b)$$

Note that $r^2 = X^2 + Y^2$ is the envelope of the signal and $\varphi = \tan^{-1}(Y/X)$ is its phase. As shown in Section 3.4.2, $r$ is Rayleigh distributed, whereas $\varphi$ is uniformly distributed in the range $0–2\pi$ rad. The random variables $X$ and $Y$ have a Gaussian distribution for large $n$.

### 4.5.2    Joint Probability Density Function

Consider two signals $s_j(t)$, $j = 1, 2$, propagated at frequencies $\omega_j$, $j = 1, 2$. Assume that $s_1(t)$ arrives at an instant $t$ and $s_2(t)$ at $t + \tau$, where $\tau$ is a time delay. The four corresponding Gaussian random variables are

$$X_1 = E_0 \sum_{i=1}^{n} a_i \cos(\omega_i t - \omega_1 T_i) \triangleq r_1 \cos \varphi_1$$

$$Y_1 = E_0 \sum_{i=1}^{n} a_i \sin(\omega_i t - \omega_1 T_i) \triangleq r_1 \sin \varphi_i$$

$$X_2 = E_0 \sum_{i=1}^{n} a_i \cos(\omega_i t + \omega_i \tau - \omega_2 T_i) \triangleq r_2 \cos \varphi_2 \qquad (4.20)$$

$$Y_2 = E_0 \sum_{i=1}^{n} a_i \sin(\omega_i t + \omega_i \tau - \omega_2 T_i) \triangleq r_2 \cos \varphi_2$$

We want to determine the joint distribution of $X_1$, $Y_1$, $X_2$, and $Y_2$.

An $n$-dimensional Gaussian probability density function is written as follows[4]:

$$p(R_1, R_2, \ldots, R_n) = \frac{\exp\left[ -\dfrac{1}{2[\Lambda]} \sum_{j=1}^{n} \sum_{k=1}^{n} |\Lambda|_{jk}(R_j - M_j)(R_k - M_k) \right]}{(2\pi)^{n/2}|\Lambda|^{1/2}} \qquad (4.21)$$

where $R_j$, $j = 1, \ldots, n$, are the random variables,

$M_j = E[R_j]$ are their mean values,

$|\Lambda|_{jk}$ is the cofactor of the element $\lambda_{jk}$ of the determinant $|\Lambda|$.

The matrix $\Lambda$ is called the *covariance matrix* and its elements $\lambda_{jk}$ are the covariances, defined as follows:

$$\lambda_{jk} \triangleq \mathrm{Cov}(R_j, R_k) = E\big[(R_j - M_j)(R_k - M_k)\big]$$

$$= E[R_j R_k] - M_k E[R_j] - M_j E[R_k] + M_j M_k$$

$$= E[R_j R_k] - E[R_j]E[R_k] \tag{4.22}$$

For the random variables $X_1$, $Y_1$, $X_2$, and $Y_2$ we assume the covariance matrix

$$\Lambda = \begin{bmatrix} \mathrm{Cov}(X_1, X_1) & \mathrm{Cov}(X_1, Y_1) & \mathrm{Cov}(X_1, X_2) & \mathrm{Cov}(X_1, Y_2) \\ \mathrm{Cov}(Y_1, X_1) & \mathrm{Cov}(Y_1, Y_1) & \mathrm{Cov}(Y_1, X_2) & \mathrm{Cov}(Y_1, Y_2) \\ \mathrm{Cov}(X_2, X_1) & \mathrm{Cov}(X_2, Y_1) & \mathrm{Cov}(X_2, X_2) & \mathrm{Cov}(X_2, Y_2) \\ \mathrm{Cov}(Y_2, X_1) & \mathrm{Cov}(Y_2, Y_1) & \mathrm{Cov}(Y_2, X_2) & \mathrm{Cov}(Y_2, Y_2) \end{bmatrix}$$

$$\tag{4.23}$$

In order to compose the covariance matrix we must determine the various moments $E[R_j R_k]$, $E[R_j]$, and $E[R_k]$. In particular,

$$E[X_1] = \langle X_1 \rangle = E_0 \sum_{i=1}^{n} \langle a_i \cos(\omega_i t - \omega_1 T_i) \rangle = 0 \tag{4.24}$$

where

$$\langle x(t) \rangle = \lim_{T \to \infty} \frac{1}{2T} \int_{-T}^{T} x(t)\, dt$$

is the time average of $x(t)$ over the period $2T$. In the same way $E[X_1] = E[Y_1] = E[X_2] = E[Y_2] = 0$. Therefore, in this case $\mathrm{Cov}[R_j, R_k] = E[R_j R_k]$. Then

$$\mathrm{Cov}[X_1, X_1] = E[X_1^2] = \langle X_1^2 \rangle$$

$$= E_0^2 \sum_{i,j} \langle a_i a_j \cos(\omega_i t - \omega_1 T_i)\cos(\omega_j t - \omega_1 T_i) \rangle$$

It is clear that this average will vanish unless $i = j$. In this case $\langle a_i^2 \cos^2(\omega_i t - \omega_1 T_i) \rangle = a_i^2/2$. Therefore,

$$\text{Cov}[X_1, X_1] = \sigma^2 \sum_i a_i^2$$

where $\sigma^2 = E_0^2/2$ (refer to Section 3.4.2).

In the limit with $i \to \infty$, this equation takes the integral form. Then

$$\text{Cov}[X_1, X_1] = \sigma^2 \int_0^{2\pi} \int_0^{\infty} \frac{1}{2\pi\overline{T}} \exp\left(-\frac{T}{\overline{T}}\right) d\theta \, dT = \sigma^2 \qquad (4.25)$$

Similarly,

$$\text{Cov}[X_i, X_i] = \text{Cov}[Y_i, Y_i] = \sigma^2, \qquad i = 1, 2 \qquad (4.26)$$

$$\text{Cov}[X_1, Y_1] = \text{Cov}[Y_1, X_1] = \text{Cov}[X_2, Y_2] = \text{Cov}[Y_2, X_2] = 0 \quad (4.27)$$

$$\text{Cov}[X_1, X_2] = \text{Cov}[X_2, X_1] = \text{Cov}[Y_1, Y_2]$$

$$= \text{Cov}[Y_2, Y_1] \triangleq \mu_1, \quad \text{to be determined}$$

$$\text{Cov}[X_1, Y_2] = \text{Cov}[Y_2, X_1] = -\text{Cov}[Y_1, X_2]$$

$$= -\text{Cov}[X_2, Y_1] \triangleq \mu_2, \quad \text{to be determined}$$

Using the same procedure,

$$\mu_1 = E[X_1 X_2] = \langle X_1 X_2 \rangle$$

$$= E_0^2 \sum_{i,j} \langle a_i a_j \cos(\omega_i t - \omega_1 T_i) \cos(\omega_j t + \omega_j \tau - \omega_2 T_j) \rangle$$

The average will vanish unless $i = j$. For this case, using a well-known trigonometric identity, we have

$$\mu_1 = E[X_1 X_2] = \sigma^2 \sum_i a_i^2 \cos(\omega_i \tau - \Delta\omega \, T_i)$$

where $\Delta\omega = \omega_2 - \omega_1$ and $\sigma^2 = E_0^2/2$.

In the limit, with $i \to \infty$,

$$\mu_1 = \sigma^2 \int_0^{2\pi} \int_0^{\infty} \frac{1}{2\pi\overline{T}} \exp\left(-\frac{T}{\overline{T}}\right) \cos(\beta v \tau \cos\theta - \Delta\omega \, T) \, d\theta \, dT$$

Then

$$\mu_1 = \frac{\sigma^2 J_0(\omega_m \tau)}{1 + (\Delta\omega \overline{T})^2} \qquad (4.28)$$

where $\omega_m = \beta v$ is the maximum Doppler shift and

$$J_n(z) = \frac{1}{\pi} \int_0^\pi \cos(z \sin x - nx)\, dx, \quad n \text{ an integer}$$

is the Bessel function of the first kind of order $n$.

In a similar way,

$$\mu_2 = \frac{-\sigma^2 \Delta\omega\, \overline{T} J_0(\omega_m \tau)}{1 + (\Delta\omega\, \overline{T})^2} \tag{4.29}$$

Therefore, the covariance matrix is

$$\Lambda = \begin{bmatrix} \sigma^2 & 0 & \mu_1 & \mu_2 \\ 0 & \sigma^2 & -\mu_2 & \mu_1 \\ \mu_1 & -\mu_2 & \sigma^2 & 0 \\ \mu_2 & \mu_1 & 0 & \sigma^2 \end{bmatrix}$$

Its determinant is given by

$$|\Lambda| = \sigma^8 (1 - \rho^2)^2 \tag{4.30}$$

where

$$\rho^2 = \frac{\mu_1^2 + \mu_2^2}{\sigma^4} \tag{4.31a}$$

Using Equations 4.28 and 4.29 in Equation 4.31a, we obtain

$$\rho^2 = \frac{J_0^2(\omega_m \tau)}{1 + (\Delta\omega\, \overline{T})^2} \tag{4.31b}$$

After a tedious procedure to determine the cofactors $|\Lambda|_{jk}$, we arrive at the following joint distribution

$$p(X_1, Y_1, X_2, Y_2) = \frac{1}{4\pi^2 \sigma^4 (1 - \rho^2)}$$

$$\times \exp\left\{ -\frac{1}{2\sigma^8(1-\rho^2)^2}\left[\sigma^2(X_1^2 + Y_1^2 + X_2^2 + Y_2^2)\right.\right.$$

$$\left.\left. - 2\mu_1(X_1 X_2 + Y_1 Y_2) - 2\mu_2(X_1 Y_2 - X_2 Y_1)\right]\right\} \tag{4.32}$$

This density can be expressed in terms of the random variables $r_1$, $\varphi_1$, $r_2$, and $\varphi_2$. Hence

$$p(r_1, r_2, \varphi_1, \varphi_2) = |J| p(X_1, Y_1, X_2, Y_2) \tag{4.33}$$

where $|J|$ is the Jacobian of the transformation, given by

$$J = \begin{bmatrix} \dfrac{\partial X_1}{\partial r_1} & \dfrac{\partial X_1}{\partial \varphi_1} & \dfrac{\partial X_1}{\partial r_2} & \dfrac{\partial X_1}{\partial \varphi_2} \\[2mm] \dfrac{\partial Y_1}{\partial r_1} & \dfrac{\partial Y_1}{\partial \varphi_1} & \dfrac{\partial Y_1}{\partial r_2} & \dfrac{\partial Y_1}{\partial \varphi_2} \\[2mm] \dfrac{\partial X_2}{\partial r_1} & \dfrac{\partial X_2}{\partial \varphi_1} & \dfrac{\partial X_2}{\partial r_2} & \dfrac{\partial X_2}{\partial \varphi_2} \\[2mm] \dfrac{\partial Y_2}{\partial r_1} & \dfrac{\partial Y_2}{\partial \varphi_1} & \dfrac{\partial Y_2}{\partial r_2} & \dfrac{\partial Y_2}{\partial \varphi_2} \end{bmatrix} \tag{4.34}$$

After determining the required derivatives of Equation 4.20 and substituting them into Equation 4.34 we obtain (unbelievably)

$$|J| = r_1 r_2 \tag{4.35}$$

Therefore,

$$
\begin{aligned}
& p(r_1, r_2, \varphi_1, \varphi_2) \\[2mm]
&= \frac{r_1 r_2}{4\pi^2 \sigma^4 (1 - \rho^2)} \\[2mm]
& \quad \times \exp\left\{ -\frac{1}{2\sigma^8(1-\rho^2)^2} \left[ \sigma^2(r_1^2 + r_2^2) - 2r_1 r_2 \mu_1 \cos(\varphi_2 - \varphi_1) \right.\right. \\[2mm]
& \qquad\qquad\qquad\qquad\qquad\qquad \left.\left. - 2r_1 r_2 \mu_2 \sin(\varphi_2 - \varphi_1) \right] \right\}
\end{aligned} \tag{4.36}
$$

The joint density of the envelope is

$$p(r_1, r_2) = \int_0^{2\pi} \int_0^{2\pi} p(r_1, r_2, \varphi_1, \varphi_2)\, d\varphi_1\, d\varphi_2$$

Then

$$p(r_1, r_2) = \frac{r_1 r_2}{\sigma^4(1 - \rho^2)} \exp\left[ -\frac{r_1^2 + r_2^2}{2\sigma^2(1 - \rho^2)} \right] I_0\left( \frac{r_1 r_2 \rho}{\sigma^2(1 - \rho^2)} \right) \tag{4.37}$$

where $I_0(x)$ is the modified Bessel function of zero order (refer to Equation 3.90b).

In a similar way, the joint density of the phases is

$$p(\varphi_1, \varphi_2) = \int_0^\infty \int_0^\infty p(r_1, r_2, \varphi_1, \varphi_2) \, dr_1 \, dr_2$$

Then

$$p(\varphi_1, \varphi_2) = \left(\frac{1 - \rho^2}{4\pi^2}\right) \frac{(1 - U^2)^{1/2} + U \cos^{-1}(-U)}{(1 - U^2)^{3/2}} \qquad (4.38)$$

where

$$U = \rho \cos\left[\varphi_2 - \varphi_1 + \tan^{-1}(\Delta\omega \, \overline{T})\right]$$

Note that if both signals are independent, then $\mu_1 = \mu_2 = \rho = 0$ and $I_0(0) = 1$. Therefore,

$$p(r_1, r_2) = p(r_1)p(r_2) = \frac{r_1 r_2}{\sigma^4} \exp\left[-\frac{r_1^2 + r_2^2}{2\sigma^2}\right]$$

In the same way, $U = 0$ and

$$p(\varphi_1, \varphi_2) = p(\rho_1, \rho_2) = \frac{1}{4\pi^2}$$

as expected.

### 4.5.3 Envelope Correlation

The normalized envelope covariance, also known as the envelope correlation coefficient, is given by[4]

$$\rho_r = \frac{\text{Cov}(r_1, r_2)}{\sqrt{\text{Var}(r_1)} \sqrt{\text{Var}(r_2)}} = \frac{E[r_1 r_2] - E[r_1]E[r_2]}{\sqrt{E[r_1^2] - E^2[r_1]} \sqrt{E[r_2^2] - E^2[r_2]}}$$

$$(4.39)$$

From Section 3.4.2

$$E[r_1] = E[r_2] = \sqrt{\pi/2} \, \sigma$$

$$(4.40)$$

$$\text{Var}(r_1) = \text{Var}(r_2) = (2 - \pi/2)\sigma^2$$

The joint moment of the envelope is

$$E[r_1 r_2] = \int_0^\infty \int_0^\infty r_1 r_2 p(r_1, r_2) \, dr_1, dr_2 \tag{4.41}$$

where $p(r_1, r_2)$ is given by Equation 4.37. Using tables of integrals,[5] we obtain

$$E[r_1, r_2] = \frac{\pi}{2}\sigma^2 F\left(-\frac{1}{2}, -\frac{1}{2}; 1; \rho^2\right) \tag{4.42}$$

where $F(a, b; c; z)$ is the hypergeometric function. This function can be evaluated by[6]

$$F(a, b; c; z) = \sum_{n=0}^{\infty} \frac{(a)_n (b)_n}{(c)_n} \frac{z^n}{n!}$$

$$= 1 + \frac{ab}{c} z + \frac{a(a+1)b(b+1)}{c(c+1)} \frac{z^2}{2!} + \cdots \tag{4.43}$$

Expanding Equation 4.42 as in Equation 4.43 we obtain

$$E[r_1 r_2] = \frac{\pi}{2}\sigma^2 \left[1 + \left(\frac{1}{2}\right)^2 \rho^2 + \left(\frac{1}{2}\right)^6 \rho^4 + \left(\frac{1}{2}\right)^9 \rho^6 + \cdots \right]$$

If the higher-order terms are neglected,

$$E[r_1 r_2] \simeq \frac{\pi}{2}\sigma^2 \left[1 + \left(\frac{\rho}{2}\right)^2\right] \tag{4.44}$$

With Equations 4.40 and 4.44 in Equation 4.39 we have

$$\rho_r = \frac{\pi}{4(4 - \pi)}\rho^2 \simeq \rho^2 \tag{4.45}$$

Therefore, from Equation 4.31b

$$\rho_r = \frac{J_0^2(\omega_m \tau)}{1 + (\Delta \bar{\omega} \bar{T})^2} \tag{4.46}$$

Note that, from Equation 4.46 the larger the frequency separation $\Delta \omega$ and/or the larger the delay spread $\bar{T}$, the smaller the correlation coefficient. Moreover, the correlation coefficient is a quadratic function of $J_0(\omega_m \tau)$. Figure 4.3 shows some plots of the correlation coefficient.

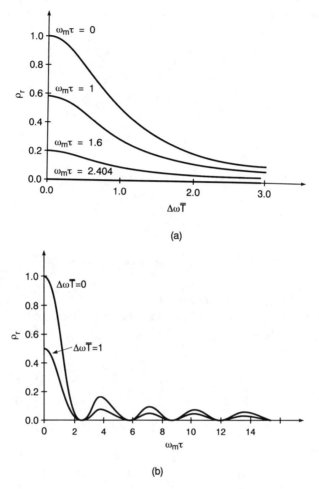

**Figure 4.3.** Envelope correlation: (a) versus frequency separation times delay spread; (b) versus Doppler shift times time delay.

### 4.5.4 Phase Correlation

In a way similar to the envelope correlation, the phase correlation is given by

$$\rho_\varphi = \frac{E[\varphi_1\varphi_2] - E[\varphi_1]E[\varphi_2]}{\sqrt{E[\varphi_1^2] - E^2[\varphi_1]}\sqrt{E[\varphi_2^2] - E^2[\varphi_2]}} \tag{4.47}$$

Both phases $\varphi_1$ and $\varphi_2$ are random variables uniformly distributed in the range $0-2\pi$ rad. In other words,

$$p(\varphi_1) = \begin{cases} 1/2\pi, & 0 \le \varphi_i \le 2\pi, i = 1,2 \\ 0, & \text{otherwise} \end{cases}$$

Therefore,

$$E[\varphi_1] = E[\varphi_2] = \frac{1}{2\pi} \int_0^{2\pi} \varphi \, d\varphi = \pi$$

$$E[\varphi_1^2] = E[\varphi_2^2] = \frac{1}{2\pi} \int_0^{2\pi} \varphi^2 \, d\varphi = \frac{4}{3}\pi^2$$

(4.48)

Hence

$$\rho_\varphi = \frac{3}{\pi^2}\left(E[\varphi_1\varphi_2] - \pi^2\right)$$

(4.49)

The expectation $E[\varphi_1\varphi_2]$ is given by

$$E[\varphi_1\varphi_2] = \int_0^{2\pi}\int_0^{2\pi}\varphi_1\varphi_2 p(\varphi_1, \varphi_2) \, d\varphi_1 \, d\varphi_2$$

(4.50)

where $p(\varphi_1, \varphi_2)$ is given by Equation 4.38.

The integral in Equation 4.50 does not lend itself to a closed-form evaluation. A good approximation to this integral is given by Jakes[7] as follows:

$$E[\varphi_1\varphi_2] = \pi^2\left[1 + \Gamma(\rho, \phi) + 2\Gamma^2(\rho, \phi) - \frac{1}{4\pi^2}\sum_{n=1}^{\infty}\frac{\rho^{2n}}{n^2}\right]$$ (4.51)

where

$$\Gamma(\rho, \phi) = \frac{1}{2\pi}\sin^{-1}(\rho\cos\phi)$$

$$\phi = -\tan^{-1}(\Delta\omega\,\bar{T})$$

and

$$\rho^2 = \frac{J_0^2(\omega_m\tau)}{1 + (\Delta\omega\,\bar{T})^2}$$

Finally, the phase correlation is

$$\rho_\varphi = 3\Gamma(\rho, \phi)[1 + 2\Gamma(\rho, \phi)] - \frac{3}{4\pi^2}\sum_{n=1}^{\infty}\frac{\rho^{2n}}{n^2}$$

(4.52)

For a time delay $\tau = 0$ we have the plot of the phase correlation as shown in Figure 4.4.

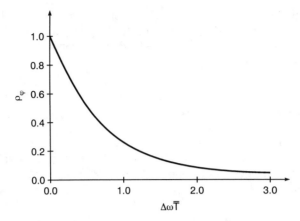

**Figure 4.4.** Phase correlation versus frequency separation times delay spread.

### 4.5.5 Coherence Bandwidth

There is not a rigid rule used to determine the coherence bandwidth. One well-accepted criterion establishes that the coherence bandwidth corresponds to the frequency separation for which the correlation factor equals 0.5.

Let $B_c$ be the coherence bandwidth. As far as envelope correlation is concerned, for a time delay $\tau = 0$, it follows that, from Equation 4.46,

$$\rho_r = \frac{J_0^2(\omega_m 0)}{1 + (\Delta\omega\,\overline{T})^2} = 0.5$$

Then,

$$B_c \triangleq \frac{\Delta\omega}{2\pi} = \frac{1}{2\pi\overline{T}} \tag{4.53}$$

As for the phase correlation, for $\rho_\varphi = 0.5$ with $\tau = 0$, it is seen that* $\Delta\omega\,\overline{T} = 0.5$. Hence

$$B_c = \frac{1}{4\pi\overline{T}} \tag{4.54}$$

Note that the coherence bandwidth given by Equation 4.53 is always greater than that given by Equation 4.54. Therefore, to be on the safe side, Equation 4.53 should preferably be used. Thus, for a delay spread $\overline{T} = 2\ \mu s$, the corresponding $B_c$ is 80 kHz. Values of delay spread exceeding 10 $\mu s$ are

---

*Refer to Figure 4.4 for a quick assessment.

comparatively rare[1]. Consider $\bar{T} = 5$ $\mu$s, such that $B_c = 32$ kHz. Consequently, a narrowband FM system of up to 30-kHz bandwidth would essentially experience nonselective fading. However, wideband systems will obviously experience selective fading.

## 4.6  LEVEL CROSSING RATE

Level crossing rate is defined as the average number of times a fading signal crosses a given signal level within a certain period. Let the time variation of the received envelope $r$ be $\dot{r}$ ($\dot{r} \triangleq dr/dt$) and let the crossing signal level be $R$. Hence, the level crossing rate $R_c$ is the mean value of $\dot{r}$ at $r = R$, i.e.,

$$R_c = E[\dot{r}|r = R] = \int_0^\infty \dot{r} p(R, \dot{r}) \, d\dot{r} \tag{4.55}$$

where $p(R, \dot{r})$ is the joint probability density function of $r$ and $\dot{r}$ at $r = R$.

### 4.6.1  Joint Probability Density Function

Consider the fading signal as described by Equations 4.19b. Their derivatives are

$$\dot{X} = E_0 \beta v \sum_{i=1}^{n} - a_i \sin(\omega_i t - \omega T_i) \cos \theta_i$$

$$\dot{Y} = E_0 \beta v \sum_{i=1}^{n} a_i \cos(\omega_i t - \omega T_i) \cos \theta_i \tag{4.56}$$

By following the same steps as in Section 4.5.2 we arrive at

$$E[\dot{X}] = E[\dot{Y}] = E[XY] = E[X\dot{X}] = E[X\dot{Y}]$$

$$= E[Y\dot{X}] = E[Y\dot{Y}] = E[\dot{X}\dot{Y}] = 0$$

$$E[X^2] = E[Y^2] = E_0^2/2 = \sigma^2 \tag{4.57}$$

$$E[\dot{X}^2] = E[\dot{Y}^2] = (E_0 \beta v/2)^2 \triangleq \dot{\sigma}^2$$

(note that $\dot{\sigma}$ is not the derivative of $\sigma$, but just a notation).

These results are then used to compose the covariance matrix $\Lambda$. Note that, from Equations 4.57 $\Lambda$ is a diagonal matrix with diagonal elements equal to $\sigma^2$, $\sigma^2$, $\dot{\sigma}^2$, and $\dot{\sigma}^2$.

Accordingly, its determinant is $|\Lambda| = \sigma^4\dot{\sigma}^4$. Using these results in Equation 4.21 we obtain

$$p(X,Y,\dot{X},\dot{Y}) = \frac{1}{4\pi^2\sigma^2\dot{\sigma}^2} \exp\left[-\frac{1}{2}\left(\frac{X^2 + Y^2}{\sigma^2} + \frac{\dot{X}^2 + \dot{Y}^2}{\dot{\sigma}^2}\right)\right] \quad (4.58)$$

With the transformation of variables we have

$$p(r,\dot{r},\varphi,\dot{\varphi}) = |J|p(X,Y,\dot{X},\dot{Y}) \quad (4.59)$$

where $J$ is the Jacobian of the transformation. From Equations 4.19b

$$X = r\cos\varphi$$

$$Y = r\sin\varphi$$

Therefore

$$\dot{X} = \dot{r}\cos\varphi - r\dot{\varphi}\sin\varphi$$

$$\dot{Y} = \dot{r}\sin\varphi + r\dot{\varphi}\cos\varphi \quad (4.60)$$

The Jacobian is easily found to be* $|J| = r^2$. Thus

$$p(r,\dot{r},\varphi,\dot{\varphi}) = \frac{r^2}{4\pi^2\sigma^2\dot{\sigma}^2} \exp\left[-\frac{1}{2}\left(\frac{r^2}{\sigma^2} + \frac{\dot{r}^2 + r^2\dot{\varphi}^2}{\dot{\sigma}^2}\right)\right] \quad (4.61)$$

The distribution $p(r,\dot{r})$ can be obtained by integrating Equation 4.61 over $\varphi$ from 0 to $2\pi$ and over $\dot{\varphi}$ from $-\infty$ to $\infty$. Hence

$$p(r,\dot{r}) = \frac{r}{\sigma^2\sqrt{2\pi\dot{\sigma}^2}} \exp\left[-\frac{1}{2}\left(\frac{r^2}{\sigma^2} + \frac{\dot{r}^2}{\dot{\sigma}^2}\right)\right] \quad (4.62)$$

Using Equation 4.62 in Equation 4.55 we have

$$R_c = \frac{1}{\sqrt{2\pi}} \frac{\dot{\sigma}}{\sigma^2} R \exp\left[-\left(\frac{R}{\sqrt{2}\sigma}\right)^2\right] \quad (4.63)$$

---

*Equation 4.34 may be used with the following adaptations: $X,Y$ in place of $X_1,Y_1$; $\dot{X},\dot{Y}$ in place of $X_2,Y_2$; $r,\varphi$ in place of $r_1,\varphi_1$; $\dot{r},\dot{\varphi}$ in place of $r_2,\varphi_2$.

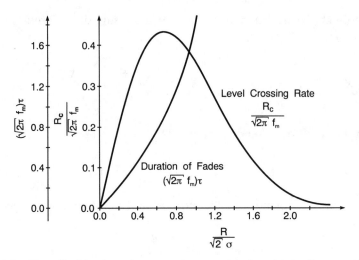

**Figure 4.5.** Normalized level crossing rate and average duration of fades (linear scale).

With $\sigma$ and $\dot{\sigma}$ given by Equations 4.57

$$R_c = \sqrt{2\pi}\, f_m \frac{R}{\sqrt{2}\,\sigma} \exp\left[-\left(\frac{R}{\sqrt{2}\,\sigma}\right)^2\right] \tag{4.64}$$

where $f_m = \omega_m/2\pi = \beta v/2\pi = v/\lambda$ is the maximum Doppler shift. The curve $R_c/\sqrt{2\pi}\,f_m$ versus $R/\sqrt{2}\,\sigma$ is plotted in Figures 4.5 and 4.6.

Note that the maximum rate occurs at $R = \sigma$ (obtained from $dR_c/d(R/\sqrt{2}\,\sigma) = 0$). In other words, because the rms value of the envelope* is $\sqrt{2}\,\sigma$, the maximum rate occurs at a level 3 dB below this value. In this case $R_c = 1.08\, f_m$. As an example, consider a mobile travelling at 72 km/h and a frequency of 900 MHz. Then $f_m = v/\lambda = 60$ Hz and $R_c = 65$ crossings per second at the rms value of the received envelope.

Let $N_c$ be the mean number of crossings within a period of time $T$. Then $N_c = R_c T$, and at the maximum $R_c$

$$N_c = 1.08 v T/\lambda = 1.08 d/\lambda$$

where $d$ is the distance travelled by the vehicle in time $T$. Therefore, the mean distance between crossings at the maximum rate is $d/N_c = 0.93\lambda$. It is shown in Appendix 5A that the overall mean distance between crossings is approximately equal to $\lambda/2$.

---

*Refer to Section 3.4.2.

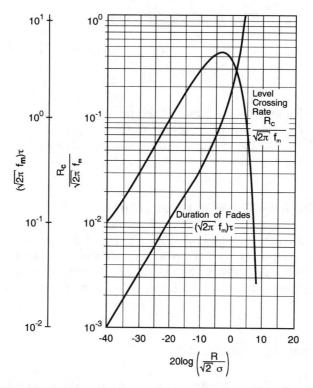

**Figure 4.6.** Normalized level crossing rate and average duration of fades (logarithmic scale).

## 4.7 AVERAGE DURATION OF FADES

Given a signal level $R$, the average duration of fades is the ratio between the total time the received signal is below $R$ and the total number of fades, both measured during a time interval $T$. Let $\tau$ be the average duration of fades and $\tau_i$ the duration of each fade. Then

$$\tau = \frac{\sum \tau_i}{R_c T} \tag{4.65}$$

The ratio $\sum \tau_i / T$ in Equation 4.65 corresponds to the probability that the signal is below $R$. Hence

$$\tau = \frac{1}{R_c} \text{prob}[r \leq R] = \frac{1}{R_c} \int_0^R p(r)\, dr \tag{4.66}$$

Using Equation 3.88 in Equation 4.66 we obtain

$$\tau = \frac{1}{\sqrt{2\pi}\, f_m(R/\sqrt{2}\,\sigma)}\left[\exp\left(\frac{R}{\sqrt{2}\,\sigma}\right)^2 - 1\right] \qquad (4.67)$$

Figures 4.5 and 4.6 show a plot of the normalized average fade duration. Note that, at the maximum crossing rate $(R = \sigma)$, the average duration of fade is $0.33/f_m$. For the example of the previous section, where $f_m = 60$ Hz, the fade duration is 5.5 ms.

## 4.8  RANDOM FREQUENCY MODULATION

The random nature of the time-varying phase of the fading signal causes a phenomenon known as *random frequency modulation* (random FM). As we shall see, the random FM adds an extra noise component to the already deteriorated signal, having a power proportional to the square of the vehicle speed.

### 4.8.1  Probability Distribution

Because the random variable $\varphi$ describes the fading signal's phase, its derivative $\dot{\varphi}$ characterizes the random FM. Its probability density function $p(\dot{\varphi})$ is obtained from Equation 4.61 as follows

$$p(\dot{\varphi}) = \int_0^\infty \int_{-\infty}^\infty \int_0^{2\pi} p(r,\dot{r},\varphi,\dot{\varphi})\, dr\, d\dot{r}\, d\varphi$$

Hence

$$p(\dot{\varphi}) = \frac{\sigma}{2\dot{\sigma}}\left[1 + \frac{\sigma^2}{\dot{\sigma}^2}\dot{\varphi}^2\right]^{-3/2} \qquad (4.68)$$

The corresponding distribution is

$$P(\dot{\Phi}) = \text{prob}(\dot{\varphi} \le \dot{\Phi}) = \int_{-\infty}^{\dot{\Phi}} p(\dot{\varphi})\, d\dot{\varphi}$$

Then

$$P(\dot{\Phi}) = \frac{1}{2} + \frac{\sigma}{2\dot{\sigma}}\dot{\Phi}\left(1 + \frac{\sigma^2}{\dot{\sigma}^2}\dot{\Phi}^2\right)^{-1/2} \qquad (4.69)$$

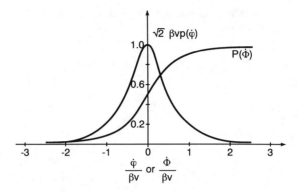

**Figure 4.7.**  Probability density and distribution functions of the random frequency modulation.

From Equations 4.57, $\sigma/\dot{\sigma} = \sqrt{2}/\beta v$. Thus

$$p(\dot{\varphi}) = \frac{1}{\sqrt{2}\,\beta v}\left[1 + \frac{2}{(\beta v)^2}\dot{\varphi}^2\right]^{-3/2} \qquad (4.70)$$

$$P(\dot{\Phi}) = \frac{1}{2} + \frac{1}{\sqrt{2}\,\beta v}\dot{\Phi}\left[1 + \frac{2}{(\beta v)^2}\dot{\Phi}^2\right]^{-3/2} \qquad (4.71)$$

Both functions are plotted in Figure 4.7

### 4.8.2  Power Spectrum

The power spectrum of a power signal is defined as the Fourier transform of its autocorrelation function (refer to Appendix 9A). Let $R_{\dot{\varphi}}(\tau)$ be the autocorrelation function of $\dot{\varphi}$. Then

$$R_{\dot{\varphi}}(\tau) = E\left[\dot{\varphi}(t)\dot{\varphi}(t-\tau)\right]$$

Rice[8] showed that

$$R_{\dot{\varphi}}(\tau) = -\frac{1}{2}\left\{\left[\frac{R_x'(\tau)}{R_x(\tau)}\right]^2 - \left[\frac{R_x''(\tau)}{R_x(\tau)}\right]^2\right\}\ln\left\{1 - \left[\frac{R_x(\tau)}{R_x(0)}\right]^2\right\} \qquad (4.72)$$

where $R_x(\tau)$ is the autocorrelation function of $X$. The random variable $X$ is given by Equations 4.19b.

We have already determined the correlation $E[X_1 X_2]$ of the signals $X_1$ and $X_2$ as given by Equation 4.28. Note that, if the frequency separation $\Delta\omega$

between $X_1$ and $X_2$ is nil, then Equation 4.28 is the autocorrelation function of $X_1$ ($= X$). Accordingly,

$$R_x(\tau) = \sigma^2 J_0(\omega_m \tau)$$

Then

$$\frac{R_x(\tau)}{R_x(0)} = J_0(\omega_m \tau)$$

$$\frac{R'_x(\tau)}{R_x(\tau)} = \frac{1}{R_x(\tau)} \frac{dR_x(\tau)}{d\tau} = -\omega_m \frac{J_1(\omega_m \tau)}{J_0(\omega_m \tau)} \tag{4.73}$$

$$\frac{R''_x(\tau)}{R_x(\tau)} = \frac{1}{R_x(\tau)} \frac{d^2 R_x(\tau)}{d\tau^2} = \omega_m^2 \left[ \frac{J_1(\omega_m \tau)}{\omega_m \tau J_0(\omega_m \tau)} - 1 \right]$$

Therefore, the power spectrum $S_{\dot{\varphi}}(f)$ is

$$S_{\dot{\varphi}}(f) = \int_{-\infty}^{\infty} R_{\dot{\varphi}}(\tau) \exp(-j\omega\tau) \, d\tau = 2\int_0^{\infty} R_{\dot{\varphi}}(\tau) \cos(\omega\tau) \, d\tau \tag{4.74}$$

This integral has been evaluated by numerical methods[7] and the result is shown in Figure 4.8.

Within the audio frequency range (300–3400 Hz), an asymptotic form, as that given by[7] Equation 4.75, can be used with fairly accurate results, for

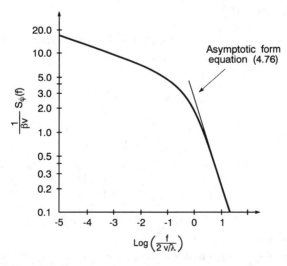

**Figure 4.8.**   Random FM power spectrum.

mobiles travelling with a speed not exceeding 96 km/h (60 mi/h). Then

$$\lim_{f \to \infty} S_{\dot\phi}(f) = \left[ \left( \frac{\dot\sigma}{\sigma} \right)^2 - \left( \frac{\dot\sigma_{xy}}{\sigma} \right)^4 \right] f^{-1} \tag{4.75}$$

where

$$\dot\sigma_{xy} = E\left[ X_1 \dot Y_2 \right] = -E\left[ X_2 \dot Y_1 \right] = 0 \quad \text{(refer to Equations 4.57)}$$

Then, from Equations 4.57 and 4.75

$$\frac{1}{\beta v} S_{\dot\phi}(f) \simeq \frac{\beta v}{2f} \tag{4.76}$$

Consequently, the noise power $N$ within the audio frequency range $(\omega_1, \omega_2)$, due to random FM, is

$$N = \int_{\omega_1}^{\omega_2} S_{\dot\phi}(f) \, df = \frac{(\beta v)^2}{2} \ln\left( \frac{\omega_2}{\omega_1} \right) \tag{4.77}$$

It can be seen that the noise power is a quadratic function of vehicle speed. Therefore, a mobile, initially running at a speed $v$ and then at $2v$ ($2v \le 98$ km/h), experiences an increase of the noise power due to the random FM of 6 dB. That is,

$$\frac{N_2}{N_1} = \left( \frac{2v}{v} \right)^2 = 4 = 6 \; dB$$

where $N_1$ is the noise power at speed $v$ and $N_2$ is this power at speed $2\ v$.

## 4.9 POWER SPECTRA OF THE RECEIVED SIGNAL

Let $\theta$ be the incidence angle of a radio wave received by the mobile, and let $\omega(\theta)$ the instantaneous angular frequency of such a received wave. If the vehicle moves at a constant speed $v$, then there will be a Doppler frequency shift equal to $\beta v \cos \theta$. Therefore,

$$\omega(\theta) = \omega_c + \omega_m \cos \theta \tag{4.78}$$

where $\omega_c$ is the carrier frequency and $\omega_m = \beta v$ is the maximum Doppler shift. Note that, from Equation 4.78, the effective bandwidth of the received wave is equal to twice the maximum Doppler shift.

Consider a mobile station with directional antenna* in the horizontal plane, having a gain equal to $G(\theta)$. If the distribution of $\theta$ is $p(\theta)$, then within a differential angle $d\theta$, the total power is $W_0 G(\theta) p(\theta)\, d\theta$, where $W_0$ is the power that would be received by an isotropic antenna. This obviously equals the differential variation of the received power with frequency $S(f)\, df$. Noting that $\omega(\theta) = \omega(-\theta)$, then

$$S(\omega)|d\omega| = 2\pi W_0 [G(\theta) p(\theta) + G(-\theta) p(-\theta)]|d\theta| \qquad (4.79)$$

where $\omega = 2\pi f$. From Equation 4.78,

$$|d\omega| = |-\beta v \sin\theta|\,|d\theta| = \left[(\beta v)^2 - (\omega - \omega_c)^2\right]^{1/2}|d\theta| \qquad (4.80)$$

where, for simplicity, we have used $\omega(\theta) \triangleq \omega$. Assume that the received power is uniformly distributed in the range $-\pi \le \theta \le \pi$. Hence

$$p(\theta) = \begin{cases} 1/2\pi, & -\pi \le \theta \le \pi \\ 0, & \text{otherwise} \end{cases} \qquad (4.81)$$

With Equations 4.80 and 4.79 we obtain

$$S(\omega) = \frac{W_0 [G(\theta) + G(-\theta)]}{\omega_m \sqrt{1 - \left(\dfrac{\omega - \omega_c}{\omega_m}\right)^2}} \qquad (4.82)$$

Suppose that the transmitted signal is vertically polarized. Then the electric field $E_z$ will be in the $z$ direction and can be sensed by a vertical monopole (whip) antenna. In the same way, small loop antennas along the $x$ axis or along the $y$ axis can be used to sense the magnetic field $H_y$ or $H_x$, respectively. The corresponding power spectra are determined as follows:

1. *Power spectrum of the electric field $E_z$—vertical monopole (whip) antenna, $G(\theta) = G(-\theta) = \frac{3}{2}$,*

$$S(\omega) = \frac{3W_0}{\omega_m}\left[1 - \left(\frac{\omega - \omega_c}{\omega_m}\right)^2\right]^{-1/2} \qquad (4.83a)$$

2. *Power spectrum of the magnetic field $H_x$—small loop antenna along the $y$ axis, $G(\theta) = G(-\theta) = \frac{3}{2}\sin^2\theta$,*

$$S(\omega) = \frac{3W_0}{\omega_m}\left[1 - \left(\frac{\omega - \omega_c}{\omega_m}\right)^2\right]^{1/2} \qquad (4.83b)$$

---

*Refer to Section 3.1.

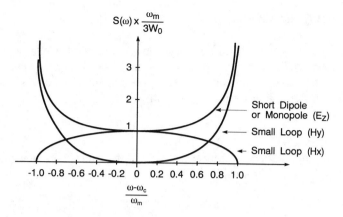

**Figure 4.9.**   Spectra of the three field components.

3. *Power spectrum of the magnetic field $H_y$*—small loop antenna along the $x$ axis, $G(\theta) = G(-\theta) = \frac{3}{2}\cos^2\theta$,

$$S(\omega) = \frac{3W_0}{\omega_m}\left(\frac{\omega - \omega_c}{\omega_m}\right)^2\left[1 - \left(\frac{\omega - \omega_c}{\omega_m}\right)^2\right]^{-1/2} \quad (4.83c)$$

These three power spectra are plotted in Figure 4.9.

## 4.10   FIELD MEASUREMENT

Having described the main mobile radio propagation phenomena and the received signal characteristics, it is now appropriate to address the main topics to be looked at when field measurements are to be carried out. In general, we are interested in characterizing the radio path between a stationary transmitter (base station) and a mobile vehicle (mobile station). The parameters to be known prior to the measurements include the power of the transmitter, the types of transmitting and receiving antennas, their gains and heights, and the carrier frequency. The mobile receiver must be equipped with data recording facilities to collect and store the equally spaced signal samples. A transducer, set up at the vehicle wheel, can be used to interrupt a processor dedicated to data acquisition. In case the vehicle is stationary, a clock generator can be used for the same purpose.

We now describe the relevant points in the field-strength measurement process.

## 4.10.1  Sampling Interval

As shown in Figure 4.9, for a vehicle moving at speed $v$, the bandwidth of the received carrier is equal to $2\omega_m$, where $\omega_m = \beta v$ is the maximum Doppler shift. If the samples are taken at regular intervals of distance $d$, the corresponding sampling frequency $\omega_s$ will be

$$\omega_s = 2\pi v/d \text{ rad/s}$$

The sampling theorem requires that the sampling frequency must be at least twice the signal bandwidth. Then

$$\omega_s = 2\pi v/d \geq 2(\text{bandwidth}) = 2(2\beta v)$$

Hence,

$$d \leq \lambda/4 \tag{4.84}$$

If, for instance, $f_c = 900$ MHz ($\lambda = \frac{1}{3}$ m), then the samples must be taken at intervals less than or equal to 8.3 cm.

## 4.10.2  Separation of Slow and Fast Fading

The received signal in a mobile radio environment experiences two types of fading: slow fading, due to topographical changes, and fast fading, due to multipath propagation. Therefore, the signal $s(t)$ has an area mean $m(t)$ (slow fading) and a local mean $r(t)$ (fast fading) modifying the area mean multiplicatively. Then $s(t) = m(t)r(t)$. If the variables $s(t)$, $m(t)$, and $r(t)$, expressed in decibels, are denoted by $S(t)$, $M(t)$, and $R(t)$, respectively, then

$$S(t) = M(t) + R(t) \tag{4.85}$$

From field measurement we obtain $S(t)$. We may wish to separate (filter out) $M(t)$ from $R(t)$ to determine, for instance, their distributions. The mean signal $M(t)$ is expected to vary slowly compared with $R(t)$. Accordingly, by conveniently low-pass filtering $S(t)$ we may be able to obtain $M(t)$ and then extract $R(t)$ by performing $R(t) = S(t) - M(t)$. The filtering process can be carried out by averaging the samples over a range of $2k + 1$ samples ($k$ integer) where the mean signal is considered to be sensibly constant. Let $S_i$ be the $i$th sample. Hence, the $i$th estimated area mean $\hat{M}_i$ is

$$\hat{M}_i = \frac{1}{2k+1} \sum_{j=-k}^{k} S_{i+j} \tag{4.86}$$

Note that the signal is continuously averaged over $2k + 1$ samples, symmetrically distributed about the time index $i$. It is shown in Appendix 4B that

Equation 4.86 performs the function of a digital low-pass filter. The estimated fast fading signal is then

$$\hat{R}_i = S_i - \hat{M}_i \tag{4.87}$$

The main concern now is with the determination of the filter length $2k + 1$ to perform an appropriate filtering. If the averaging is carried out over a sufficiently small number of neighboring samples, then the estimated mean will contain a large number of high-frequency components. In this case the filter is considered to have a large bandwidth. On the other hand, if the averaging is over a sufficiently large number of neighboring samples, the mean signal may not remain "constant" within such an interval so that the slow variation of the signal can be lost. In this case the filter is considered to have a small bandwidth and the estimated signal will be mostly constituted of low-frequency components. It is demonstrated in Appendix 4B that the cutoff frequency $f_{co}$ of the digital filter shown in Equation 4.86 is

$$f_{co} = \frac{f_s}{2k + 1} \tag{4.88}$$

where $f_s = \omega_s/2\pi$ is the sampling frequency. From Equation 4.84 we see that $f_s \geq 4\beta v/(2\pi) = 4f_m$. Consider that

$$f_s = 4\gamma f_m = 4\gamma v/\lambda \tag{4.89}$$

where $\gamma \geq 1$.

The averaging interval may be chosen so as to have the filter cutoff frequency equal to

$$f_{co} = \alpha f_m \tag{4.90}$$

where $0 \leq \alpha \leq 1$.

From Equations 4.88 through 4.90, we obtain

$$2k + 1 = 4\gamma/\alpha \tag{4.91}$$

The $2k + 1$ samples are collected within a time interval equal to $(2k + 1)/f_s$. If the vehicle moves at a speed $v$, then the distance $L$, where the signal is considered to be "sensibly constant", is

$$L = \frac{2k + 1}{f_s} v \tag{4.92}$$

With Equations 4.89 and 4.91 in Equation 4.92 we obtain

$$L = \lambda/\alpha \tag{4.93}$$

As an example, consider a carrier $f_c$ at 900 MHz. Assume that $\gamma = 2$, corresponding to a sampling frequency equal to four times the bandwidth (or eight times the maximum Doppler shift, which equals twice the minimum required sampling rate). Moreover, let $\alpha = 0.04$, corresponding to a cutoff frequency equal to 4% of the maximum Doppler shift. Hence, the number of samples $2k + 1$ in the filtering process (filter's length) is equal to 200. This corresponds to a distance of $L = 25\lambda = 8.33$ m. If the car is moving at $v = 72$ km/h, then the sampling rate will be 480 samples per second, with each sample being taken at 4-cm intervals.

### 4.10.3   Validation of the Measurements

One set of measurements may not be necessarily enough to consider the collected data as representative of the signal strength in that geographical region. It may be necessary to take several runs at different conditions (time of the day, weather, etc.) before we can validate our results. Even if we keep the "same" measuring conditions, another set of measurements cannot be expected to coincide with the previous set. From Equation 4.46 we understand that two same-frequency signals ($\Delta\omega = 0$) are considered to be uncorrelated if $J_0^2(2\pi l/\lambda) = 0$, occurring for $2\pi l/\lambda = 2.404$. Accordingly, $l = 0.38\lambda \simeq 0.5\lambda$. For a 900-MHz carrier, this implies a distance of 13 cm. Thus, it is very unlikely that any other set of runs will be taken within this distance. Moreover, the surrounding objects may not be exactly the same anymore.

Having collected the data we need to assess the accuracy of our results, let $p = \text{prob}(R \leq V)$ be the true cumulative distribution of the Rayleigh signal $R(t)$. We estimate $p$ by $\hat{p}$, where $\hat{p}$ is the number of occurrences of the event $R_i \leq V$, averaged over the total number $N$ of samples. Then

$$\hat{p} = \frac{1}{N} \sum_{j=1}^{N} p_i \tag{4.94}$$

where

$$p_i = \begin{cases} 1, & R_i \leq V \\ 0, & \text{otherwise*} \end{cases}$$

The expected value of $\hat{p}$ and its variance $\hat{\sigma}^2$ are (see Appendix 4C)

$$E[\hat{p}] = p \tag{4.95}$$

$$\hat{\sigma}^2 = \frac{1}{N} p(1 - p) \tag{4.96}$$

---

*This can be written as $p_i = 1 - \frac{1}{2}[\text{sgn}(R_i - V) + |\text{sgn}(R_i - V)|]$, where $\text{sgn}(x) = 1, 0, -1$ for $x > 0$, $x = 0$, $x < 0$, respectively.

Note that, from Equation 4.96 the larger the number of samples, the less the estimated probability will deviate from the true probability. In Equation 4.96 we are assuming the samples to be uncorrelated. However, as we have seen, uncorrelated samples occur at distances $l$ equal to or greater than $0.38\lambda$.

The $N$ samples are taken within a distance of $L_T = N(1/f_s)v$. With Equation 4.89 in Equation 4.92, where we replace $L$ by $L_T$ and $2k + 1$ by $N$, we obtain

$$L_T \triangleq \text{total distance} = N\lambda/4\gamma$$

Hence, the number of uncorrelated samples within $N$ is $L_T/l = 0.65\,N/\gamma$. We now modify Equation 4.96 by introducing this perturbation factor. Then

$$\hat{\sigma}^2 \simeq \frac{\gamma}{0.65N}p(1-p) \qquad (4.97)$$

If the distribution of $\hat{R}$ is plotted on Rayleigh-scaled axes, the resulting curve will be a straight line. A straight line is also obtained if the distribution of $\hat{M}$ is plotted on Gaussian-scaled axes.

### 4.10.4  Final Considerations

Many other parameters, such as street orientation, tunnels, etc., may affect the area mean, but not the multipath statistics.

Observations at stationary conditions also constitute an interesting investigation. Ideally, in this condition, we would not expect the received signal to experience a time-varying fading. However, although in a reduced way, multipath effects will still be present due to motion of the scatterers.

## 4.11  RADIO CHANNEL SIMULATION

The testing of equipment or techniques using the mobile radio channel can be carried out on the field over real HF circuits or by means of simulation. The first approach is usually time-consuming, costly, and inexhaustive because the parameters influencing the radio channel are, in general, out of control. The second approach is more attractive, but requires the development of both a proper theoretical model and an apparatus to meet certain specifications.

In this section we shall examine two solutions to the simulation problem: one using analog techniques and another using digital signal processing (DSP) techniques. The theory behind these simulators has already been developed in Section 4.5. Accordingly, a fading signal can be written as in Equation 4.19a. Note that, in Section 4.5, we considered the transmission of an unmodulated carrier represented by $\cos \omega t$ and $\sin \omega t$ in Equation 4.19a. For a generic signal, say $x(t)$, the fading signal is given by

$$s(t) = Xx(t) - Y\hat{x}(t) \qquad (4.98)$$

where

$$X = E_0 \sum_{i=1}^{M} a_i \cos(\omega_i t + \psi_i)$$

$$Y = E_0 \sum_{i=1}^{M} a_i \sin(\omega_i t + \psi_i)$$

$$a_i^2 = p(\theta_i) \, d\theta \, p(\psi_i) \, d\psi$$

$$\omega_i = \beta v \cos \theta_i$$

and

$$\hat{x}(t) = \frac{1}{\pi} \int_{-\infty}^{\infty} \frac{x(\tau)}{t - \tau} \, d\tau$$

is the Hilbert transform* of $x(t)$.

Note that $X$ and $Y$ are independent zero-mean Gaussian random processes

### 4.11.1 Analog Solution

The random processes $X$ and $Y$ can be generated by waves-sum or Gaussian-noise filtering as follows.

#### 4.11.1.1 *Waves Sum Solution*

Because $\theta_i$ is assumed to be uniformly distributed in the range $0-2\pi$ rad., then $p(\theta_i) = 1/2\pi$, $d\theta = 2\pi/n$, and $\theta_i = 2\pi i/n$, $i = 1, \ldots, n$. The same reasoning applies to the phase $\psi_i$. By using some simple trigonometric identities it is straightforward to show that Figure 4.10 implements such a simulator. Some works[9, 10] have shown that good results can be obtained with $n \geq 6$.

---

*The Hilbert transform imposes a phase shift of 90° to all of the frequency components of $x(t)$.

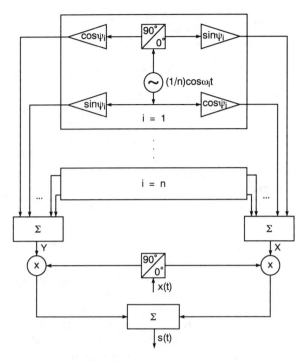

**Figure 4.10.**  Wave-sum Rayleigh fading simulator.

Jakes[7] has shown that the simulator is greatly simplified if the signal is represented in "terms of waves whose frequencies do not overlap". Then[7]

$$X = \frac{2E_0}{\sqrt{n}} \left[ \sqrt{2} \sum_{i=1}^{n_0} \cos \psi_i \cos \omega_i t + \cos \psi_n \cos \beta vt \right]$$

$$Y = \frac{2E_0}{\sqrt{n}} \left[ \sqrt{2} \sum_{i=1}^{n_0} \sin \psi_i \cos \omega_i t + \sin \psi_n \cos \beta vt \right]$$

where $n_0 = (n/2 - 1)/2$. This approach substantially reduces the number of required amplifiers.

Because the resultant signal is the sum of several cosine waves, this approach yields a discrete output power spectrum.

### 4.11.1.2 Gaussian-Noise Filtering Solution

Instead of summing shifted waves to obtain the Gaussian processes $X$ and $Y$, a Gaussian noise generator (e.g., a Zener diode) can be used. The Doppler shift can be introduced if the output of the noise generator is conveniently filtered so as to produce an output power spectrum with a shape

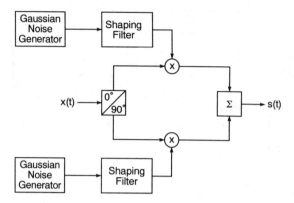

**Figure 4.11.** Gaussian noise filtering Rayleigh fading simulator.

similar to that shown in Figure 4.9 for the electric field (monopole antenna). The block diagram showing this solution is depicted in Figure 4.11.

Note that in this case the Doppler shift is directly dependent on the shaping filter's cutoff frequency (refer to Figure 4.9). Because we have used a Gaussian noise generator to obtain the random processes $X$ and $Y$, this approach yields a continuous output power spectrum.

### 4.11.2 DSP Solution

A fading simulator implemented with digital signal processing (DSP) techniques has the usual advantages of computer (processor) supported equipment: adaptability, modularity, wider variety of features, compactness, reliability, etc. The functioning principles are exactly the same as seen before, but in this case digital techniques are used.

Basically, the mobile radio channel is modelled as a tapped delay line. Each delayed signal is modulated in amplitude and phase by an independent complex-Gaussian process with variable center frequency and bandwidth[12]. Figure 4.12 shows a fading simulator using DSP techniques[11].

The input interface block performs the A/D conversion of the input $x(t)$. The $N$th sample of $x(t)$ is $x(NT)$ and its Hilbert transform is $\hat{x}(NT)$, where $T$ is the sampling period. Both in-phase $x(NT)$ and quadrature components $\hat{x}(NT)$ are delayed by integer steps of $T$ in the complex tapped delay line block to simulate the radio paths. The in-phase and quadrature components in each path $i$ have their amplitudes modulated by the low-pass complex Gaussian processes $W_i$ and $\hat{W}_i$, respectively. The amplitude modulation simulates the Rayleigh fading. Moreover, these amplitude-modulated signals are then frequency-modulated by the complex exponentials $e^{j\theta i}$ and $e^{j\hat{\theta} i}$, imposing a Doppler shift on these signals.

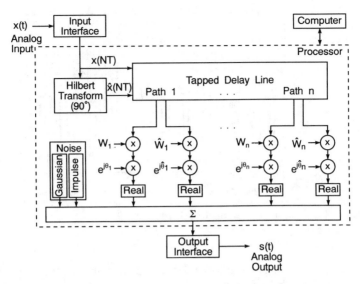

**Figure 4.12.**   Mobile radio channel simulator (DSP solution). (*Source*: After L. Ehrman, L. B. Bates, and J. M. Kates, Real time software simulation of the H.F. radio channel, *IEEE Trans. Comm.*, COM-30(8), pp. 1809–1817, August 1982.)

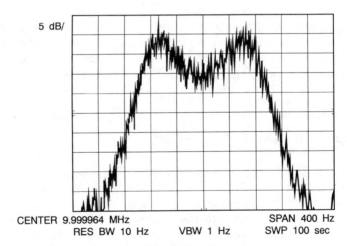

**Figure 4.13.**   Output power spectrum of the fading simulator.

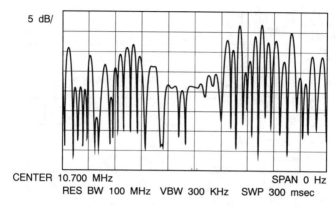

5 dB/

CENTER 10.700 MHz                                           SPAN 0 Hz
RES BW 100 MHz  VBW 300 KHz  SWP 300 msec

**Figure 4.14.**  Detected envelope of the fading simulator.

The real part of these signals are then added up to give the resultant digital signal. Gaussian and impulsive noise are also added to this signal. The output interface block performs the D/A conversion. Three paths ($n = 3$) as used in Reference 11 seemed to be enough to yield very good results.

### 4.11.3  Some Results

In this section we show some results[21] of a fading simulator using the approach as sketched in Figure 4.11. The simulated transmission frequency and the vehicle speed are 880 MHz and 86 km/h, respectively, corresponding to a Doppler shift of 70 Hz. The output power spectrum and the detected envelope are shown in Figures 4.13 and 4.14, respectively.

### 4.12  SUMMARY AND CONCLUSIONS

In this chapter we have analyzed the main phenomena concerning multipath propagation in a mobile radio environment. The first main topic dealt with the frequency shift experienced by the radio signal when the vehicle is in motion: the Doppler effect. We conjectured on a fairly simple manner of estimating the Doppler shift, but we also showed that this matter is rather complex, having a precise solution given by the relativity theory. However, the precise solution reduces to the approximate one when the ratio $v/c$ ([speed of the vehicle]/[speed of light]) is small, and this is obviously our case.

Due to multipath propagation, a single transmitted wave is received at the mobile as a sum of multiple waves each of which has its own time delay. One of the best accepted models establishes that the time delay follows a negative exponential distribution. Field measurements confirm this model and show that the delay spread (corresponding to the standard deviation of such a distribution) varies from fractions of microseconds (in rural areas) to several microseconds (in urban areas).

Time delay imposes a phase shift on the received signal and the Doppler effect provokes a frequency shift. Moreover, the multiple paths produce waves with different amplitudes. The resultant signal is then given by the sum of waves having different amplitudes, phases, and frequencies. Accordingly, one matter of interest is the determination of the correlation between two signals arriving at different time instants and having two different frequencies. We showed that signals with no frequency separation between them but received at antennas distant 0.5λ from one another are considered to be uncorrelated. In the same way, if the frequency separation is such that the signal correlation is equal to 0.5, these signals are uncorrelated. This frequency separation defines the coherence bandwidth. Systems having channel bandwidth smaller than the coherence bandwidth are known as narrowband systems. In the wideband systems the channel bandwidth exceeds the coherence bandwidth. In the first case the signal experiences nonselective fading whereas in the second case we have selective fading.

The fading signal was then characterized by the level crossing rate and the average duration of fades. It was clear that, if the crossing (threshold) level is set to be sufficiently low or sufficiently high, the crossing rate will be small. The maximum rate occurs at 3 dB below the rms signal value, and the mean distance between crossings was proved to be approximately equal to 0.5λ. It was shown that the average duration of fades increases exponentially with the threshold level and is on the order of some milliseconds at the maximum crossing rate.

It was seen that an unmodulated carrier transmitted by the base station arrives at the mobile station as a frequency modulated wave due to the time variation of the signal's phase. Because this variation is random, the phenomenon is known as random FM. Within the voice-frequency band, the noise introduced by the random FM is a quadratic function of the mobile's speed. The power spectrum of the received signal is confined within a frequency band equal to twice the maximum Doppler shift. The shape of the spectrum is different for each one of the electromagnetic field components.

As far as field measurements are concerned, the usual procedure is to have the transmitter at the base station and a receiving setup at the mobile station. This receiving setup includes, among others, data recording facilities to store the signal samples. Slow fading is separated from fast fading by averaging the received signal over a distance of tens of wavelengths. The area mean is considered to be representative of that geographical area after several runs under the "same" conditions.

Field measurements are important for a better characterization of the service area. However, if equipments or techniques are to be tested for mobile radio applications, a wise strategy is to use a laboratory simulation apparatus. The advantage of using a mobile radio simulator is that it provides means of controlling the relevant parameters, which, in a field test, are usually out of our reach. The functioning principles of these simulators are very simple and the latest implementations use the digital signal processing approach.

## APPENDIX 4A   MEAN DISTANCES BETWEEN FADINGS

Let $r$ be the envelope of the Rayleigh fading signal. Assume that each maximum of $r$ is followed by a minimum of $r$ such that, on average, the occurrence rate $R_M$ of maxima equals that of the minima occurrence $R_m$.

Let $p(r, \dot{r}, \ddot{r})$, be the joint probability density function of $r$, $\dot{r}$, and $\ddot{r}$, where the dots denote differentiation with respect to time, $t$. Following Rice[22] the mean maxima (or minima) occurrence rate is

$$R_m = R_M = E[\ddot{r}|\dot{r} = 0] = -\int_0^\infty dr \int_{-\infty}^0 \ddot{r} p(r, \dot{r} = 0, \ddot{r})\, d\ddot{r} \quad (4A.1)$$

Assuming that $R_m$ is known, the mean distance $d$ between minima is

$$d = v/R_m \quad (4A.2)$$

where $v$ is the vehicle speed.

### 4A.1   Joint Density $p(r, \dot{r}, \ddot{r})$ of $r$, $\dot{r}$, and $\ddot{r}$

In order to find the joint distribution $p(r, \dot{r}, \ddot{r})$ of $r$, $\dot{r}$, and $\ddot{r}$, we follow the same procedure as that described in Section 4.6, as appropriate.

First we find the joint density $p(x, \dot{x}, \ddot{x}, y, \dot{y}, \ddot{y})$ of $x$, $\dot{x}$, $\ddot{x}$, $y$, $\dot{y}$, and $\ddot{y}$, where $x$ and $y$ are given by Equations 4.19b and the dots denote differentiation with respect to $t$. Then we change variables to find $p(r, \dot{r}, \ddot{r}, \varphi, \dot{\varphi}, \ddot{\varphi})$, where $r^2 = x^2 + y^2$ and $\varphi = \tan^{-1} y/x$. The expression for $p(r, 0, \ddot{r})$ is obtained by setting $\dot{r} = 0$ and integrating $p(r, \dot{r} = 0, \ddot{r}, \varphi, \dot{\varphi}, \ddot{\varphi})$ over the ranges of $0 \leq \varphi \leq 2\pi$, $-\infty \leq \dot{\varphi} \leq \infty$, and $-\infty \leq \ddot{\varphi} \leq \infty$.

### 4A.2   Minima Occurrence Rate

Following the (tedious) procedure described in the previous section, it is found that[22]

$$R_m = \frac{(a^2 - 1)^2}{(2a)^{5/2}} \left(\frac{b_2}{\pi b_0}\right)^{1/2} \sum_{n=0}^\infty \frac{\Gamma(n/2 + \frac{5}{4})}{\Gamma(n/2 + \frac{7}{4})} \frac{A_n}{a^n} \quad (4A.3)$$

where

$$b_n = (2\pi)^n \int_0^\infty S(f)(f - f_c)^n \, df \tag{4A.4}$$

$S(f)$    is the power spectrum of the envelope $r$

$f_c$    is the midband frequency

$$a^2 = \frac{b_0 b_4}{b_2^2} \tag{4A.5}$$

$$A_n = \sum_{i=0}^n \frac{(\frac{1}{2})(\frac{3}{2}) \cdots (i - \frac{1}{2})}{i!}(n - i + 1)b^i \tag{4A.6}$$

$$b = \frac{1}{2}(3 - a^2) \tag{4A.7}$$

and

$$\Gamma(z) = \int_0^\infty t^{z-1} e^{-t} \, dt \tag{4A.8}$$

is the Gamma (factorial) function.


## 4A.3   Mean Distance Between Fadings

Using Equation 4.83a for the power spectrum of the electric field in Equation 4A.4, it is found that

$$b_n = \frac{3}{2} W_0(\omega_m)^n \frac{1 \cdot 3 \cdot 5 \cdots (n - 1)}{2 \cdot 4 \cdot 6 \cdots n} \tag{4A.9}$$

where $\omega_m = 2\pi f_m$.
Using Equation 4A.9 as required in Equations 4A.5 through 4A.7 and the results in Equation 4A.3, it is found that

$$R_m \simeq 2.6 f_m \tag{4A.10}$$

Finally, with Equation 4A.10 in Equation 4A.2,

$$d \simeq 0.383\lambda \simeq 0.5\lambda \quad \text{Q.E.D.}$$

## APPENDIX 4B   DIGITAL LOW-PASS FILTER

Let $x(n)$ be the $n$th sample of a discrete signal. We want to prove that the $n$th sample of the low-pass filtered signal may be estimated as

$$y(n) = \frac{1}{2k + 1} \sum_{j=-k}^{k} x(n + j) \tag{4B.1}$$

We introduce the delay operator $z^{-1}$ to represent a delay of $j$ sampling periods. Consequently, Expression 4B.1 may be rewritten as

$$H(z) = \frac{Y(z)}{X(z)} = \frac{1}{2k + 1} \sum_{j=-k}^{k} z^{-j} \tag{4B.2}$$

Hence, using the summation properties of a geometric series

$$H(z) = \frac{z^{-k}}{2k + 1} \left( \frac{1 - z^{2k+1}}{1 - z} \right) \tag{4B.3}$$

### 4B.1   Frequency Response

The transfer function $H(z)$ describes a filter having a unique pole located at $z = 1$, for $\omega = 0$, and $2k + 1$ zeros, located at $z = 1$, equally spaced at $2\pi/2k + 1$. Moreover, this filter has $k$ zeros at $z = 0$. The frequency response of the filter can be determined by using $z = e^{j\omega}$ in Equation 4B.3. Then

$$H(e^{j\omega}) = \frac{e^{-jk\omega}}{2k + 1} \left( \frac{1 - e^{j(2k+1)\omega}}{1 - e^{j\omega}} \right) \tag{4B.4}$$

or, equivalently,

$$H(e^{j\omega}) = \frac{e^{-jk\omega} \exp\left( j\dfrac{2k + 1}{2}\omega \right)}{(2k + 1)e^{j\omega/2}} \left( \frac{\exp\left( j\dfrac{2k + 1}{2}\omega \right) - \exp\left( -j\dfrac{2k + 1}{2}\omega \right)}{e^{j\omega/2} - e^{-j\omega/2}} \right)$$

Then

$$H(e^{j\omega}) = \frac{\sin[(2k + 1)\omega/2]}{(2k + 1)\sin(\omega/2)} \tag{4B.5}$$

The modulus $|H(e^{j\omega})|$ is plotted in Figure 4B.1, for $2k + 1 = 5$

Note that, from Equation 4B.5 the phase is constant and equal to zero. Note also that the frequency is normalized, i.e., $\pi$ corresponds to half of the

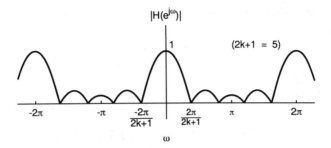

**Figure 4B.1.**   Magnitude of the frequency response of a digital filter.

sampling frequency ($f_s/2$). In other words, in Figure 4B.1 the first null occurs for $f_s/(2k + 1)$, the second for $2f_s/(2k + 1)$ and so on. (Nulls will occur for $(2k + 1)\omega/2 = n\pi$, where $n$ is an integer. Hence $\omega = nf_s/(2k + 1)$.)

## 4B.2   Cutoff Frequency

Let us estimate the value of the second peak of $|H(e^{j\omega})|$ (the first occurs for $\omega = 0$, as we can see). Such peak lies within the range $2\pi/(2k + 1) < \omega < 4\pi/(2k + 1)$. Assume, for simplicity, that the second peak occurs at $\omega \simeq 3\pi/(2k + 1)$ (see Figure 4B.1). In this case

$$\left| H\left[ \exp\left( j\frac{3\pi}{2k + 1}\omega \right) \right] \right| = \frac{1}{(2k + 1)\sin\left( \dfrac{3\pi/2}{2k + 1} \right)} \qquad (4\text{B.6})$$

Using the 3-dB criterion for the cutoff frequency, we have

$$\left| H\left[ \exp\left( j\frac{3\pi}{2k + 1}\omega \right) \right] \right| \leq \frac{1}{\sqrt{2}} \qquad (4\text{B.7})$$

With Equation 4B.6 in Equation 4B.7 we see that the inequality is always true for $2k + 1 \geq 2$. For larger values of $2k + 1$ the second peak will be even smaller, and we may consider that the effective cutoff will occur at the first null ($2\pi/(2k + 1)$). Then, the cutoff frequency is

$$f_{co} = \frac{f_s}{2k + 1} \qquad (4\text{B.8})$$

## APPENDIX 4C   SAMPLING DISTRIBUTIONS

Let $X$ be a random variable having a probability distribution $P(X) = \text{prob}(x \leq X)$, with a mean $\mu$ and variance $\sigma^2$. Let $X_1, X_2, \ldots, X_N$ be the observed samples of such a distribution. Then the sample mean $\bar{X}$ is

$$\bar{X} = \frac{1}{N}\sum_{i=1}^{N} X_i$$

The expected value of $\bar{X}$ is

$$E[\bar{X}] = E\left[\frac{1}{N}\sum_{i=1}^{N} X_i\right] = \frac{1}{N}\sum_{i=1}^{N} E[X_i] = E[X_i] = \mu$$

The variance of $\bar{X}$ is

$$\hat{\sigma}^2 = E\left[(\bar{X}-\mu)^2\right] = E\left[\left(\frac{1}{N}\sum_{i=1}^{N} X_i - \mu\right)^2\right] = E\left[\left(\frac{1}{N}\sum_{i=1}^{N}(X_i - \mu)\right)^2\right]$$

$$= \frac{1}{N^2}E\left[\sum_{i=1}^{N}(X_i - \mu)^2 + 2\sum_{i=1}^{N-1}\sum_{j=i+1}^{N}(X_i - \mu)(X_j - \mu)\right]$$

$$= \frac{1}{N^2}\left\{\sum_{i=1}^{N} E\left[(X_i - \mu)^2\right] + 2\sum_{i=1}^{N-1}\sum_{j=i+1}^{N} E\left[(X_i - \mu)(X_j - \mu)\right]\right\}$$

$$= \frac{1}{N^2}\left\{NE\left[(X_i - \mu)^2\right] + N(N-1)E\left[(X_i - \mu)(X_j - \mu)\right]\right\}$$

If the samples are independent, then the cross product will have zero expectation. Hence

$$\hat{\sigma}^2 = \frac{1}{N}E\left[(X_i - \mu)^2\right] = \frac{\sigma^2}{N}$$

Now consider the following distribution:

$$X_i = \begin{cases} 0 & \text{with probability } P[X_i = 0] = 1 - p = \text{prob}(R_i > V) \\ 1 & \text{with probability } P[X_i = 1] = p = \text{prob}(R_1 \le V) \end{cases}$$

Then

$$E[X] = \sum_{i=0}^{\infty} X_i P[X = X_i] = 0(1 - p) + 1p = p$$

$$E[X^2] = \sum_{i=0}^{\infty} X_i^2 P[X = X_i] = 0^2(1 - p) + 1^2 p = p$$

$$E^2[X_i] = p^2$$

Hence

$$\sigma^2 = E[X^2] - E^2[X] = p(1-p)$$

Then

$$\hat{\sigma} = \frac{p(1-p)}{\sqrt{N}}$$

## REFERENCES

1. Davis, B. R. and Bogner, R. E., Propagation at 500 MHz for mobile radio, *IEEE Proc.*, 132(5), Part F, August 1985.
2. Halliday, D. and Resnik, R., *Physics—Part II*, 2nd ed., John Wiley & Sons, New York, 1966.
3. Ramo, S., Whinnery, J. R., and Duzer, T. V., *Fields and Waves in Communication Electronics*, John Wiley & Sons, New York, 1965.
4. Davenport, W. B., Jr., *Probability and Random Processes*, International Student Edition, McGraw-Hill KogaKusha, Ltd., Tokyo, 1970.
5. Gradshteyn, I. S. and Ryahik, I. W., *Table of Integrals, Series and Products*, Academic, New York, 1965.
6. Abramowitz, M. and Stegun I. A., Eds., *Handbook of Mathematical Functions with Formulas, Graphs and Mathematical Tables*, National Bureau of Standards, Applied Mathematical Series 55, June 1964.
7. Jakes, W. C., *Microwave Mobile Communications*, John Wiley & Sons, New York, 1974.
8. Rice, S. O., Statistical properties of a sine wave plus random noise, *Bell Syst. Tech. J.*, 27, 109–157, January 1948.
9. Bennet, W. R., Distribution of the sum of randomly phased components, *Quart. Appl. Math.*, 385–393, 5 January 1948.
10. Slack, M., The probability of sinusoidal oscillations combined in random phase, *J. IEEE*, 93, Part III, 76–86, 1946.
11. Ehrman, L., Bates, L. B., and Kates, J. M., Real time software simulation of the H.F. radio channel, *IEEE Trans. Communications*, COM-30(8), August 1982.
12. Goldberg, B., *Communication Channels, Characterization and Behavior*, IEEE Press, New York, 1975.
13. Turin, G. L., Simulation of urban radio propagation and of urban radio communication system, in *Proc. Int. Symp. Antennas and Propagation*, 543–546, Sendai, Japan.
14. Hashemi, H., Simulation of the urban radio propagation channel, *IEEE Trans. Vehicular Technology*, VT-28, 213–225, August 1979.
15. Caples, E. L., Massad, K. E., and Minor, T. R., A UHF channel simulator for digital mobile radio, *IEEE Trans. Vehicular Technology*, VT-29, 281–289, May 1980.
16. Aulin, T., Modified model for the fading signal at a mobile radio channel, *IEEE Trans. Vehicular Technology*, VT-28, 182–203, August 1979.

17. Arredondo, G. A., Chriss, W. H., and Walker, E. H., A multipath fading simulator for mobile radio, *IEEE Trans. Vehicular Technology*, VT-22, 241–244, November 1973.

18. Arnold, H. W., and Bodtman, W. F., A hybrid multi-channel hardware simulator for frequency-selective mobile radio paths, *IEEE GTC 82*, p. A.3.1, November–December 1982.

19. Ralphs, J. D. and Sladen, F. M. E., An HF channel simulator using a new Rayleigh fading method. *The Radio and Electronic Engineer*, 46, 579, December 1976.

20. Oppenheim, A. V. and Schafer, R. W., *Digital Signal Processing*, Prentice-Hall, Englewood Cliffs, N.J., 1975.

21. Branquinho, O. C. and Yacoub, M. D., Multipath effects simulator (in Portuguese), 9th Brazilian Telecommunications Symposium, São Paulo, S.P., Brazil, 2–5 September 1991.

22. Rice, S. O., Statistical properties of random noise currents, in *Selected Papers on Noise and Stochastic Processes*, N. Wax, Ed., Dover, New York, 1954.

# PART III
# Diversity-Combining Methods

# CHAPTER 5

# Fading Counteractions

Fading is considered to be one of the main causes of performance degradation in a mobile radio system. Not only can it be annoying but also disastrously harmful if we consider, for instance, its effects on data transmission. Fortunately, this is not an undefeatable enemy and, in fact, many techniques can be used to counteract its effects. This chapter aims at describing the principles, functioning, effectiveness, advantages, and disadvantages of these techniques. The problem is first approached from the macroscopic side, where it will be seen that only one method is effective to combat slow (lognormal) fading, namely, space diversity. On the microscopic side, we describe six techniques to combat fast (Rayleigh) fading—namely, space, polarization, angle, frequency, time, and hopping diversities. Adaptive equalization and coding also constitute effective means of counteracting fading. However, because these techniques are essentially applicable to digital communications, they will be explored in Chapter 6.

Four combining methods—pure selection, threshold selection, maximal ratio, and equal gain—are then investigated. The performance of each combining technique is assessed by a measure of the signal-to-noise ratio (SNR) obtained at the output of the combiner. Accordingly, the probability distributions of the various SNRs as functions of the number of diversity branches are determined and analyzed.

It is shown that, taking into account the cost-effectiveness factor, the two-branch diversity scheme constitutes the best option to combat fading.

## 5.1  INTRODUCTION

The increasing demand for mobile radio services impels the system designers and operators to find means of not only satisfying such demand but also of improving the performance of the system. One of the main causes of performance degradation in mobile radio systems is the occurrence of fading. A vehicle moving in a typical urban environment may experience extreme variations in signal levels, and 40 dB below the mean signal strength is not uncommon. The envelope of the signal fades according to a Rayleigh distribution, with successive minima occurring about every half-wavelength of the carrier frequency.

As far as voice transmission is concerned, the effects of multipath propagation, although annoying, may not be essentially critical. The main concern is when data or signalling are involved: a loss of communication at certain crucial instants may be extremely harmful (imagine, for instance, the system loosing control of a long-duration international call at the billing instant). It is obvious that the general consensus is toward providing a high-performance system where both voice and data transmissions meet established standards of the grade of service.

However, overcoming the effects of multipath propagation (fading) is not an easy task and, in fact, is considered to be one of the difficult problems in the engineering design of a mobile radio system.

Fading counteraction is usually carried out by means of diversity methods, the principles of which are based on the fact that the instances of fading occurring on independent channels constitute independent events. Therefore, if certain information is redundantly available on two or more channels (known as diversity branches), the probability that this information is affected by a deep fade, occurring simultaneously on all of the branches, is very low. Accordingly, with a convenient algorithm (known as a combining method) it is possible to obtain a resultant signal where the effects of fading are minimized.

The use of some diversity techniques in communication systems dates back to the 1920s.[5] Since then several diversity schemes have emerged. Although not widely known as such, the hand-off from one base station to another also constitutes a form of diversity. In this chapter we shall concentrate on the alternative methods, other than hand-off. Initially we describe the methods used to obtain the independent branches, i.e., how the information can be repeated over independent paths. Then we examine the methods of combining (processing) the received signals in order to obtain the best output signal.

There are basically two types of fading in a mobile radio environment: long-term fading and fast fading. The methods of counteracting their effects are different and will be considered separately.

## 5.2  LONG-TERM FADING COUNTERACTION

The methods of counteracting the lognormal fading use *macroscopic* diversity. The shadowing of a transmitted signal is provoked by large obstructions such as hills, mountains, and others, positioned between the base station and the mobile. A vehicle in this scenario may experience a significantly bad communication, sometimes with a complete loss of the signal.

The basic method used to combat this type of fading is by avoiding the obstruction. A straightforward way of achieving this is by providing more than one base station strategically positioned so that the mobiles always have a clear radio path to at least one base station. The mobile unit establishes

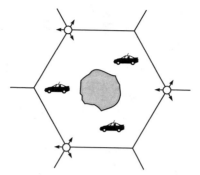

**Figure 5.1.**   Corner-excited cells to defeat long-term fading.

communication with all of the base stations. The selection procedure is a dynamic process so that the mobile station always has the best communication.

Note that this type of diversity involves geographical separation between the base stations. Therefore, it is considered as a kind of space diversity. Moreover, because the distances involved are usually long, it is classified as macroscopic diversity. Jakes[9] proposes corner-excited cells for the mobile radio system as shown in Figure 5.1.

In Figure 5.1 the circles represent the base stations, using 120° directional antennas. At first sight it looks as if the number of base stations has been tripled. In fact, the number is exactly the same because each site serves three cells. Note that the mobiles will always have a clear radio path for communication. This is a very interesting configuration because it also minimizes the cochannel interference problem, as will be seen in Chapter 7.

Shadowing appears not only in cellular systems but also in different mobile services. Portable communications integrated into local networks as described by Cox and co-workers[17, 18] constitute one of these services. In this system "high-quality voice and data service via radio must be provided to areas with varying subscriber densities, while making efficient use of the radio spectrum".[5] In order to achieve this, a fixed grid of radio access ports (base stations) is used. The layout of these ports in the grid greatly influences system performance. The selection of the best-quality radio link between a given user and the closest group of ports is supposed to be very rapid, carried out in tens of milliseconds. Figure 5.2 shows one possible configuration of macroscopic diversity for portable communication systems.

In the example of Figure 5.2 a three-branch diversity is depicted. The three ports numbered 1, 2, 3 compose a group serving the mobile users within and near the corresponding equilateral triangle. The selected signal is the best among those received from the ports 1, 2, 3. It can be seen that each port serves six different diversity groups. Other configurations using square or hexagonal grids with the ports located at different points of the grid are proposed and analyzed by Bernhardt.[5]

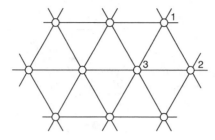

**Figure 5.2.** Example of macroscopic diversity in a triangular-grid portable communication system.

## 5.3 SHORT-TERM (FAST) FADING COUNTERACTION

The methods of counteracting fast fading use *microscopic* diversity. The name microscopic is related to the distances involved in obtaining independent radio paths, corresponding to fractions of a wavelength to several wavelengths. Microscopic diversity relies on the fact that independent signals have a low probability of experiencing deep fades at the same time instant. Consequently, the basic idea is to repeat the information through independent radio paths and then to associate them in a convenient way so that the information is recovered with the effects of the fading greatly attenuated. Section 5.4 describes the various methods used to obtain the independent radio paths, and Section 5.5 examines the schemes used to combine these radio paths.

## 5.4 DIVERSITY SCHEMES

In this section we describe the space, polarization, angle, frequency, time, and hopping diversity techniques.

### 5.4.1 Space Diversity

The space diversity technique is the precursor of all of the other diversity techniques and was first applied in 1929 by the use of spaced receiving antennas for HF sky-wave reception. Due to its simple and economical implementation, in connection with the fact that no extra frequency spectrum is required, this diversity scheme is widely used.

When space diversity is mentioned it is usually implicit that the spacing is between receiving antennas. Although spaced transmitting antennas also provide diversity, the ease of implementation is not quite the same. The transmitting antennas cannot use the same frequency and same polarization since the signals would not be separable into independent receivers.[15] If different frequencies or polarization are used, then we recognize this as another technique, as described later. Consequently, our analysis here refers to receiving antennas, spaced and located at the base station and/or at the mobile. In this analysis we assume that the information is redundantly transmitted over independent branches. We recall that, although the signal is transmitted by one single antenna, due to the scattering effects of the obstructions in the environment, the signal arrives at the receiver in the form of various (infinite) signals (multipath propagation). Among the "infinitude" of received signals we must discriminate those experiencing independent fadings. In other words, a correlation function between signals should be determined.

### 5.4.1.1 *Spaced Antennas at the Base Station*

One would intuitively expect that there should be reciprocity between signals received at the mobile unit and those received at the base station. As a consequence, the conclusions drawn for one case could be immediately applied to the other case. As far as signal correlation factor is concerned, this is not quite true. The explanation for this is as follows (refer to Figure 5.3). Scatterers are assumed to be randomly distributed around a circle of radius $R$ surrounding the mobile. This radius is usually very small compared to the distance $D$ between base station and mobile. Consequently, the beamwidth $\Delta\varphi = 2R/D$ incident on the base station is also very small. Hence, the base station antennas should be placed far away from one another so that each antenna receives a radio signal propagated through independent paths. On the other hand, the mobile unit receives the scattered signals from all of the angles. Therefore, uncorrelated signals can be received at very close distances from each other.

Our aim now is to determine the distance between the antennas at the base station receiving independent radio signals from a given mobile. This can be directly extracted from the cross-correlation expression of the envelope of two signals $r_1, r_2$ received at the base station by two antennas horizontally spaced by a distance $d$, as shown in Figure 5.4. Lee[8, 19] has

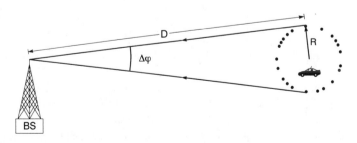

**Figure 5.3.** Scatterers in the vicinity of the mobile.

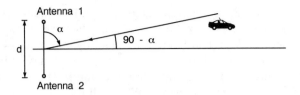

**Figure 5.4.** Two horizontally spaced antennas.

demonstrated that this correlation factor $\rho_r$ is given by (see Appendix 5A)

$$\rho_r = E^2\left\{\cos\left[2\pi\frac{d}{\lambda}\cos(\theta_i - \alpha)\right]\right\} + E^2\left\{\sin\left[2\pi\frac{d}{\lambda}\cos(\theta_i - \alpha)\right]\right\} \quad (5.1)$$

where $d = v\tau$;

    $v$ is the speed of the vehicle;

    $\tau$ is the time separation between the signals;

    $\lambda$ is the wavelength;

    $\theta_i$ is the incident angle of the $i$th wave, $\alpha - \pi/2 \le \theta_i \le \alpha + \pi/2$;

    $\alpha$ is the mean incident angle;

  $E[x]$ is the expectation of $x$.

The expectation $E[f(\theta_i)]$ shown in Equation 5.1 is calculated as

$$E[f(\theta_i)] = \int_{\alpha-\pi/2}^{\alpha+\pi/2} f(\theta_i)p(\theta_i)\, d\theta_i \quad (5.2)$$

where the probability density $p(\theta_i)$ is given by

$$p(\theta_i) \simeq \frac{Q}{\pi}\cos^n(\theta_i - \alpha) \quad (5.3)$$

In Equation 5.3, $Q$ is a normalization factor, obtained as follows:

$$\int_{\alpha-\pi/2}^{\alpha+\pi/2} p(\theta_i)\, d\theta_i = 1 \quad (5.4)$$

As an example, consider an electric field pattern $E_r$ given by

$$E_r = \cos^n(\Delta\varphi/2) \quad (5.5)$$

The 3-dB beamwidth is the value of $\Delta\varphi$ for which $E_r = 1/\sqrt{2}$. If $n = 1$, then the 3-dB beamwidth is $\Delta\varphi = 90°$. Conversely, if the 3-dB beamwidth is $\Delta\varphi = 2°$, then $n \simeq 2275$. For a given $\alpha$, the value of $Q$ can be found by using Equation 5.3 in Equation 5.4. Finally, the correlation factor $\rho_r$ in Equation 5.1 can be estimated as a function of $\lambda$ and $d$.

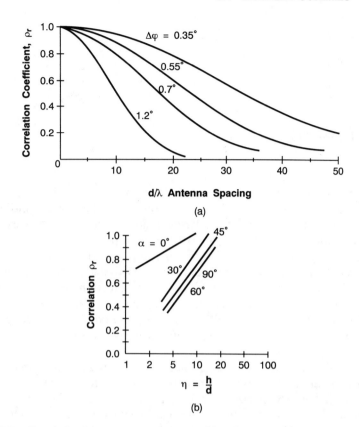

**Figure 5.5.** Correlation factor at the base station: (a) incidence angle $\alpha = 90°$ (theoretical curves); (b) frequency of 850 MHz (empirical curves). (*Source:* W. C. Y. Lee, *Mobile Communications Design Fundamentals*, Howard W. Sams, Indianapolis, IN, 1986. Reprinted with permission of Prentice-Hall Computer Publishing.)

Using this procedure Lee[8, 19] determined, by means of numerical computation, various curves for the correlation factor. A set of such theoretical curves is shown here together with some empirical curves. Figure 5.5a shows the correlation factor for an incident angle $\alpha = 90°$ (broadside propagation), having the beamwidth as a parameter. Figure 5.5b shows some empirical curves of the signal correlation as a function of the ratio $h/d$, where $h$ is the antenna height and $d$ is the distance between antennas. The empirical curves were obtained for a frequency of 850 MHz. For a different frequency, say $f$, Lee[8, 19] suggests the following correction factor:

$$d' = \left(\frac{850}{f}\right)d \tag{5.6}$$

where $d'$ is the new distance separation obtained at the frequency $f$ MHz.

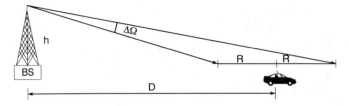

**Figure 5.6.**  Vertical separation of the antennas.

It is intuitively expected that the correlation between antennas 1 and 2 increases as the incidence angle $\alpha$ decreases. Note that for the in-line case ($\alpha = 0°$) the correlation factor is larger than that for $\alpha \neq 0°$ because the signals will have travelled along almost the same path.

The receiving antennas can also be vertically spaced. This configuration is easy to implement because the same pole can be used for the antennas. Height diversity is mainly used to provide protection against destructive interference due to ground reflections, a characteristic of the microwave line-of-sight links. Moreover, this has application in high-capacity digital radios, to provide protection against refractive ray-splitting multipath propagation.

Horizontal space diversity gives better performance than that of the vertical separation, because the decorrelation of the received signals increases faster with the horizontal rather than with the vertical separation of the antennas. This is because the vertical beamwidth $\Delta\Omega$ is much smaller than the horizontal beamwidth $\Delta\varphi$[20], as shown later. Referring to Figure 5.6 the vertical beamwidth $\Delta\Omega$ is approximately given by $\Delta\Omega \simeq 2Rh/D^2$. Because $\Delta\varphi \simeq 2R/D$, then $\Delta\Omega/\Delta\varphi \simeq h/D$. Usually the antenna height $h$ is much smaller than the distance $D$ from the mobile to the base station. Therefore, $\Delta\varphi \gg \Delta\Omega$. The great advantage of the vertical diversity is that the incidence angle of the radio wave does not affect the correlation factor.

### 5.4.1.2 *Spaced Antennas at the Mobile*

The separation distance between the antennas at the mobile can be found from the correlation factor $\rho_r$ of the two signals $r_1$ and $r_2$ arriving at these antennas. This correlation factor has already been determined in Chapter 4 and is given by Equation 4.46. Because we are interested in the correlation between the same signal arriving via two distinct paths, we set their frequency separation $\Delta\omega$ in Equation 4.46 to be equal to zero. Therefore

$$\rho_r = J_0^2(\omega_m \tau) \tag{5.7}$$

where $\omega_m = \beta v = (2\pi/\lambda)v$ is the maximum Doppler shift and $\tau$ is the time delay between the arrival of the first and the second signals. The argument

$\omega_m \tau$ of the Bessel function in Equation 5.7 can be expressed as

$$\omega_m \tau = 2\pi \frac{v\tau}{\lambda} = 2\pi \frac{d}{\lambda} \qquad (5.8a)$$

where $d = v\tau$ is the separation distance between the two signals. Because we require these signals to be uncorrelated, we must set $\rho_r$ to be equal to zero.

The first null of the Bessel function $J_0(\omega_m \tau)$ occurs for $\omega_m \tau = 2.404$, corresponding to a ratio $d/\lambda \simeq 0.38$. Consequently, a separation distance $d \simeq 0.38\lambda$ between the receiving antennas guarantees independence between the received signals. However, from Figure 4.3b we notice that, after this first null, the correlation factor starts increasing and reaches a maximum value of 0.16, obtained for $d \simeq 0.6\lambda$. Nevertheless, a correlation less than 0.2 can be neglected, and, in practice, a separation distance $d = 0.5\lambda$ has been found to be quite reasonable. As an example, with a carrier frequency of 900 MHz ($\lambda = \frac{1}{3}$ m) the distance between the receiving antennas would be $d \simeq 17$ cm. Compare this result with that from Figure 5.5a, where the distances are usually of the order of tens of wavelengths.

### 5.4.2 Polarization Diversity

The polarization diversity takes advantage of the orthogonality of the polarized wave components of the travelling radio wave. Although the two components of the polarized wave travel through similar paths, the obstacles encountered by these waves scatter each one of them differently. This implies that they reach their destination in the form of a large number of scattered waves with amplitudes and phases varying randomly. The orthogonality between the waves plus the randomness of amplitudes and phases greatly contribute to strengthening the uncorrelation between the two components of the polarized wave.

Polarization diversity can be realized by means of antennas reacting individually to the $E$ field and to the $H$ field. This is achieved by the vertical electric dipole (whip antenna) and the vertical magnetic dipole (loop antenna), respectively. It is interesting to mention that the electric field produced by a loop antenna is exactly the same as the magnetic field developed by a whip antenna.

There are two great disadvantages of using this technique: (1) only two diversity branches can be used; (2) 3 dB (half of the power) is effectively lost because the signal is divided into two transmitting antennas.

### 5.4.3 Angle Diversity

A fading signal is the resultant of the sum of an "infinite" number of scattered signals arriving at the antenna through multiple paths. If the antenna's beamwidth is restricted to a small angle, only a few of these waves would be detected and this would improve the fading characteristics of the received signal. Consequently, a directional antenna can be used to improve the fading characteristics of the radio signal. However, it is essential to keep the received signal above an acceptable threshold. This can be accomplished by providing a number of directional antennas constantly monitoring the signal. Another method is to use an adaptive antenna keeping track of the best signal by adaptively positioning itself to point to the correct angle.

### 5.4.4  Frequency Diversity

Another way of providing uncorrelation between two (or more) signals carrying the same information is by transmitting the signals on two (or more) different carriers. The independence between the diversity branches is achieved if the frequencies of the carriers are sufficiently separated from one another so that fadings occurring in these signals are uncorrelated.

One possible measure of frequency separation is the coherence bandwidth, studied in Section 4.5. The signals are considered to fade independently if the frequency separation between them exceeds the coherence bandwidth by several times.[9] As defined in Section 4.5, the coherence bandwidth $B_c$ is the frequency separation for which the correlation between the signals is equal to 0.5. Accordingly, $B_c = 1/2\pi\overline{T}$ for the envelope and $B_c = 1/4\pi\overline{T}$ for the phase correlations, respectively, where $\overline{T}$ is the delay spread. In suburban areas, $\overline{T} \simeq 0.5\ \mu s$, corresponding to a coherence bandwidth of $B_c = 320$ kHz or $B_c = 160$ kHz for the envelope or phase correlations, respectively. Therefore, two fading signals are considered to be independent if the frequency separation between them is $kB_c$, where $k$ is a constant.* In the given example, if we consider $k = 6$, then this separation would be around 2 MHz ($\simeq 6 \times 320$ kHz).

This obviously requires a wider frequency spectrum, not usually available. A way of saving spectrum is to employ the "$1 : N$ protection switching".[15] In this technique one spare frequency is used as a means of providing diversity to $N$ other carriers. These $N$ frequencies carry $N$ independent traffic, and, when diversity is needed for a given specific carrier, its traffic is switched to the spare carrier. Another way of saving spectrum is described in Section 5.4.6.

### 5.4.5  Time Diversity

As well as frequency, time separation can also provide diversity. As far as data transmission is concerned this technique is widely used and consists in

---

*The constant $k$ is usually assumed to be greater than 5.

repeating the message (or part of it) according to some criterion. If the intertransmission time is conveniently determined, the received signals may experience independent fadings.

Similarly to the data transmission, speech signals can be sampled and appropriately delayed to be used as a diversity branch. At the reception, the samples (or the data) are stored for a period of time equal to the required time delay. Then, the correct (or the best) information can be selected.

The time interval between the samples (messages) must exceed the coherence time of the channel. This can be determined from the expression of the envelope correlation between two fading signals given by Equation 5.7. From Equation 5.7 the first null of the Bessel function occurs for $\omega_m \tau = 2.404$, or $d \simeq 0.38\lambda$. Dividing both terms by $v$ (the vehicle speed), we have $\tau \simeq 0.38\lambda/v = 0.38/f_m$, where $f_m$ is the maximum Doppler frequency and $\tau$ is the time interval between independent signals. Using the same approximation as in the case of space diversity, we can assume that if

$$\tau \geq 1/(2f_m) \tag{5.8b}$$

the signals arriving with a time delay equal to $\tau$ fade independently. For a mobile at 72 km/h, operating at 900 MHz, the delay between the signals must exceed 8.3 ms.

If an $M$-branch diversity is used, then a minimum time delay of $M/2f_m$ is obviously required. In the case of voice transmission the sampling rate must be greater than $8 \times M$ kHz.

### 5.4.6 Diversity by Hopping

Frequency hopping and time (slot) hopping constitute efficient ways of using frequency or time diversities, respectively. The aim is for the information to hop from one frequency or from one time slot to another so that the information experiences independent fadings. If only the resources (in time or frequency) of each base station are used, there is no need for synchronization between base stations. This subject is more deeply explored in Chapter 10.

## 5.5 COMBINING SCHEMES

There are basically four combining methods divided in two groups, namely, (1) switched combining and (2) gain combining.

In the switched combining group, the aim is to pick (choose) one out of the $M$ received signals according to some criteria. In other words,

$$r = \text{one out of } \{r_1, r_2, \ldots, r_M\} \tag{5.9}$$

where $r$ is the resultant envelope and $r_i$, $i = 1, \ldots, M$, are the received envelopes of the signals. There are two techniques in this group, namely, pure selection combining and threshold selection combining.

In gain combining, the resultant signal is a linear combination of the received signals, i.e.,

$$r = \sum_{i=1}^{M} a_i r_i \qquad (5.10)$$

where $a_i$ are the gains at each branch. The two techniques here are maximal-ratio combining and equal-gain combining.

### 5.5.1  Switched Combining

### Pure Selection

In pure selection combining, the received signals are continuously monitored so that the best signal can be selected. In theory, the selection criterion should be based on the best signal-to-noise ratio. In practice, however, because this is difficult to obtain, the strongest signal + noise is selected. This technique requires each branch to have its own receiver. Moreover, the signals must be monitored at a rate faster than that of the fading occurrence. A block diagram showing the principles of this technique is sketched in Figure 5.7a.

### Threshold Selection

One of the major problems of pure selection combining is the number of receivers required, one for each branch. Another constraint is the high monitoring rate to select the best signal. This obviously increases both the cost of the equipment and its complexity.

A clever alternative to this approach is *threshold selection combining*. In this case the received signals are scanned in a sequential order, and the first signal with a power level above a certain threshold is selected. While above the threshold, the selected signal remains at the combiner's output; otherwise, a scanning process is reinitiated.

The threshold may be either fixed or variable. Setting its level is a task involving the knowledge of the mean signal strength in the geographical area. Consequently, the fixed approach may work well within the region for which the fixed threshold was found to be appropriate. On the other hand, if conditions change, this approach can give bad results. For instance, if the mobile moves into another area where the mean signal level is lower than its fixed threshold, instability and unnecessary switchings may occur. Accordingly, a variable threshold is preferred, giving better results at the expense of the inclusion of a circuit to estimate the mean signal level. The block diagram of both approaches are sketched in Figure 5.7b and 5.7c.

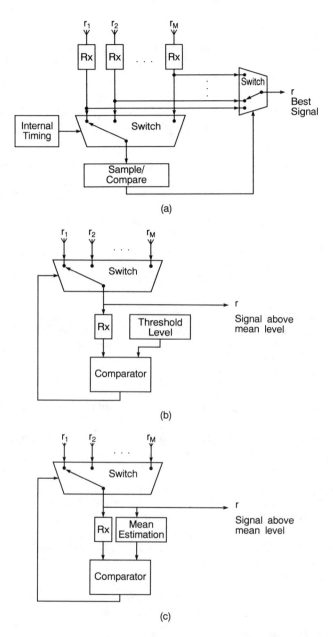

**Figure 5.7.** Switched combining block diagrams: (a) pure-selection combining; (b) threshold selection—fixed threshold; (c) threshold selection—variable threshold.

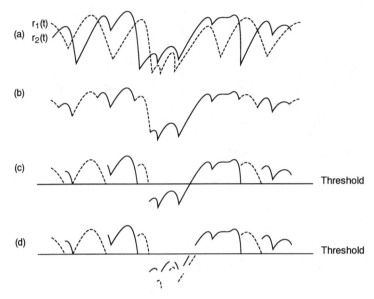

(a) $r_1(t)$
$r_2(t)$

(b)

(c)     Threshold

(d)     Threshold

**Figure 5.8.** Switched combining techniques: (a) two fading signals; (b) resultant signal using pure selection; (c) resultant signal using switch-and-stay threshold selection; (d) resultant signal using switch-and-examine threshold selection.

In the case of two branches, when the signal falls below the threshold, two strategies can be used to select the signal: (1) switch and stay or (2) switch and examine. In the first strategy the other signal is always selected regardless of its power level. In the second strategy the other signal is not selected unless its power exceeds the threshold.

The switch-and-stay approach has the disadvantage that the output signal can stay longer below the threshold. On the other hand, the switch-and-examine approach has the inconvenience of the noise bursts provoked by the rapid switchings between the two signals when both are below the threshold. As a consequence, the switch-and-stay technique is preferable. These two techniques are illustrated in Figure 5.8.

### 5.5.2 Gain Combining

#### Maximal Ratio

In this technique each one of the $M$ signals has a gain proportional to its own signal-to-noise ratio. Moreover, because the resultant signal is an overall sum of the $M$ signals, cophasing circuitry is required. The block diagram of maximal-ratio combining is shown in Figure 5.9a.

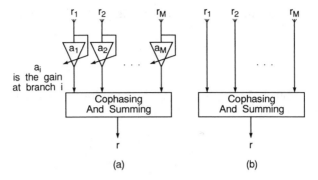

**Figure 5.9.**   Gain-combining block diagrams: (a) maximal-ratio combining; (b) equal-gain combining.

## Equal Gain

This technique differs from the previous one in the sense that all of the signals have a gain equal to 1. In other words, the variable gain amplifiers are eliminated as shown in Figure 5.9b.

## 5.6   STATISTICAL PROPERTIES AND PERFORMANCE MEASURE

Due to the random characteristic of the mobile radio signal, the performance of the diversity combining techniques is assessed on a statistical basis. The performance measure of most interest is the signal-to-noise ratio (SNR) $\gamma_i$, defined for each branch $i$ as

$$\gamma_i \triangleq \frac{\text{local mean signal power}}{\text{mean noise power}}$$

For a Rayleigh fading signal having an envelope $r_i$, the local mean power is equal to $r_i^2/2$, corresponding to the signal power averaged over one RF cycle. In the presence of Gaussian noise with mean power equal to $N_i$, the SNR is then

$$\gamma_i = \frac{r_i^2}{2N_i} = \frac{r_i^2}{2N} \tag{5.11}$$

where we assume that the mean noise power is equal to $N$ in all of the branches.

Let $p(r_i)$ be the probability density function of $r_i$. The density $p(\gamma_i)$ can be easily found as follows

$$p(\gamma_i)|d\gamma_i| = p(r_i)|dr_i| \tag{5.12}$$

We recall from Section 3.4.2 that the Rayleigh distribution is

$$p(r_i) = \frac{r_i}{\sigma^2} \exp\left(-\frac{r_i^2}{2\sigma^2}\right) \qquad (5.13)$$

From Equation 5.11, $d\gamma_i = (r_i/N)\, dr_i$. Therefore, using this result with Equation 5.13 in Equation 5.12, we obtain

$$p(\gamma_i) = \frac{N}{\sigma^2} \exp\left(-\frac{r_i^2}{2\sigma^2}\right) \qquad (5.14)$$

Define $\gamma_0$ as the ratio of the mean signal power and the mean noise power, i.e.,

$$\gamma_0 = \frac{E_0^2/2}{N} \qquad (5.15)$$

where $E_0$ is the amplitude of the radio wave. From Equation 4.57, $\sigma^2 = E_0^2/2$. Then

$$\gamma_0 = \frac{\sigma^2}{N} \qquad (5.16)$$

Consequently, the probability density function of $\gamma_i$ is

$$p(\gamma_i) = \frac{1}{\gamma_0} \exp\left(-\frac{\gamma_i}{\gamma_0}\right) \qquad (5.17)$$

The probability that $\gamma_i$ is less than or equal to a given SNR, $\Gamma$, is the probability distribution function $P(\Gamma)$. Hence

$$P(\Gamma) = \mathrm{prob}(\gamma_i \leq \Gamma) = \int_0^\Gamma p(\gamma_i)\, d\gamma_i = 1 - \exp\left(-\frac{\Gamma}{\gamma_0}\right) \qquad (5.18)$$

In the analyses that follow we assume that the signals in all of the diversity branches are uncorrelated and distributed as in Equation 5.18. Our aim now is to determine the performance of each diversity technique. This is achieved by obtaining and assessing the corresponding probability distribution function of the SNR available at each combiner's output.

### 5.6.1  Switched Combining

#### Pure Selection

We assume that the selector is ideal and that the best signal is always present at the output. Using Equation 5.18, the probability that the SNRs in all of the $M$ branches are simultaneously less than or equal to a given SNR $\Gamma_s$ is

$$P_{\text{SEL}}(\Gamma_s) = \text{prob}(\gamma_i, \ldots, \gamma_M \leq \Gamma_s) = \left[1 - \exp\left(-\frac{\Gamma_s}{\gamma_0}\right)\right]^M \quad (5.19)$$

The probability that at least one branch has an SNR exceeding $\Gamma_s$ is $1 - P_{\text{SEL}}(\Gamma_s)$, known as the *reliability*. The corresponding curves for $M = 1, 2, 3, 4, 6, 8,$ and 10 branches are shown in Figure 5.10.

These curves are plotted on Rayleigh paper, where the Rayleigh distribution (curve for $M = 1$) appears as a straight line. It can be readily seen that there is a substantial improvement when diversity is used. The larger the number of branches, the better the SNR performance. However, although increasing the number of diversity branches will always improve the SNR, the

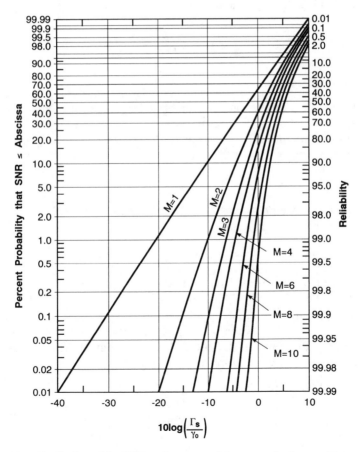

**Figure 5.10.** Distribution of the SNR at the output of the pure-selection combiner.

advantages of implementing more branches are not quite obvious, because the relative amelioration tends to be smaller. For example, a 99.9% reliability is achieved for an SNR equal to $-30$ dB for $M = 1$ (no diversity), $-15$ dB for $M = 2$, $-9.8$ dB for $M = 3$, or $-7.1$ dB for $M = 4$. The gains attained when changing from $M = 1$ to $M = 2$ or from $M = 2$ to $M = 3$ to $M = 4$ are, respectively, 50, 35, and 28%.

Another statistic of interest is the average SNR at the output of the selector. Let $\bar{\Gamma}_s$ be this average. Then

$$\bar{\Gamma}_s = \int_0^\infty \Gamma_s p_{\text{SEL}}(\Gamma_s) \, d\Gamma_s$$

where

$$p_{\text{SEL}}(\Gamma_s) = \frac{dP_{\text{SEL}}(\Gamma_s)}{d\Gamma_s} = \frac{M}{\gamma_0}\left[1 - \exp\left(-\frac{\Gamma_s}{\gamma_0}\right)\right]^{M-1} \exp\left(-\frac{\Gamma_s}{\gamma_0}\right)$$

is the probability density function of $\Gamma_s$. Hence

$$\bar{\Gamma}_s = \gamma_0 \sum_{i=1}^{M} \frac{1}{i} \tag{5.20}$$

**Threshold Selection**

We assume that the system uses the switch-and-stay strategy and that the switching occurs at an SNR threshold level $\Gamma_T$. Let the SNR at the output of the combining circuit be $\gamma$. We want to determine the probability that $\gamma$ is less than or equal to a given SNR level $\Gamma_s$, i.e.,

$$P_{\text{THS}}(\Gamma_s) = \text{prob}(\gamma \leq \Gamma_s)$$

Because the Rayleigh signals $r_1$ and $r_2$ (refer to Figure 5.8) are statistically indistinguishable we may write

$$P_{\text{THS}}(\Gamma_s) = \text{prob}(\gamma \leq \Gamma_s) = \text{prob}(\gamma \leq \Gamma_s | \gamma = \gamma_1) = \text{prob}(\gamma \leq \Gamma_s | \gamma = \gamma_2)$$

where $\gamma_1$ and $\gamma_2$ are the SNRs at branch 1 (signal $r_1$) and 2 (signal $r_2$), respectively. Let $\gamma_x$ and $\gamma_y$ be the portion of $\gamma$ above and below the threshold $\Gamma_T$, respectively, i.e., $\gamma = \gamma_x + \gamma_y$. The condition $\gamma = \gamma_i$ ($i = 1$ or $i = 2$) will occur either when $\gamma > \Gamma_T$ or $\gamma \leq \Gamma_T$, given that $\gamma = \gamma_x + \gamma_y$. Consequently, the overall distribution is the sum of the corresponding weighted conditional distributions:

$$\text{prob}(\gamma \leq \Gamma_s) = \text{prob}(\gamma \leq \Gamma_s | \gamma = \gamma_i)$$

$$= \text{prob}(\gamma \leq \Gamma_s | \gamma > \Gamma_T)\text{prob}(\gamma > \Gamma_T | \gamma = \gamma_x + \gamma_y)$$

$$+ \text{prob}(\gamma \leq \Gamma_s | \gamma \leq \Gamma_T)\text{prob}(\gamma \leq \Gamma_T | \gamma = \gamma_x + \gamma_y) \tag{5.21}$$

The probabilities $\mathrm{prob}(\gamma > \Gamma_T | \gamma = \gamma_x + \gamma_y)$ and $\mathrm{prob}(\gamma \le \Gamma_T | \gamma = \gamma_x + \gamma_y)$ correspond to the proportion of the time that $\gamma > \Gamma_T$ and $\gamma \le \Gamma_T$, respectively.

Define a successful transition as that occurring when the resultant SNR $\gamma$ is above the threshold $\Gamma_T$, and an unsuccessful transition otherwise. Given that a transition has occurred, it will be unsuccessful with probability $q$ such that

$$q = P(\Gamma_T) = \mathrm{prob}(\gamma \le \Gamma_T) = 1 - \exp\left(-\frac{\Gamma_T}{\gamma_0}\right) \tag{5.22}$$

A successful transition occurs with probability $p$, where

$$p = 1 - q = \exp\left(-\frac{\Gamma_T}{\gamma_0}\right) \tag{5.23}$$

Let $\tau_a$ be the average time that the SNR $\gamma_i$ is above the threshold. Similarly, $\tau_b$ is defined for $\gamma_i$ below the threshold. Because the switching instants occur at random, the average time that $\gamma$ exceeds $\Gamma_T$, given that a successful transition has occurred, is $\tau_a/2$. In the same way, $\tau_b/2$ is the average time that $\gamma$ is below $\Gamma_T$, given that an unsuccessful switching has occurred. Following an unsuccessful transition, the next transition will only occur after $\gamma$ has exceeded $\Gamma_T$ and fallen below it again. Therefore, the average time between the occurrence of an unsuccessful transition and the next transition is $\tau_b/2 + \tau_a$. On the other hand, the time span between a successful transition and the next transition is $\tau_a/2$. We notice that the condition $\gamma > \Gamma_T$ occurs in the successful transitions with probability $(1 - q)$ and in the unsuccessful transitions with probability $q$. In the first case $\gamma$ exceeds $\Gamma_T$ $(\gamma > \Gamma_T)$ for an average time of $\tau_a/2$ whereas in the second case the average time is $\tau_a$. Accordingly, the average time that the SNR $\gamma$ exceeds the threshold SNR $\Gamma_T$ is $\tau_x$ such that

$$\tau_x = (1 - q)\frac{\tau_a}{2} + q\tau_a = (1 + q)\frac{\tau_a}{2} \tag{5.24}$$

The same reasoning applies to $\tau_y$ (the average time that $\gamma \le \Gamma_T$). Hence

$$\tau_y = q\frac{\tau_b}{2} \tag{5.25}$$

The ratio $\tau_a/\tau_b$ is obviously equal to the ratio between the occurrence of a successful and an unsuccessful transition, i.e.,

$$\frac{\tau_a}{\tau_b} = \frac{p}{q} = \frac{1 - q}{q} \tag{5.26}$$

Similarly, the ratio between the probabilities $\text{prob}(\gamma > \Gamma_T | \gamma = \gamma_x + \gamma_y)$ and $\text{prob}(\gamma \le \Gamma_T | \gamma = \gamma_x + \gamma_y)$ is equal to the ratio between $\tau_x$ and $\tau_y$. Hence, using Equations 5.24 through 5.26,

$$\frac{\text{prob}(\gamma > \Gamma_T | \gamma = \gamma_x + \gamma_y)}{\text{prob}(\gamma \le \Gamma_T | \gamma = \gamma_x + \gamma_y)} = \frac{\tau_x}{\tau_y} = \frac{1 - q^2}{q^2} \tag{5.27}$$

The two remaining probabilities in Equation 5.21, namely, $\text{prob}(\gamma \le \Gamma_s | \gamma > \Gamma_T)$ and $\text{prob}(\gamma \le \Gamma_s | \gamma \le \Gamma_T)$, are determined as follows:

$$\text{prob}(\gamma \le \Gamma_s | \gamma > \Gamma_T) = \frac{\text{prob}(\Gamma_T < \gamma \le \Gamma_s)}{\text{prob}(\gamma > \Gamma_T)} = \frac{\int_{\Gamma_T}^{\Gamma_s} p(\gamma)\, d\gamma}{\int_{\Gamma_T}^{\infty} p(\gamma)\, d\gamma}$$

$$= \begin{cases} \dfrac{P(\Gamma_s) - P(\Gamma_T)}{1 - P(\Gamma_T)} = \dfrac{P(\Gamma_s) - q}{1 - q}, & \Gamma_s \ge \Gamma_T \\ 0, & \Gamma_s < \Gamma_T \end{cases} \tag{5.28}$$

In the same way,

$$\text{prob}(\gamma \le \Gamma_s | \gamma \le \Gamma_T) = \frac{\text{prob}(\gamma \le \Gamma_s \text{ and } \gamma \le \Gamma_T)}{\text{prob}(\gamma \le \Gamma_T)}$$

$$= \begin{cases} \dfrac{P(\Gamma_T)}{P(\Gamma_T)} = 1, & \Gamma_s \ge \Gamma_T \\ \dfrac{P(\Gamma_s)}{P(\Gamma_T)} = \dfrac{P(\Gamma_s)}{1 - q}, & \Gamma_s < \Gamma_T \end{cases} \tag{5.29}$$

By combining Equations 5.27 through 5.29 into Equation 5.21 we obtain the overall distribution function

$$P_{\text{THS}}(\Gamma_s) = \begin{cases} (1 + q)P(\Gamma_s) - q, & \Gamma_s \ge \Gamma_T \\ qP(\Gamma_s), & \Gamma_s < \Gamma_T \end{cases} \tag{5.30}$$

The probability density function is

$$p_{\text{THS}}(\Gamma_s) = \frac{dP_{\text{THS}}(\Gamma_s)}{d\Gamma_s} = \begin{cases} (1 + q)p(\Gamma_s), & \Gamma_s \ge \Gamma_T \\ qp(\Gamma_s), & \Gamma_s < \Gamma_T \end{cases} \tag{5.31}$$

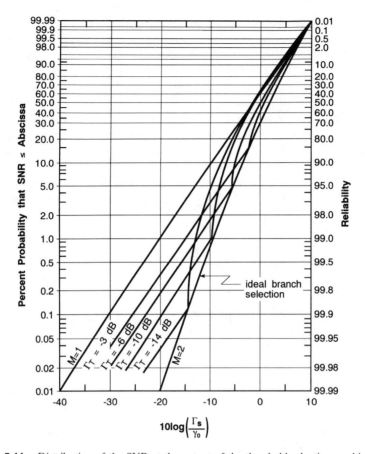

**Figure 5.11.** Distribution of the SNR at the output of the threshold-selection combiner.

The average SNR at the output of the selector is $\overline{\Gamma}_s$,

$$\overline{\Gamma}_s = \int_0^\infty \Gamma_s \, p_{\mathrm{THS}}(\Gamma_s) \, d\Gamma_s = \int_0^{\Gamma_T} \Gamma_s \, p_{\mathrm{THS}}(\Gamma_s) \, d\Gamma_s + \int_{\Gamma_T}^\infty \Gamma_s \, p_{\mathrm{THS}}(\Gamma_s) \, d\Gamma_s$$

$$\overline{\Gamma}_s = \gamma_0 + \Gamma_T \exp\left(-\frac{\Gamma_T}{\gamma_0}\right) \tag{5.32}$$

Figure 5.11 shows the distribution of $\Gamma_s$ given by Equation 5.30 for some values of threshold $\Gamma_T$. We also plot, for comparison, the distribution of $\Gamma_s$ for an ideal selector, corresponding to the distribution of $\Gamma_s$ for $M = 2$ in pure-selection combining. Also shown is the Rayleigh distribution ($M = 1$). Note that the curves for threshold selection are confined between the curves for $M = 1$ (no diversity) and $M = 2$ (pure selection). For values of SNR below the threshold, the curves have approximately a Rayleigh slope. Above the threshold they merge very quickly with the Rayleigh curve ($M = 1$). For an SNR equal to the threshold, the curves touch that of the ideal selection ($M = 2$). This is expected because the ideal selector works as a scanning circuit having the threshold dynamically switched to be the input signal currently in use. It can be seen that the largest gains are obtainable at lower levels (below the threshold). At higher levels (above the threshold) the gains are negligible. Accordingly, the threshold level must be set to the lowest acceptable SNR. As an example, suppose that in a certain geographical area a signal of 6 dB below its mean value is considered to be acceptable. Then the reliability goes from 77.8% (no diversity) to 95% if a threshold of −6 dB is used.

## 5.6.2   Gain Combining

### Maximal Ratio

Refer to Figure 5.9a for the block diagram of the maximal-ratio combiner. The envelope of the combined signal is

$$r = \sum_{i=1}^{M} a_i r_i \tag{5.33}$$

where $a_i$ is the gain at branch $i$. Assuming a mean noise power equal to $N$ for each branch, the total noise power $\mathcal{N}$ at the combiner's output is

$$\mathcal{N} = N \sum_{i=1}^{M} a_i^2 \tag{5.34}$$

Consequently, the resultant SNR $\gamma$ is

$$\gamma = \frac{r^2/2}{\mathcal{N}} = \frac{1}{2} \frac{\left(\sum_{i=1}^{M} a_i r_i\right)^2}{N \sum_{i=1}^{M} a_i^2} \tag{5.35}$$

It is shown in Appendix 5B that the SNR $\gamma$ is maximized if the gains $a_i$ are chosen to be equal to the ratio of the signal voltage to noise power of the respective branch. Thus

$$a_i = \frac{r_i}{N} \tag{5.36}$$

Consequently, using Equation 5.36 in Equation 5.35, we obtain

$$
\gamma = \frac{1}{2} \frac{\left(\sum_{i=1}^{M} r_i^2/N\right)^2}{N\sum_{i=1}^{M}(r_i/N)^2} = \sum_{i=1}^{M} \frac{r_i^2}{2N} = \sum_{i=1}^{M} \gamma_i
\qquad (5.37)
$$

Equation 5.37 shows that the resultant SNR at the output of the combiners is the sum of the SNRs of the $M$ branches. It was shown in Section 3.4.2 that the envelope $r_i$ is a function of two independent Gaussian variates, namely, $x_i$ and $y_i$, having means equal to zero and variances $\sigma^2$ equal to $E_0^2/2$. Recall that $r_i^2 = x_i^2 + y_i^2$. Then

$$
\gamma = \sum_{i=1}^{M} \gamma_i = \sum_{i=1}^{M} \frac{1}{2N} r_i^2 = \sum_{i=1}^{M} \frac{x_i^2}{2N} + \sum_{i=1}^{M} \frac{y_i^2}{2N}
\qquad (5.38)
$$

It is shown in Appendix 5C that the sum of the squares of independent standard normal random variables has a chi-square distribution with degrees of freedom equal to the number of terms in the sum. Because in Equation 5.38 we have two $M$-term sums, the degree of freedom is $2M$. The variance is equal to $\sigma^2/2N = \frac{1}{2}\gamma_0$. Thus the probability density function of $\gamma$ is

$$
p_{\mathrm{MAX}}(\gamma) = \frac{\gamma^{M-1} \exp(-\gamma/\gamma_0)}{\gamma_0^M (M-1)!}, \qquad \gamma \geq 0
\qquad (5.39a)
$$

For low SNR,

$$
p_{\mathrm{MAX}}(\gamma) \simeq \frac{\gamma^{M-1}}{\gamma_0^M (M-1)!}
\qquad (5.39b)
$$

The distribution function $P_{\mathrm{MAX}}(\Gamma_s)$ is

$$
P_{\mathrm{MAX}}(\Gamma_s) = \mathrm{prob}(\gamma \leq \Gamma_s) = \int_0^{\Gamma_s} p_{\mathrm{MAX}}(\gamma)\, d\gamma
$$

Using the density given by Equation 5.39a,

$$
P_{\mathrm{MAX}}(\Gamma_s) = 1 - \exp\left(\frac{-\Gamma_s}{\gamma_0}\right) \sum_{i=1}^{M} \frac{(\Gamma_s/\gamma_0)^{i-1}}{(i-1)!}
\qquad (5.40a)
$$

For low SNR (Equation 5.39b),

$$
P_{\mathrm{MAX}}(\Gamma_s) \simeq \frac{(\Gamma_s/\gamma_0)^M}{M!}
\qquad (5.40b)
$$

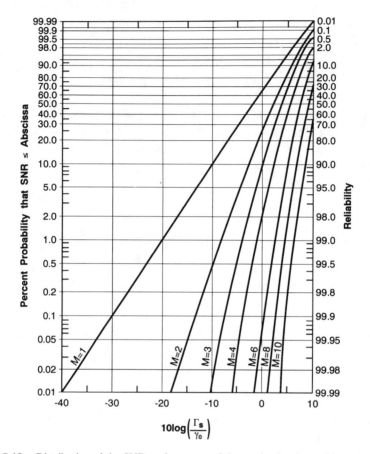

**Figure 5.12.** Distribution of the SNR at the output of the maximal-ratio combiner.

The corresponding (exact) distribution is plotted in Figure 5.12. Comparing the curves of Figure 5.10 with those of Figure 5.12, we note that maximal-ratio combining gives better results than pure-selection combining. We observe that, as the number of branches increases ($M \to \infty$), the sum in Equation 5.40a tends to the exponential $\exp(\Gamma_s/\gamma_0)$. Accordingly, the distribution $P_{MAX}(\Gamma_s)$ tends to zero.

The mean SNR at the output of the combiner is

$$\bar{\Gamma}_s = \langle \gamma \rangle = \left\langle \sum_{i=1}^{M} \gamma_i \right\rangle = \sum_{i=1}^{M} \langle \gamma_i \rangle = M\gamma_0 \qquad (5.41)$$

### Equal Gain

The signal output of the combiner is the direct sum of the received signals. The resultant envelope and the corresponding SNR are directly obtained from Equations 5.33 and 5.34 by setting the gains $a_i = 1$. Hence

$$r = \sum_{i=1}^{M} r_i \qquad (5.42)$$

and

$$\gamma = \frac{r^2/2}{\mathcal{N}} = \frac{\left(\sum_{i=1}^{M} r_i\right)^2}{2NM} = \frac{r^2}{2NM} \qquad (5.43)$$

The envelope $r$ is a random variable, resulting from the sum of $M$ independent Rayleigh variates. Let $p(r)$ be the density of $r$. Therefore, the density $p_{EQU}(\gamma)$ of $\gamma$ can be found as

$$p_{EQU}(\gamma)|d\gamma| = p(r)|dr|$$

Using Equation 5.43

$$p_{EQU}(\gamma) = NM\frac{p(r)}{r} = NM\frac{p\left(\sqrt{2\gamma NM}\right)}{\sqrt{2\gamma NM}} \qquad (5.44)$$

The question is how to find $p(r)$. For the case where $M = 2$ (two branches), the problem can be easily tackled. The joint distribution $p(r_1, r_2)$ of two Rayleigh signals has already been determined in Chapter 4 and is given by Equation 4.37.

For $M = 2$,

$$r = r_1 + r_2 \qquad \text{or} \qquad r_2 = r - r_1$$

Then

$$p(r) = \int_0^r p(r_1, r_2)\bigg|_{r_2 = r - r_1} dr_1 = \int_0^r p(r_1, r - r_1)\, dr_1 \qquad (5.45)$$

Such an integral can be solved in terms of tabulated functions, and the final distribution is given by

$$P_{EQU}(\Gamma_s) = 1 - \exp(-2\Gamma_s) - \sqrt{\pi\Gamma_s}\,\exp(-\Gamma_s)\mathrm{erf}\left(\sqrt{\Gamma_s}\right) \qquad (5.46)$$

where $\mathrm{erf}(\cdot)$ is the error function.

For $M > 2$ branches, the problem is rather complicated but can be tackled with the help of computer methods, using, for instance, simulation techniques. An approximate solution for low SNR is presented by Schwartz,[31] where it is shown that

$$p_{\text{EQU}}(\gamma) = \frac{2^{M-1}M^M}{(2M-1)!}\frac{\gamma^{M-1}}{\gamma^M} \tag{5.47}$$

$$P_{\text{EQU}}(\Gamma_s) \simeq \frac{(M/2)^M\sqrt{\pi}}{(M-1/2)!}\frac{1}{M!}\left(\frac{\Gamma_s}{\gamma_0}\right)^M \tag{5.48}$$

where $(\cdot)!$ is the gamma function. In particular,

$$\left(M - \frac{1}{2}\right)! = \frac{1\cdot 3\cdot 5\cdot 7\cdots(2M-1)!}{2^M}$$

Note that Equation 5.48 differs from Equation 5.40b by a factor equal to $(M/2)^M\sqrt{\pi}/[(M-1/2)!]$. The distribution of $\Gamma_s$ is shown in Figure 5.13.

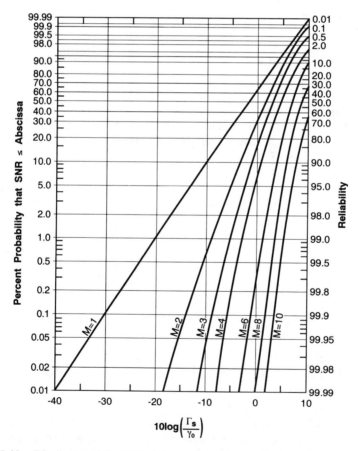

**Figure 5.13.** Distribution of the SNR at the output of the equal-gain combiner.

It can be seen that the performance of this technique is slightly poorer than that of the maximal ratio.

The mean value of $\gamma$ is easily obtained from Equation 5.43 as follows:

$$\bar{\Gamma}_s = \langle \gamma \rangle = \frac{1}{2NM} \left\langle \left( \sum_{i=1}^{M} r_i \right)^2 \right\rangle = \frac{1}{2NM} \sum_{i,j}^{M} \langle r_i r_j \rangle \qquad (5.49)$$

In Equation 5.49 we notice that there are $M$ elements equal to $\langle r_i^2 \rangle$ and $M(M-1)$ elements equal to $\langle r_i r_j \rangle$, $i \neq j$. From the Rayleigh distribution (refer to Chapter 3), $\langle r_i^2 \rangle = E(r_i^2) = 2\sigma^2$. Because the signals are assumed to be independent, $\langle r_i r_j \rangle = \langle r_i \rangle \langle r_j \rangle = E(r_i)E(r_j)$, $i \neq j$. However, $E(r_i) = E(r_j) = \sqrt{\pi/2}\,\sigma$. Therefore, $\langle r_i r_j \rangle = (\pi/2)\sigma^2$. Using these results in Equation 5.49, we obtain

$$\bar{\Gamma}_s = \frac{1}{2NM} \left[ 2M\sigma^2 + M(M-1)\pi \frac{\sigma^2}{2} \right]$$

$$= \gamma_0 \left[ 1 + (M-1)\frac{\pi}{4} \right] \qquad (5.50)$$

where $\gamma_0 = \sigma^2/N$ (see Equation 5.16).

## 5.7 COMPARATIVE PERFORMANCE OF COMBINING TECHNIQUES

A quick look at the graphs of Section 5.6 shows that the maximal ratio combining is the best technique followed by equal gain, pure selection and threshold selection, in this order. The distributions of the SNR at the output of the various combiners using $M = 2$ branches are shown in Figure 5.14. The performance of these techniques can also be shown as functions of the number of branches for a given SNR. However, it is questionable whether the improvement achieved for a number of branches greater than 2 justifies the implementation costs of the hardware required.

From the plots of Figure 5.14 it can be seen that a substantial improvement is accomplished when the combining techniques are used. As an example, in the conditions where the use of no diversity gives a 90% reliability, maximal ratio, equal gain, pure selection, and threshold selection for $\Gamma_T = -3$ dB would yield 99.6, 99.5, 99.1, and 97% reliabilities, respectively. Note that the equal gain technique gives a slightly poorer performance than that of the maximal ratio combining, even though it is much simpler to be implemented.

The performance of these techniques can also be assessed by the average SNR at the output of the respective combiners. We rewrite Equations 5.20, 5.32, 5.41 and 5.50 as the ratio $\bar{\Gamma}_s/\gamma_0$ between the average SNR of each

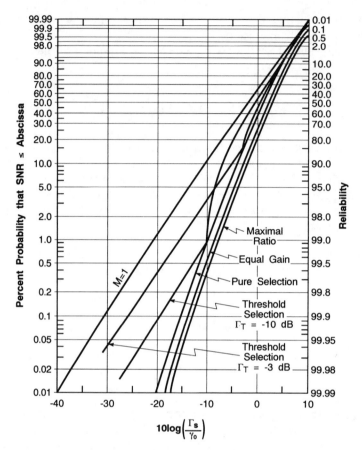

**Figure 5.14.** Distribution of the SNR at the output of the various combiners using two branches.

technique and the average SNR for $M = 1$ branch

$$\frac{\bar{\Gamma}_s}{\gamma_0} = \sum_{i=1}^{M} \frac{1}{i} \qquad \text{for pure selection} \qquad (5.51a)$$

$$\frac{\bar{\Gamma}_s}{\gamma_0} = 1 + \frac{\Gamma_T}{\gamma_0} \exp\left(-\frac{\Gamma_T}{\gamma_0}\right) \qquad \text{for threshold selection} \qquad (5.51b)$$

$$\frac{\bar{\Gamma}_s}{\gamma_0} = M \qquad \text{for maximal ratio} \qquad (5.51c)$$

$$\frac{\bar{\Gamma}_s}{\gamma_0} = 1 + (M - 1)\frac{\pi}{4} \qquad \text{for equal gain} \qquad (5.51d)$$

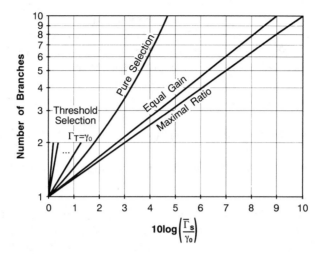

**Figure 5.15.**   Mean SNR of each combining technique compared to one branch as a function of the number of branches.

Note that in our derivations we have assumed $M = 2$ for the threshold selection. Moreover, its performance depends on the threshold $\Gamma_T$. In this case the best performance (maximum of Equation 5.51b) is obtained when $\Gamma_T = \gamma_0$, for which $\overline{\Gamma}_s = 1.38\gamma_0$. The worst results are obtained by setting $\Gamma_T$ to be either very low ($\Gamma_T \ll \gamma_0$) or very high ($\Gamma_T \gg \gamma_0$). In these situations the distribution of $\Gamma_s$ approximates the Rayleigh distribution. The curves of the mean SNR versus the number of branches are shown in Figure 5.15. It can be seen that, for $M = 2$, the difference between the average SNR of the best performance technique (maximal ratio) and that of the "worst" performance technique (threshold selection) does not exceed 2 dB (in case $\Gamma_T$ is set equal to $\gamma_0$).

## 5.8   OTHER RELEVANT POINTS

In this section we shall comment on some diversity-related topics, exploring more the conceptual ideas and avoiding mathematical derivations.

### 5.8.1   Combining Correlated Signals

The performance analyses of the combining techniques have been entirely based on the assumption of independent fading signals, i.e., correlation factor equal to zero. If the signals are considered to be correlated, the performance of the techniques will be different from that previously obtained. The extreme situation occurs when the correlation factor is equal to

1, which corresponds to having exactly the same signal in all of the branches. In this case the resultant distribution is the well-known Rayleigh distribution ($M = 1$), as expected. Therefore, the corresponding curves for these extreme cases, namely, correlation factor equal to 0 and equal to 1, determine the performance bounds of the diversity-combining techniques.

## 5.8.2  Level Crossing Rate (LCR)

The use of diversity-combining techniques smooths out the deep fades, increasing the mean level of the received signal. Accordingly, lower signal levels will be crossed at lower rates compared with the crossing rates obtained when no diversity is used. In a similar way, higher signal levels will be crossed at higher rates. On the other hand, because the resultant signal is a combination of two or more signals, the number of ripples at high levels will increase.

As far as correlation factor is concerned, its increase also leads to an increase of the LCR at low levels and to a reduction of the LCR at high levels. This is intuitively explained by the fact that, because the correlation factor gives a degree of "similarity" between the signals, a high correlation implies that the signals can be treated as only one signal.

## 5.8.3  Average Duration of Fades

Because the diversity-combining techniques smooth out the fades, the average fade duration is, obviously, reduced. It can be proved that, if $\tau$ is the average fade duration for $M = 1$ branch and $\tau_M$ is this average for $M$ branches, then $\tau_M = \tau/M$.[9] With the increase of the correlation between signals the average fade duration also increases.

## 5.8.4  Random FM

Diversity-combining techniques, if properly used, can substantially reduce the random frequency modulation. The cophasing process is one of the factors playing a significant role in this. For instance, if the reference signal is chosen to be equal to one of the $M$ received signals, then no improvement is achieved. In other words, the random FM of the resultant signal will be exactly the same as that of the reference signal. On the other hand, if the reference signal is chosen to be equal to the resultant signal itself (the sum of the $M$ received signals), then the resultant random FM can be reduced. Moreover, the random FM can be eliminated (or drastically reduced) if a pilot tone, having a frequency close to the carrier frequency, is transmitted along with the signal.

### 5.8.5 Predetection and Postdetection

The diversity receivers performing predetection have the signals detected before combining. In the receivers using postdetection, the signals are first combined and then detected. Predetection systems are usually more complex and costly. Because coherent combining is performed at an intermediate frequency (IF), the use of individual down-converters, amplifiers, and cophasing circuitry is required. On the other hand, the combined signals produce an output with less random FM and better statistics than that produced by the postdetection systems.

There are two basic methods of cophasing the received signals, as shown in Figure 5.16. In one method (switches in the positions 1) a phase shift is imposed on each one of the received signals, whereas in the other method (switches in the positions 2) the phase shifts are applied to the local oscillator.

The reference signal for phase comparison can be either one of the incoming signals $(r_1, r_2, \ldots, r_n)$ or the combined (sum) signal $r$ itself. In the former case, if the signal $r_i$ is chosen to be the reference signal, the corresponding phase shifter and phase comparator elements can be eliminated (bypassed). Maximal-ratio and equal-gain combining techniques require cophasing circuitry. In Figure 5.16, the block diagram corresponds to the equal-gain combining scheme. Maximal ratio can be obtained by adding a gain control element in each branch before the summing.

Postdetection receivers are usually very simple to implement. They are the precursors of all the diversity-combining systems. Initially, they were manually controlled, requiring an operator to select the signal that sounded best.

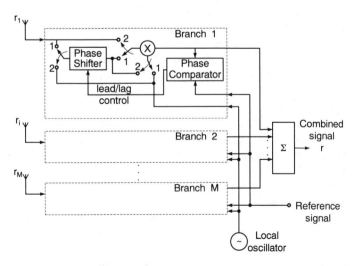

**Figure 5.16.** Cophasing block diagram. (With the switches at the positions 1 we have the variable-phase shifter; at the positions 2, the variable-phase local oscillator.)

In postdetection receivers, cophasing is not necessary because demodulation (detection) is performed before combining. Once demodulation is accomplished, the baseband signals are all in phase. Pure selection and threshold selection use postdetection receivers.

A great number of practical diversity systems can be found in the literature,[8, 9, 11, 32] with appropriate description and analysis.

## 5.9  SUMMARY AND CONCLUSIONS

One of the main causes of performance degradation in a mobile radio system is the occurrence of fading. The fading may be annoying for voice but perhaps critical for data communication. Many methods for counteracting the effects of fading have been proposed, and they greatly contribute to minimize such effects, consequently improving the performance of the system. The basic principle of these methods (diversity methods) is redundant repetition of the information on two or more independently fading paths (diversity branches) so that the probability of simultaneous occurrence of fade on all of the paths is negligible. At reception, a convenient processing of such information (diversity combining) can lead to a better received signal.

The precursor of all of the diversity techniques is space diversity, where transmitting and/or receiving antennas are spaced far enough from each other so that the fadings occur independently in each branch. Therefore, the probability that two or more diversity branches are simultaneously in deep fade is substantially reduced. The distances between the receiving antennas depend on several factors, including the kind of fading (lognormal or Rayleigh) and the location of the antennas (base station or mobile), among others. When antennas are located at the base station, the antenna heights, the beamwidth, and the wave arrival angle also affect these distances.

Slow (lognormal) fading is usually counteracted by means of space diversity where base stations are strategically positioned so that the mobile will always have a clear path for communication. Because this involves long distances the slow fading counteraction methods are classified into the macroscopic diversity group.

Fast (Rayleigh) fading is counteracted by means of microscopic diversity. Space diversity is considered to be one of the simplest and most effective means of combatting fading. Its great advantage is that it does not require extra frequency spectrum. Although spacing between antennas at the mobile involves distances of about half a wavelength, at the base station some tens of wavelengths may be necessary (but still the distances involved are nothing compared to those used in macroscopic diversity).

Polarization diversity may also be regarded as a special case of space diversity, because it requires the use of spaced antennas. Nevertheless, it takes advantage of the independence between orthogonally polarized waves to obtain independent fading signals. This technique has two limitations: Because there are only orthogonal polarized waves, only two diversity branches can be used. Moreover, half of the signal power is lost, because the signal is used to feed two antennas.

Angle diversity uses directional antenna(s) to restrict the detection of the signal to a small number of scattered waves. It may employ either several antennas to implement the diversity branches or an adaptive antenna to track the best signal.

Frequency diversity requires one antenna at each end of the radio path but a separate transmitter for each frequency. One of its main disadvantages is the need for extra frequency spectrum.

In an $M$-branch time-diversity system the same information is repeated $M$ times in order to obtain $M$ independent signals. These signals are assumed to be independent if the intertransmission time exceeds the channel coherence time. If voice transmission is involved, an increase of the sampling rate by a factor of $M$ is required. This implies a proportional increase of the channel bandwidth. Moreover, because the diversity branches are obtained by delaying and repeating the information, storage elements must be provided at both ends of the path. The great limitation of this technique, however, is that the intertransmission time $\tau$ is a function of the maximum Doppler shift $f_m$. In other words, the signals are considered to be uncorrelated if $\tau \geq 1/(2f_m) = 1/[2(v/\lambda)]$. Accordingly, if the vehicle is stationary ($v = 0$), time diversity does not apply ($\tau \to \infty$).

Time-slot hopping or frequency hopping are alternative ways of achieving diversity using exactly the same available resources. On the other hand they require a complex control (see Chapter 10).

As far as combining techniques are concerned, maximal ratio yields the best performance, followed by equal gain, pure selection, and threshold selection, in this order. Maximal-ratio and equal-gain combining give almost the same performance. Accordingly, the latter is preferred due to its simplicity. Pure selection is not feasible because all of the $M$ signals are required to be continuously monitored, demanding a considerable amount of signal processing. An alternative to pure selection is threshold selection, where the threshold level can be conveniently set for the best performance.

It was shown that the larger the number of branches, the better the system performance. However, the percentage of improvement in the SNR for a given reliability decreases with the increase of the number of diversity branches. In practice, the use of more than two diversity branches at the mobile is not feasible due to the costs and complexity. Moreover, two diversity branches already provide a substantial improvement in the received signal. For instance, at a 99% reliability the simple threshold selection technique provides a 9-dB improvement in the SNR. This, in fact, represents the power to be saved at the transmitter if the initial grade of service is to be maintained. In the same way, we can say that the effects of fading can be reduced with the increase of the transmitter power, hence avoiding diversity. In the example given, a 9-dB increase of the transmitter power would lead to a reliability of 99% without the need for diversity. However, the effects of this on cochannel interference can be disastrous.

## APPENDIX 5A   CORRELATION FACTOR OF TWO SIGNALS AT THE BASE STATION

Let $r_1$ and $r_2$ be the envelopes of two Rayleigh fading signals. It has been demonstrated in Chapter 4* that the cross-correlation of these signals is given by

$$\rho_r \simeq \frac{\pi}{4(4 - \pi)} \rho^2 \simeq \rho^2 \tag{5A.1}$$

where

$$\rho^2 = \frac{\mu_1^2 + \mu_2^2}{\sigma^4} \tag{5A.2}$$

and

$$\mu_1 = E[X_1 X_2] \tag{5A.3}$$

$$\mu_2 = E[X_1 Y_2] \tag{5A.4}$$

$$\sigma^2 = E[X_1^2] \tag{5A.5}$$

If the frequency difference $\Delta\omega$ between the signals is nil, then

$$E[X_1 X_2] = \frac{E_0^2}{2} \sum_i a_i^2 \cos(\omega_i \tau) = \sigma^2 \sum_i a_i^2 \cos(\omega_i \tau) \tag{5A.6}$$

$$E[X_1 Y_2] = \frac{E_0^2}{2} \sum_i a_i^2 \sin(\omega_i \tau) = \sigma^2 \sum_i a_i^2 \sin(\omega_i \tau) \tag{5A.7}$$

where

$$a_i^2 = p(\theta_i)\, d\theta \tag{5A.8}$$

Let the $i$th incoming wave arrive at an angle $\theta_i - \alpha$. Hence, when in the limit $i \to \infty$, Equations 5A.6 and 5A.7 will assume the integral form. Thus

$$\mu_1 = \sigma^2 \int_{-\pi/2+\alpha}^{\pi/2+\alpha} p(\theta)\cos[\beta v \tau \cos(\theta - \alpha)]\, d\theta \tag{5A.9}$$

$$\mu_2 = \sigma^2 \int_{-\pi/2+\alpha}^{\pi/2+\alpha} p(\theta)\sin[\beta v \tau \cos(\theta - \alpha)]\, d\theta \tag{5A.10}$$

where we have used the Doppler frequency $\omega_i = \beta v \cos(\theta_i - \alpha)$.

---

*The reader is strongly urged to review Section 4.5

With Equations 5A.9 and 5A.10 in Equation 5A.2, we obtain

$$\rho_r = \left( \int_{-\pi/2+\alpha}^{\pi/2+\alpha} p(\theta)\cos[\beta v \tau \cos(\theta - \alpha)] \, d\theta \right)^2$$

$$+ \left( \int_{-\pi/2+\alpha}^{\pi/2+\alpha} p(\theta)\sin[\beta v \tau \cos(\theta - \alpha)] \, d\theta \right)^2 \qquad (5A.11)$$

Equation 5A.11 is the correlation factor at the base station. Assume that the probability density function of the $i$th incoming wave incident at an angle $\alpha$ is given by

$$p(\theta_i) = \frac{Q}{\pi}\cos^n(\theta_i - \alpha) \qquad (5A.12)$$

In Equation 5A.12 $Q$ is a normalization factor obtained from

$$\int_{-\pi/2+\alpha}^{\pi/2+\alpha} p(\theta_i) \, d\theta_i = 1 \qquad (5A.13)$$

and $n$ is related to the beamwidth ($\Delta\varphi/2$) of the incoming wave.

If the electric field pattern is given by $\cos^n(\Delta\varphi/2)$, then $n$ is obtained at its half-power, i.e.,

$$\cos^n\left( \frac{\Delta\varphi}{2} \right) = \frac{1}{\sqrt{2}}$$

Note that the larger the value of $n$, the smaller the beamwidth and, consequently, the more likely the angle of arrival is closer to $0°$. Figure 5A.1 shows the function $\cos^n x$ for $n = 1, 10, 50$.

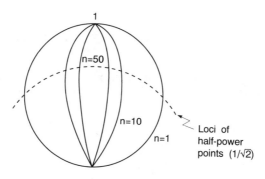

**Figure 5A.1.** Examples of the function $\cos^n x$ for $n = 1, 10,$ and $50$.

## APPENDIX 5B  OPTIMUM BRANCH GAIN FOR MAXIMAL-RATIO COMBINING

This appendix aims at determining the optimum branch gain for maximal-ratio combining. This is easily done by means of the Schwarz inequality.

### 5B.1  Schwarz Inequality

Let $R(\omega)$ and $S(\omega)$ be two complex-valued functions having complex conjugates equal to $R^*(\omega)$ and $S^*(\omega)$, respectively. These functions are related to each other by the Schwarz inequality as follows:

$$\left| \int_{-\infty}^{\infty} R(\omega)S(\omega)\,d\omega \right|^2 \leq \int_{-\infty}^{\infty} |R(\omega)|^2\,d\omega \int_{-\infty}^{\infty} |S(\omega)|^2\,d\omega \qquad (5B.1)$$

Such an inequality also holds for real functions $f(t)$ and $g(t)$,

$$\left[ \int_{-\infty}^{\infty} f(t)g(t) \right]^2 \leq \int_{-\infty}^{\infty} g^2(t)\,dt \int_{-\infty}^{\infty} f^2(t)\,dt \qquad (5B.2)$$

and for real random variables $X$ and $Y$ with finite second moments,

$$E^2[XY] \leq E[X^2]E[Y^2] \qquad (5B.3)$$

It can be seen that the *equality* in the three respective equations is satisfied if and only if

$$R(\omega) = KS^*(\omega)$$

$$f(t) = Kg(t)$$

$$X = KY$$

where $K$ is a real-valued number. Let us start by proving Equation 5B.1. Consider the following integral:

$$I = \int_{-\infty}^{\infty} d\omega \int_{-\infty}^{\infty} |R(\omega)S(z) - S(\omega)R(z)|^2\,dz \geq 0$$

where $R$ and $S$ are complex-valued functions. The integral $I$ is necessarily real and positive. Expanding the functions inside the integral, we have

$$|R(\omega)S(z) - S(\omega)R(z)|^2 = |R(\omega)|^2|S(z)|^2 + |S(\omega)|^2|R(z)|^2$$
$$- R(\omega)S(z)S^*(\omega)R^*(z)$$
$$- R^*(\omega)S^*(z)S(\omega)R(z)$$

Integrating term by term and observing that the first two terms after the equality sign are identical, as are the last two, we have

$$I = 2\int_{-\infty}^{\infty} |R(\omega)|^2 d\omega \int_{-\infty}^{\infty} |S(\omega)|^2 d\omega$$

$$- 2\int_{-\infty}^{\infty} R^*(\omega)S(\omega) \, d\omega \int_{-\infty}^{\infty} R(z)S^*(z) \, dz$$

However,

$$\left| \int_{-\infty}^{\infty} R(\omega)S(\omega) \, d\omega \right|^2 = \int_{-\infty}^{\infty} R(\omega)S(\omega) \, d\omega \int_{-\infty}^{\infty} R^*(z)S^*(z) \, dz$$

Then

$$\frac{I}{2} = \int_{-\infty}^{\infty} |R(\omega)|^2 d\omega \int_{-\infty}^{\infty} |S(\omega)|^2 d\omega - \left| \int_{-\infty}^{\infty} R(\omega)S(\omega) \, d\omega \right|^2 \geq 0$$

proving the inequality.

The proof of Equation 5B.2 follows the same reasoning as that of Equation 5B.1. For the case of Equation 5B.3 the proof is also simple. Consider the function

$$Q(k) \triangleq E\left[(X - KY)^2\right] \tag{5B.4}$$

The right-hand side of Equation 5B.4 is always nonnegative. Then

$$Q(K) = E[X^2] - 2KE[XY] + K^2E[Y^2] \geq 0$$

$Q(K)$ is a nonnegative quadratic function of $K$. Its minimum is determined by equating $dQ(K)/dK$ to 0. Let $K_0$ be such a minimum. Hence,

$$K_0 = E[XY]/E[Y^2]$$

It is straightforward to show that

$$Q(K_0) > 0 \quad \text{if } E^2[XY] < E[X^2]E[Y^2]$$

and

$$Q(K_0) = 0 \quad \text{if } E^2[XY] = E[X^2]E[Y^2]$$

proving the inequality.

## 5B.2  Optimum Branch Gain

Now let us return to the problem of determining the branch gain of maximal-ratio combining. From the corresponding section in Chapter 5 we have that the envelope of the combined signal is

$$r = \sum_{i=1}^{M} a_i r_i \tag{5B.5}$$

The total power noise at the output of the combiner is

$$\mathcal{N} = \sum_{i=1}^{M} a_i^2 N_i \tag{5B.6}$$

where here we are assuming that each branch $i$ has a noise power equal to $N_i$. The resultant SNR $\gamma$ is

$$\gamma = \frac{1}{2} \frac{\left(\sum_{i=1}^{M} a_i r_i\right)^2}{\sum_{i=1}^{M} a_i^2 N_i} \tag{5B.7}$$

From the Schwarz inequality we can write

$$\left(\sum_{i=1}^{M} a_i r_i\right)^2 = \left(\sum_{i=1}^{M} a_i \sqrt{N_i} \frac{r_i}{\sqrt{N_i}}\right)^2 \le \sum_{i=1}^{M} a_i^2 N_i \sum_{i=1}^{M} \frac{r_i^2}{N_i} \tag{5B.8}$$

Consequently, from Equations 5B.7 and 5B.8

$$\gamma = \frac{1}{2} \frac{\left(\sum_{i=1}^{M} a_i r_i\right)^2}{\sum_{i=1}^{M} a_i^2 N_i} \le \sum_{i=1}^{M} \frac{r_i^2}{2 N_i} \tag{5B.9}$$

The maximum of Equation 5B.9 is obtained when the "less than or equal to" sign is transformed into an "equal to" sign. In this case

$$a_i \sqrt{N_i} = K \frac{r_i}{\sqrt{N_i}}$$

Thus

$$a_i = K \frac{r_i}{N_i}$$

where $K$ is a real-valued number. Hence it has been proved that the optimum gain for each branch is proportional to its respective signal-voltage–to–noise-power ratio.

## APPENDIX 5C   THE CHI-SQUARE DISTRIBUTION[30]

Let $X_i$, $i = 1, 2, \ldots, k$, be independent and normal-Gaussian-distributed random variables with means $\mu_i$ and variances $\sigma_i^2$. We want to prove that the random variable

$$R = \sum_{i=1}^{k} \left( \frac{X_i - \mu_i}{\sigma_i} \right)^2 \tag{5C.1}$$

has a chi-square distribution with $k$ degrees of freedom.

Define $V_i$ as

$$V_i \triangleq \frac{X_i - \mu_i}{\sigma_i} \tag{5C.2}$$

Hence $V_i$ has a standard normal distribution (zero mean value and unity standard deviation). The moment generating function denoted by $M_R(s)$ is given by

$$M_R(s) \triangleq E[\exp(sR)] = \int_{-\infty}^{\infty} \exp(sr)p(r)\, dr \tag{5C.3}$$

where $p(r)$ is the probability density function of $R$ to be determined. However,

$$E[\exp(sR)] = E\left[\exp\left(s\sum_{i=1}^{k} V_i^2\right)\right] = E\left[\prod_{i=1}^{k} \exp(sV_i^2)\right] = \prod_{i=1}^{k} E[\exp(sV_i^2)] \tag{5C.4}$$

and

$$E[\exp(sV_i^2)] = \int_{-\infty}^{\infty} \exp(sv^2)p(v)\, dv \tag{5C.5}$$

where $p(v)$ is the density of $V$. As previously shown, $p(v)$ has a standard normal (Gaussian) distribution. Then

$$
\begin{aligned}
E[\exp(sV_i^2)] &= \int_{-\infty}^{\infty} \exp(sv^2) \frac{1}{\sqrt{2\pi}} \exp\left(-\frac{1}{2}v^2\right) dv \\
&= \int_{-\infty}^{\infty} \frac{1}{\sqrt{2\pi}} \exp\left[-\frac{1}{2}(1 - 2s)v^2\right] dv \\
&= \frac{1}{\sqrt{1 - 2s}} \int_{-\infty}^{\infty} \frac{\sqrt{1 - 2s}}{\sqrt{2\pi}} \exp\left[-\frac{1}{2}(1 - 2s)v^2\right] dv \\
&= \frac{1}{\sqrt{1 - 2s}}, \quad \text{for } s < \frac{1}{2} \tag{5C.6}
\end{aligned}
$$

because the integral is equal to unity (normal curve with variance $1/(1 - 2s)$). Consequently,

$$M_R(s) = \prod_{i=1}^{k} \frac{1}{\sqrt{1 - 2s}} = \left(\frac{1}{1 - 2s}\right)^{k/2}, \quad \text{for } s < \frac{1}{2} \quad (5C.7)$$

Equation 5C.7 is the moment generating function of a chi-square distribution with $k$ degrees of freedom. The density may be found by means of the inverse function or directly from tables. Hence, the chi-square density with $k$ degrees of freedom is given by

$$p(r) = \frac{1}{\Gamma(k/2)} \left(\frac{1}{2}\right)^{k/2} r^{k/2-1} \exp\left(-\frac{1}{2}r\right), \quad r \geq 0 \quad (5C.8)$$

where $\Gamma(\cdot)$ is the gamma function defined by

$$\Gamma(t) = \int_{0}^{\infty} x^{t-1} \exp(-x)\, dx \quad (5C.9)$$

Using this definition and then integrating by parts, we obtain

$$\Gamma(t + 1) = -x^t \exp(-x)\Big|_{0}^{\infty} + t\int_{0}^{\infty} x^{t-1} \exp(-x)\, dx$$

Then

$$\Gamma(t + 1) = t\Gamma(t) \quad (5C.10)$$

If $t = n$, where $n$ is an integer,

$$\Gamma(n + 1) = n! \quad (5C.11)$$

Now, returning to the problem of determining the distribution of the SNR $\gamma$ given by Equation 5.38, we have

$$\gamma = \sum_{i=1}^{M} \gamma_i = \sum_{i=1}^{M} \frac{1}{2N} r_i^2 = \sum_{i=1}^{M} \frac{x_i^2}{2N} + \sum_{i=1}^{M} \frac{y_i^2}{2N} \quad (5C.12)$$

Assume that $\sigma^2$ is the variance of the Gaussian random variables $x_i$ and $y_i$. Then the SNR $\gamma'$, defined by Equation 5C.13,

$$\gamma' = \sum_{i=1}^{M} \frac{x_i^2}{\sigma^2} + \sum_{i=1}^{M} \frac{y_i^2}{\sigma^2} \quad (5C.13)$$

has a chi-square distribution with $k = 2M$ degrees of freedom. Its density

$p(\gamma')$ is given by Equation 5C.8 with the appropriate substitution of variables. Hence

$$p(\gamma') = \frac{1}{2(M-1)!}\left(\frac{\gamma'}{2}\right)^{M-1}\exp\left(-\frac{1}{2}\gamma'\right) \qquad (5C.14)$$

From Equations 5C.12 and 5C.13 we have

$$\gamma = \frac{\sigma^2}{2N}\gamma' = \frac{1}{2}\gamma_0\gamma' \qquad (5C.15)$$

Then, changing variables, we obtain the distribution $p(\gamma)$

$$p(\gamma)|d\gamma| = p(\gamma')|d\gamma'| \qquad (5C.16)$$

but, from Equation 5C.15

$$d\gamma = \tfrac{1}{2}\gamma_0\,d\gamma' \qquad (5C.17)$$

Hence, using Equations 5C.14 through 5C.17, we have

$$p(\gamma) = \frac{\gamma^{M-1}}{\gamma_0^M}\frac{\exp(-\gamma/\gamma_0)}{(M-1)!}, \qquad \gamma \geq 0$$

as used in the corresponding section of Chapter 5 (Equation 5.39a).

## REFERENCES

1. Yeh, Y. S. and Reudink, D. O., Efficient spectrum utilization for mobile radio systems using space diversity, *IEEE Trans. Commun.*, COM-30(3), March 1982.
2. Wong, W. C. (L.), Steele, R., Glance, B., and Horn, D., Time diversity with adaptive error detection to combat Rayleigh fading in digital mobile radio, *IEEE Trans. Commun.*, COM-31(3), March 1983.
3. Darnell, M., Problems of mobile H.F. communications and techniques for performance improvement, *IEE Proc.*, 132(5), Part F, August 1985.
4. Frullone, M., Riva, G., Sentinelli, M., and Serra, A. M., Performance of digital mobile systems suitable for pan-European operations, in 2nd Nordic Seminar on Digital Land Mobile Radio Communication, Stockholm, 14–16 October 1986.
5. Bernhardt, R. C., Macroscopic diversity in frequency reuse radio systems, *IEEE J. Selected Areas Commun.*, SAC-5(5), June 1987.
6. Baghdady, E. J., Novel techniques for counteracting multipath interference effect in receiving systems, *IEEE J. Selected Areas Commun.*, SAC-5(2), February 1987.
7. Turkmani, A. M. D., Performance evaluation of *M*-branch diversity with continuous phase modulation systems in Rayleigh and log-normal fading, in Int. Symp. Signals, Systems and Electronics, ISSSE'89, Erlangen, September 18–20, 1989.
8. Lee, W. C. Y., *Mobile Communications Engineering*, McGraw-Hill, New York, 1982.
9. Jakes, W. C., Jr., *Microwave Mobile Communications*, John Wiley & Sons, New York, 1974.

10. Brennan, D. G., Linear diversity combining techniques, *Proc. IRE*, 47, June 1959.

11. Rsutako, A. J., Yeh, Y. S., and Murray, R. R., Performance of feedback and switch space diversity 900 MHz mobile radio systems with Rayleigh fading, *IEEE Trans. Commun. Tech.* COM-21, 1257–1268, November 1973.

12. Adachi, F. T., Hirade, K., and Kamata, T., A periodic switching diversity technique for a digital FM land mobile radio, *IEEE Trans. Vehicular Tech.*, VT-27, November 1978.

13. Adachi, F., Selection and scanning diversity effects in a digital FM land mobile radio with discriminator and differential detections, *IECE Japan*, 64-E, June 1981.

14. Adachi, F., Postdetection selection diversity effects on digital FM land mobile radio, *IEEE Trans. Vehicular Tech.*, VT-31, November 1982.

15. Stein, S., Fading channel issues in system engineering, *IEEE J. Selected Areas Commun.*, SAC-5(2), February 1987.

16. Fluhr, Z. C. and Poster, P. T., Control architecture, *Bell Syst. Tech. J.*, 58(1), January 1979.

17. Cox, D. C., Universal portable communications, in *N. Commun. For.*, NCF'84, Chicago, Illinois, September 24–26, pp. 169–174, 1984 and in *IEEE Trans. Vehicular Tech.*, VT-34, 117–121, August 1985.

18. Cox, D. C., Arnold, H. W., and Poster, P. T., Universal digital portable communications: a system perspective, *IEEE J. Selected Areas Commun.*, SAC-5(5), 764–773, June 1987.

19. Lee, W. C. Y., Effects on correlation between two mobile radio base stations antennas, *IEEE Trans. Commun.* COM-21, 1214–1224, November 1973.

20. Camwell, P., Propagation and Diversity, unpublished notes on mobile communications, 1988.

21. Alisouskas, V. F. and Tomasi, W., *Digital and Data Communications*, Prentice-Hall, Englewood Cliffs, NJ, 1985.

22. Haykin, S., *Adaptive Filter Theory*, Prentice-Hall, Englweood Cliffs, NJ, 1986.

23. Oureshi, S. U. H., Adaptive equalization, *Proc IEEE*, 73(9), September 1985.

24. Oppenheim, A. V. and Schafer, R. W., *Digital Signal Processing*, Prentice-Hall, Englewood Cliffs, NJ, 1975.

25. Mota, J. C. M., Romano, J. M. T., and Souza, R. F., Blind deconvolution in data transmission systems (in Portuguese), RT-179, FEE, UNICAMP, Brazil, November 1989.

26. Foschini, G. V., Equalizing without altering or detecting data, *AT & T Tech. J.*, 64(8), October 1985.

27. Macchi, O. and Eweda, E., Convergence analysis of self-adaptive equalizers, *IEEE Trans. Inform. Theory*, IT-30 (2), March 1984.

28. Schwartz, M., *Information Transmission, Modulation and Noise*, McGraw-Hill, New York, 1970.

29. Davenport, W. B., Jr., *Probability and Random Processes*, McGraw-Hill KogaKusha Ltd., Tokyo, 1970.

30. Mood, A. M., Graybell, F. A., and Boes, D. C., *Introduction to the Theory of Statistics*, McGraw-Hill, New York, 1974.

31. Schwartz, M., Bennet W. R., and Seymour, S., *Communications Systems and Techniques*, McGraw-Hill, New York, 1966.

32. Parsons, J. D. and Gardiner, J. G., *Mobile Communications Systems*, Blackie and Son, Ltd., Glasgow, 1989.

# CHAPTER 6

# Data Transmission and Signalling

This chapter examines the performance of digital transmission over a fading environment. The performance measure is the bit error rate (or, equivalently, the probability of bit error) for a given average signal-to-noise ratio per bit. We first review the bit error rate of some binary modulation schemes, namely, FSK and PSK; later we examine how their performance can be improved. The aim is to gain an insight into the qualitative improvement achieved when diversity, coding, or multiple-transmission techniques are applied. Accordingly, the simple binary modulation techniques were chosen in order to ease our analysis. However, the basic principles described here can be directly applied to more elaborate modulation schemes such as those studied in Chapter 9. Some other improvement techniques such as interleaving, ARQ, and adaptive equalization are also briefly investigated. An appendix on channel coding is included, with the objective of introducing the basic concepts and main codes.

## 6.1 INTRODUCTION

In a mobile radio system, data or control signalling can be transmitted over speech channels or separate signalling channels, depending on the control task to be performed. Examples of control tasks include the phases of a mobile call setup, hand-off between base stations, and call disconnect. Message exchange between mobile and base station is carried out over radio channels, whereas messages between base station and mobile switching centers usually flow through land links.

There are two types of signalling messages transmitted over the radio channels: (1) those sent as a continuous stream of bits and (2) those sent in bursts. The former include mobile pages, systems status reports, and overhead messages and are transmitted over separate signalling channels. The latter comprise (1) call release and hand-off, making use of speech channels, as well as (2) requests from mobile to base station, using separate signalling channels.

In analog systems, special attention is given to problems arising from the association of voice, using analog transmission, and data, using digital transmission. In digital systems, data and digitized voice transmission can use the same digital techniques.

One very important issue in digital transmission is synchronization. Time jitter or frequency drift of the clock may provoke an increase of the bit or word error rate, degrading the performance of the system. Techniques to improve bit or frame synchronization can be used to minimize the effects of both time jitter and frequency drift. A sequence of alternate 0s and 1s can be used to achieve bit synchronization. In the same way, the inclusion of one or more known word-patterns at the beginning and/or at the end of the frame can be used to achieve frame synchronization. At the reception, the word-patterns are stored to be compared with the incoming sequence.*

Despite synchronization, bit or word errors can still occur due to many other factors, including fading. This chapter aims at examining the occurrence of errors in data transmission of the mobile radio system as well as investigating the usual methods employed to minimize error occurrence. We shall restrict our attention to some binary modulation systems, due to their simplicity and universal applicability. Moreover, these are the digital modulation schemes used in analog mobile radio systems. It is understood that digital systems use higher-order modulation schemes. Our aim in this chapter, however, is not to compare modulation techniques but to investigate the qualitative effects of the techniques used to improve bit error rate performance. Furthermore, all of the procedures here described can be directly applied to higher-order modulation schemes.†

## 6.2  DIGITAL MODULATION SCHEMES

The transmission of binary data over a band-pass channel requires the modulation of the data onto a carrier wave. The carrier is usually a sinusoid, conveniently keyed by the data according to one of the three basic techniques: amplitude-shift keying (ASK), frequency-shift keying (FSK), and phase-shift keying (PSK). The PSK and FSK present a constant envelope, leading to a lower sensitivity to amplitude nonlinearities. Consequently, use of these digital modulation schemes is preferable to using ASK.

At the reception side, the detection can be carried out by either coherent or noncoherent techniques. Consequently, we may have coherent FSK or coherent PSK and noncoherent FSK or "noncoherent" PSK. In fact, it is not possible to have noncoherent PSK, because PSK requires the knowledge of the phase (contradicting the name noncoherent). A variation of PSK, based on a differential encoding of the binary data, provides the means for

---

*For more on synchronization and the definition of the TDMA structure, refer to Chapter 10.

†For the description and analysis of the digital modulation schemes used in digital cellular systems, refer to Chapter 9.

noncoherent detection as required. This modulation technique is named differential phase-shift keying (DPSK).

## 6.3 ERROR RATES FOR BINARY SYSTEMS

As we shall see in Chapter 9, the probability of a symbol error in the presence of stationary additive Gaussian noise depends only on the SNR associated with each symbol. This is true if we consider a receiver with a linear filter processing an undistorted waveform. In other words, the probability of error occurrence depends on the ratio between the energy and noise power per bit. For each one of the binary systems, given the SNR ($\gamma_b$) per bit, these probabilities are

$$\text{prob}(\text{error}|\gamma_b) = \tfrac{1}{2}\exp(-\alpha\gamma_b),$$

$$\alpha = \begin{cases} \tfrac{1}{2} & \text{for noncoherent FSK} \\ 1 & \text{for "noncoherent" PSK (DPSK)} \end{cases} \tag{6.1}$$

$$\text{prob}(\text{error}|\gamma_b) = \tfrac{1}{2}\text{erfc}\left(\sqrt{\alpha\gamma_b}\right), \qquad \alpha = \begin{cases} \tfrac{1}{2} & \text{for coherent FSK} \\ 1 & \text{for ideal PSK} \end{cases} \tag{6.2}$$

It was shown in Chapter 5 that the distribution of $\gamma_b$ in a Rayleigh fading environment is

$$p(\gamma_b) = \frac{1}{\gamma_{b0}}\exp\left(-\frac{\gamma_b}{\gamma_{b0}}\right) \tag{6.3}$$

where $\gamma_{b0}$ is the average SNR per bit. Consequently, the unconditional probability of error is

$$\text{prob}(\text{error}) \triangleq p = \int_0^\infty \text{prob}(\text{error}|\gamma_b)p(\gamma_b)\,d\gamma_b \tag{6.4}$$

Using Equations 6.1 and 6.3 in Equation 6.4, we obtain

$$p = \frac{1}{2(1 + \alpha\gamma_{b0})} \tag{6.5}$$

for noncoherent detection. In a similar way, with Equations 6.2 and 6.3 in

Equation 6.4, we have

$$p = \frac{1}{2\left(1 - \sqrt{1 + \dfrac{1}{\alpha \gamma_{b0}}}\right)} \tag{6.6}$$

for coherent detection.

## 6.4  PROBABILITY OF ERRORS IN A DATA STREAM

Let $p$ be the probability of bit error. The probability of having exactly $m$ bits in error in a message of $N$ bits is given by the binomial expansion

$$p(N, m) = \binom{N}{m}(1 - p)^{N-m}p^m \tag{6.7}$$

where

$$\binom{N}{m} = \frac{N!}{(N - m)!m!} \tag{6.8}$$

The probability of error ($p_{eM}$) in a message containing $N$ bits is

$$P_{eM} = 1 - p(N, 0) = 1 - (1 - p)^N \tag{6.9}$$

We can also calculate the probability of receiving a falsely recognizable word. This corresponds to the event that the receiver recognizes a signal as one message, given that the transmitter had sent another message. If the two messages differ from each other by $d$ bits, then the probability of this event is

$$p_f = (1 - p)^{N-d}p^d \tag{6.10}$$

## 6.5  IMPROVING THE PERFORMANCE OF DIGITAL TRANSMISSION

Several techniques can be applied to improve the performance of digital transmission. The fading of channels has an important effect on the bit error rate, degrading the performance of any modulation scheme as shown in Chapters 8 and 9. In other to counteract fading and, consequently, to diminish the bit error rate, the following techniques can be used (individually or combined):

- Diversity
- Error detecting and correcting codes
- Multiple transmission
- Interleaving
- Automatic repeat request
- Adaptive equalization

Bit error rate performance enhancement is accomplished at the expense of increasing system complexity and cost. Some techniques can yield better results than others, but can also be more costly. Therefore, the decision for one or another technique depends on the analysis of cost versus effectiveness. In the following sections we shall investigate these techniques.

## 6.6  DIVERSITY AND DIGITAL TRANSMISSION

The use of diversity techniques for analog transmission has proved to be very effective. In this case we use the signal-to-noise ratio as the effectiveness criterion (refer to Chapter 5). The effectiveness of diversity on data (digital) transmission can be assessed by the reduction in the error rate achieved when these techniques are used. The probability of bit error can be evaluated by averaging the conditional probability prob(error$|\gamma_b$) over the distribution $p(\gamma_b)$ of $\gamma_b$ as shown in Equation 6.4. The distribution of $\gamma_b$ depends on the diversity-combining technique as described in Chapter 5. For convenience, we reproduce these distributions and then estimate the probability of error, $p$.

### 6.6.1  Switched Combining: Pure Selection

The distribution of $\gamma_b$ is

$$p(\gamma_b) = \frac{M}{\gamma_{b0}} \left[ 1 - \exp\left( -\frac{\gamma_b}{\gamma_{b0}} \right) \right]^{M-1} \exp\left( -\frac{\gamma_b}{\gamma_{b0}} \right) \qquad (6.11)$$

#### 6.6.1.1 *Noncoherent Detection*

With Equations 6.1 and 6.11 in Equation 6.4, we obtain

$$\text{prob(error)} \triangleq p = \frac{M!}{2\prod_{i=1}^{M}(i + \alpha\gamma_{b0})} \qquad (6.12)$$

### 6.6.1.2 Coherent Detection

Switched combining does not require any coherent phase reference for detection. Consequently, this kind of technique is of no use for coherent detection.

### 6.6.2 Switched Combining: Threshold Selection

The distribution of $\gamma_b$ is

$$
p(\gamma_b) = \begin{cases} \dfrac{1+q}{\gamma_{b0}} \exp\left(-\dfrac{\gamma_b}{\gamma_{b0}}\right), & \gamma_b \geq \gamma_T \qquad (6.13a) \\[3mm] \dfrac{q}{\gamma_{b0}} \exp\left(-\dfrac{\gamma_b}{\gamma_{b0}}\right), & \gamma_b < \gamma_T \qquad (6.13b) \end{cases}
$$

where $q = 1 - \exp(-\gamma_T/\gamma_{b0})$ and $\gamma_T$ is the SNR threshold level.

### 6.6.2.1 Noncoherent Detection

The probability of error is determined as follows:

$$
\text{prob(error)} \triangleq p = \int_0^{\gamma_T} \text{prob(error}|\gamma_b)p(\gamma_b)\, d\gamma_b
$$

$$
+ \int_{\gamma_T}^{\infty} \text{prob(error}|\gamma_b)p(\gamma_b)\, d\gamma_b
$$

In the first integral we use Equation 6.13a for $p(\gamma_b)$, whereas in the second integral we use Equation 6.13b. In both of them the conditional probability prob(error$|\gamma_b$) is given by Equation 6.1. Hence

$$
p = \frac{q + (1-q)\exp(-\alpha\gamma_T)}{2(1+\alpha\gamma_{b0})} \tag{6.14}
$$

### 6.6.2.2 Coherent Detection

Not applicable (see Section 6.6.1.2).

### 6.6.3 Gain Combining: Maximal Ratio

The distribution of $\gamma_b$ is

$$
p(\gamma_b) = \frac{\gamma_b^{M-1} \exp(-\gamma_b/\gamma_{b0})}{\gamma_{b0}^M(M-1)!} \tag{6.15}
$$

### 6.6.3.1 *Noncoherent Detection*

With Equations 6.1 and 6.15 in Equation 6.4 we have

$$\text{prob(error)} \triangleq p = \frac{1}{2(1 + \alpha\gamma_{b0})^M} \tag{6.16}$$

### 6.6.3.2 *Coherent Detection*

In a similar way, using Equations (6.2) and (6.15) in Equation (6.4) gives the probability of error for maximal-ratio coherent combining. However, no convenient closed-form solution seems to be available. We can approximate Equation 6.15 by using the leading term of its series expansion or, equivalently, we may consider a low SNR. Both approximations correspond to the case of low bit error rates. Hence, the approximate distribution of $\gamma_b$ is

$$p(\gamma_b) \simeq \frac{\gamma_b^{M-1}}{\gamma_{b0}^M (M-1)!} \tag{6.17}$$

With Equations 6.2 and 6.17 in Equation 6.4, we obtain

$$\text{prob(error)} \triangleq p = \frac{(M - \frac{1}{2})!}{2\sqrt{\pi}\,(\alpha\gamma_{b0})^M M!} \tag{6.18}$$

where $(\cdot)!$ is the gamma function. In this particular case

$$\left(M - \frac{1}{2}\right)! = \frac{1 \cdot 3 \cdot 5 \cdot 7 \cdots (2M-1)}{2^M}\sqrt{\pi}$$

### 6.6.4 Gain Combining: Equal Gain

A closed-form solution for equal-gain combining can only be given by an approximate expression. Moreover, as shown in Chapter 5, this approximate solution differs from that of maximal-ratio combining by a factor of $(M/2)^M\sqrt{\pi}/(M - \frac{1}{2})!$. Therefore, we may use this same factor, modifying the probability of error of the maximal ratio, to obtain the probability of error of the equal-gain case.

### 6.6.4.1 *Noncoherent Detection*

$$\text{prob(error)} \triangleq p = \frac{(M/2)^M\sqrt{\pi}}{2(M - \frac{1}{2})!} \frac{1}{(1 + \alpha\gamma_{b0})^M} \tag{6.19}$$

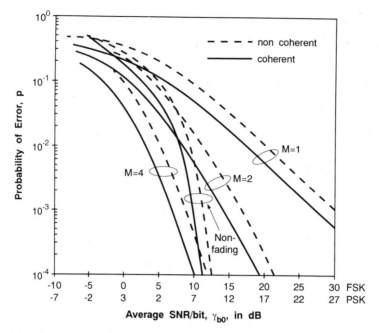

**Figure 6.1.**    Error rates for FSK and PSK with maximal-ratio combining.

### 6.6.4.2 *Coherent Detection*

$$\text{prob(error)} \triangleq p = \frac{(M/2)^M}{2(\alpha\gamma_{b0})^M M!} \tag{6.20}$$

A plot of the error rates with maximal-ratio combining for both coherent and noncoherent detections is shown in Figure 6.1. We also show the probability of error in a Gaussian (nonfading) environment, for comparison. Note that the curves for PSK ($\alpha = 1$) are obtained from the corresponding curves for FSK ($\alpha = \frac{1}{2}$), at an SNR 3 dB lower.

## 6.7  ERROR DETECTING AND CORRECTING CODES

The ratio of signal energy per bit to noise-power density ($E_b/N_0 = \gamma_b$) is intimately related to the bit error rate. Practical implementation of a digital transmission system imposes limits on the power ratio $\gamma_b$. This, in connection with the chosen modulation scheme, dictates the system performance. The bit error rate performance can be substantially improved by means of coding. Coding is a rather extensive subject, covered in many textbooks (e.g., References 3 through 6). For a summary on the subject, refer to Appendix 6A.

This section is essentially concerned with the forward error correction (FEC) technique as a means of error control. As far as mobile radio application is concerned, the codes were initially developed for efficient correction of burst errors, a situation likely to occur on a fading channel. However, it was verified that the bursts usually occur during an extremely long period of time, forcing the codes to have proportionally long constraint length.* A new class of codes, able to correct random errors, was then developed to be more efficiently used for mobile radio applications.

There are several types of error correcting code available: Hamming, Golay, Reed–Muller, BCH, Reed–Solomon, etc. A linear block code $A$ having $k$ information bits and $n - k$ redundancy bits is described as $A(n, k)$. Moreover, if its minimum distance is $d$, this code is able to correct up to $t = (d - 1)/2$ bits in $n$ bits. The ratio $r = k/n$ is called the rate. As an example, the Golay $(23, 12)$ has a total length of 23 bits with 12 information bits (11 redundancy bits). Its minimum distance is known to be equal to 7. Therefore, this code is able to correct up to 3 in 23 bits. Its code rate is $\frac{12}{23} \simeq \frac{1}{2}$.

It is easy to see that, for a code that can correct up to $t$ bits, the probability of message error is

$$p_{eM} = 1 - \sum_{m=0}^{t} p(N, m) \qquad (6.21)$$

where $p(N, m)$, given by Equation 6.7, is the probability of having exactly $m$ bits in error in an $N$-bit message. It is important to note that, in order to keep the same transmitted power, the SNR per bit in the encoded message is multiplied by the code rate, $r$. In other words the average SNR per bit of the encoded message is $(k/n)\gamma_{b0}$.

In order to get an insight into the improvement achieved when coding is used, we shall use a 12-bit message and two codes, namely, shortened Hamming $(17, 12)$ and Golay $(23, 12)$. The aim is to estimate the probability of error in the encoded message, given by Equation 6.21. For the Hamming $(17, 12)$ we have $N = 17$ and $t = 1$, whereas for the Golay $(23, 12)$ we have $N = 23$ and $t = 3$. As for the uncoded message, $N = 12$ and $t = 0$ (no error can be corrected). The bit error probability $p$ used in Equation 6.7 varies according to the modulation technique. It is given by Equation 6.5 for noncoherent detection or by Equation 6.6 for coherent detection. The resultant curves for these cases are shown in Figure 6.2.

In Figure 6.2 only the curves for noncoherent FSK are shown. Those for coherent FSK can be obtained at an SNR approximately 2 dB lower. Furthermore, as seen in Figure 6.1, the curves for PSK can be obtained at an SNR approximately 3 dB lower. Bearing this in mind, hereinafter, for simplicity, only the curves for noncoherent FSK will be shown. Note that the improvement is really substantial and this is even more noticeable at higher SNR.

---

*Refer to Section 6A.4 for the definition of constraint length.

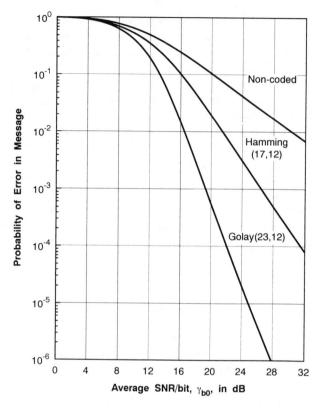

**Figure 6.2.**   Error rates for noncoherent FSK using coding.

## 6.8   MULTIPLE TRANSMISSION

In this technique each message is sent an odd number of times and, at the reception, a bit-by-bit majority decision is applied. If $s$ is the number of repeats, an $(s + 1)/2$-out-of-$s$ majority voting process is used to determine each valid bit in the message.

The probability of bit error, $p'$, after an $(s + 1)/2$-out-of-$s$ majority voting is the probability of at least $(s + 1)/2$ errors occurring, i.e.,

$$p' = \sum_{i=(s+1)/2}^{s} \binom{s}{i} p^i (1 - p)^{s-i} \qquad (6.22)$$

where $p$ is the bit error probability of a single transmission. The probability of having exactly $m$ corrupted bits in an $N$-bit message is, as in Equation 6.7,

$$p(N, m) = \binom{N}{m} (1 - p')^{N-m} p'^m \qquad (6.23)$$

The probability that more than $t$ bits are in error is, as in Equation 6.21,

$$P_{eM} = 1 - \sum_{m=0}^{t} p(N, m) \tag{6.24}$$

where $p(N, m)$ is given by Equation 6.23.

Multiple transmission is also a type of code, known as repetition code. The repetition code is represented as $(s, 1)$, having a minimum distance equal to $s$. Therefore, it is able to correct up to $(s - 1)/2$ errors in an $s$-bit message, having a rate $r = 1/s$.

As an example, consider the transmission of a 12-bit message, using the repetition code $(3, 1)$, corresponding to a 2-out-of-3 majority voting. Then consider this same message, using the repetition code $(5, 1)$, corresponding to a 3-out-of-5 majority voting. We shall determine the probability of at least one error occurring in the message. This can be done by setting $t = 0$ in Equation 6.24 and using the appropriate equations as required. Note that, in this case, $N = 12$ and $s = 3$ or $s = 5$, depending on the repetition code. The resultant curves are shown in Figure 6.3.

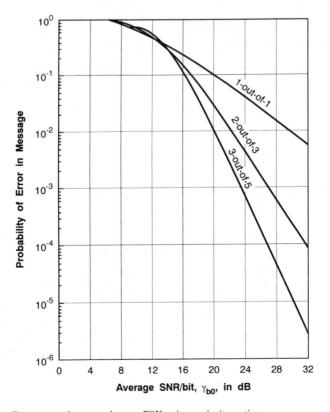

**Figure 6.3.**   Error rates for noncoherent FSK using majority voting.

## 6.9 INTERLEAVING

Interleaving is a simple and efficient way of using coding to combat error bursts occurring on a fading channel. The basic principle is to spread the code word, conveniently positioning the bits one away from another, so that they experience independent fading. In this case the error bursts affect several clustered bits belonging to several code words. Therefore, the effect of the error burst is spread over the message so that it may be possible to recover the data with the nonaffected bits.

Interleaving is easily implemented with the use of memories, where the code words are written row-by-row and read out column-by-column.

## 6.10 AUTOMATIC REPEAT REQUEST

Automatic repeat request (ARQ) systems require error-detecting code only. If an error is detected at the receiver end, a request for repeat is sent over the return channel. ARQ systems can be very efficient, with a relatively small amount of added redundancy. However, the message throughput* tends to be reduced, due to the obvious reasons. It can be emphasized that even when a message is repeated, the repetition may still fail. Note that the ARQ message may also fail (be in error). This can be overcome by considering any incoming message in error as an ARQ message. Hence the most recent outgoing message is repeated. It is obvious that interleaving and ARQ are not effective if combined. Let $p_{eM}$ be the probability of error in a message. Then the time to transmit a set of messages is increased by $1/(1 - p_{eM})$, whereas the throughput is reduced by $(1 - p_{eM})$.

## 6.11 ADAPTIVE EQUALIZATION

*Equalization* is the compensation for phase and amplitude distortions of signals using a telephone channel. This term is now widespread and is applied to any type of channel. The use of equalizers as a technique to combat fading is a rather challenging problem. If the channel characteristics are considered to be known, the equalizers can be easily (sometimes manually) adjusted to obtain the best performance. If, however, the channel characteristics are time-varying, the equalizers are given adaptive properties. Adaptive equalizers are also used in a stationary environment, where their internal settings remain unchanged after the steady-state condition is achieved. In a nonstationary condition, as in the mobile radio environment, adaptive equalizers should adjust their coefficients in real time so as to track the changes on the channel. Therefore, a conveniently designed adaptive

---

*Refer to Chapter 11 for the concept of throughput.

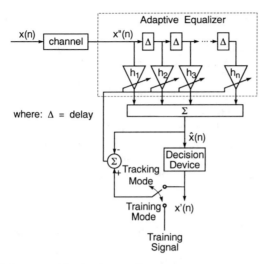

**Figure 6.4.**  Digital system with adaptive equalizer.

equalizer can be used to combat fading. It may also be effective against intersymbol interference (ISI).*

A simplified model of a digital system using adaptive equalizer is sketched in Figure 6.4. The transmitted digital sequence $x(n)$ arrives at the receiver as the sequence $x''(n)$, a distorted form of $x(n)$. The aim of the adaptive equalizer is to restore the signal as close as possible to its original form. An adaptive equalizer is a transversal filter having as an output the weighted sum of the input $x''(n)$ delayed by a set of delaying elements $(\Delta)$.† The estimate of $x(n)$ is $x'(n)$, obtained by conveniently sampling the equalizer's output $\hat{x}(n)$. The sampling is carried out by the decision device block. After a training period the estimate $x'(n)$ closely approximates $x(n)$ and the adaptive equalizer starts its "tracking mode". In this mode the difference $x'(n) - \hat{x}(n)$ (the error) is used for continuous adjustment of the filter coefficients $h_i$. If the variations of the received signal are slow enough, these perturbations can be tracked and corrected.

Note that we have assumed that there is an initial training period or, equivalently, a training sequence can be interleaved with the data packets to update the equalizer's settings. If the changes in the channel characteristics are slow, these assumptions are quite reasonable. However, in digital radio

---

*Due to the time-dispersive characteristics of the channel the transmitted symbol, ideally occupying a certain time interval, is distorted and extends to the next interval used by another symbol. This type of interference is known as intersymbol interference.

†Refer to Appendix 4B for a quick look at a digital filter. This subject is fully explored in Reference 21.

systems the conditions are exactly the opposite, and this approach cannot be used.

Instead, a new class of adaptive equalizers, using "blind" strategies to adjust the filter coefficients, are preferred. "Blind equalization" or "self-adaptive equalization" does not require a training sequence. Therefore, the equalizer operates only on the tracking mode with a convergence rate slower than conventional adaptive equalizers. Many blind equalization strategies have been proposed but this is still a subject of investigation .[16–18]

An alternative adaptive equalizer, known as a decision feedback equalizer, uses a feedback configuration. In this case, from the detected symbols the ISI is determined and is subtracted out from the new incoming symbols.[19, 20]

## 6.12 COMPARATIVE PERFORMANCE AND COMBINED TECHNIQUES

The curves obtained in Figures 6.1, 6.2 and 6.3 are now combined in Figure 6.5 for comparison.

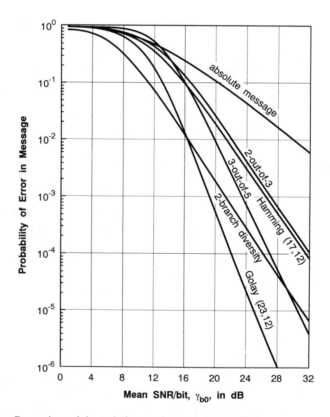

**Figure 6.5.** Comparison of the techniques using noncoherent FSK.

Note that the use of coding substantially improves the bit error rate performance. Moreover, the Golay $(23, 12)$ code performs remarkably better than the 3-out-of-5 majority voting. In a similar way, the Hamming $(17, 12)$ gives better results than the 2-out-of-3 majority voting. The use of coding implies an increase of the transmission rate, corresponding to a proportional increase of the necessary bandwidth. In this sense both Golay $(23, 12)$ and Hamming $(17, 12)$ codes require less bandwidth than the 3-out-of-5 and 2-out-of-3 majority voting, respectively.

If, however, transmission rate or (equivalently) bandwidth are critical, the use of diversity can be an excellent option. Nevertheless, it can be seen that although a two-branch diversity and a half-rate code present the same degree of redundancy, the latter gives better results. In particular, the Golay $(23, 12)$ has a minimum distance of 7, roughly corresponding to a fourth-order diversity if hard-decision decoding is used.

Diversity, coding, and majority voting can be combined to improve even further the bit error performance. Because a great deal of combination is possible, as far as diversity is concerned, we shall restrict to the maximal-ratio combining technique.

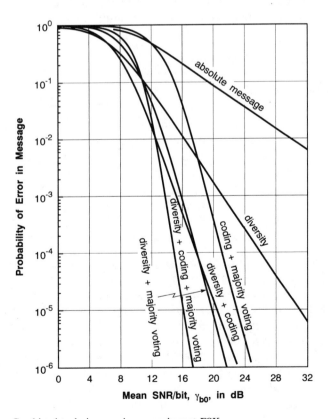

**Figure 6.6.**  Combined techniques using noncoherent FSK.

1. *Diversity and coding*—The procedure for estimating the bit error rate is the same as that described in Section 6.7. We notice, however, that instead of using Equation 6.5 or 6.6 for $p$ we use Equation 6.16 or 6.18, respectively.
2. *Diversity and majority voting*—The procedure is the same as that described in Section 6.8. In a similar way we use Equation 6.16 and 6.18 for $p$ instead of Equation 6.5 or 6.6, respectively.
3. *Coding and majority voting*—The procedure is similar to that described in Section 6.8. However, in Equations 6.5 and 6.6 the average SNR per bit is reduced to $(k/sn)\gamma_{b0}$. Furthermore, in Equation 6.24, the parameter $t$ is set to be equal to the number of correctable bits of the used code.
4. *Diversity and coding and majority voting*—The procedure is similar to that used for the combination of coding and majority voting. The difference is that instead of using Equation 6.5 or 6.6, we use Equation 6.16 or 6.18, respectively.

In the examples shown here we have assumed a 12-bit message, 2-branch diversity, Golay (23, 12) code, and the 3-out-of-5 majority voting. The results are depicted in Figure 6.6. Note that the combination of diversity plus coding is an excellent and simple option for a substantial increase in bit error rate performance.

## 6.13   CHOICE OF CODE

Consider a mobile receiver working with a capture effect such that it detects only the signals found to be above the rms value by a certain threshold. Let $R$ be the received signal, $\sqrt{2}\,\sigma$ the rms value, and $T$ the threshold. Capture occurs if

$$20\log(R/\sqrt{2}\,\sigma) \geq T \qquad (6.25)$$

Due to the occurrence of fading, a data stream transmitted over a mobile radio channel will carry a proportion of corrupted bits, erroneously detected at the receiver. We want to determine the appropriate code, able to minimize the effects of the fades. In other words we want to correct the corrupted bits. The first step is to determine the mean proportion of corrupted bits in the message and then to apply an appropriate code able to correct this proportion. Constraints on power and bandwidth are not considered here. Moreover, a perfect interleaving is assumed so as to minimize error-burst effects. Given the threshold as in Equation 6.25, it is possible to estimate the average duration of fades $\tau$ and the level crossing rate $R_c$ in a straightforward manner (refer to Sections 4.6 and 4.7),

$$R_c = \sqrt{2\pi} f_m \left(\frac{R}{\sqrt{2}\sigma}\right) \exp\left[-\left(\frac{R}{\sqrt{2}\sigma}\right)^2\right] \tag{6.26}$$

$$\tau = \frac{1}{\sqrt{2\pi} f_m (R/\sqrt{2}\sigma)} \left\{\exp\left[\left(\frac{R}{\sqrt{2}\sigma}\right)^2\right] - 1\right\} \tag{6.27}$$

Consider a transmission rate of $z$ bits per second. With $R_c$ crossings per second, the mean time between crossings is equal to $1/R_c$ and the corresponding mean number of bits within this time interval is $z/R_c$. Similarly, if the mean duration of the fades is $\tau$, the mean number of faded (corrupted) bits is $z\tau$.

Therefore, the mean proportion of corrupted bits is $z\tau/(z/R_c) = \tau R_c$ in the fading interval. The proportion $\tau R_c$ is the mean proportion of bits in the message to be corrected. Using Equations 6.26 and 6.27, we obtain

$$\tau R_c = 1 - \exp\left[-\left(R/\sqrt{2}\sigma\right)^2\right] \tag{6.28}$$

Consider a code able to correct $t$ out of $n$ bits. Accordingly, the ratio $t/n$ must exceed $\tau R_c$ so as to achieve the minimum error correction requirement. Thus

$$t/n \geq 1 - \exp\left[-\left(R/\sqrt{2}\sigma\right)^2\right] = \tau R_c \tag{6.29}$$

Because Equation 6.25 gives the minimum requirement for the receiver to capture the signal, we may use an equality sign to obtain $R/\sqrt{2}\sigma$ as a function of $T$ and replace it in Equation 6.29. Hence

$$t/n \geq 1 - \exp(-10^{T/10}) \tag{6.30}$$

As an example, let capture occur for $T = -9$ dB. Then, using Equation 6.30, we obtain $t/n \geq 0.118$. The Golay $(23, 12)$ corrects up to $t = 3$ bits in an $n = 23$-bit message, corresponding to a proportion of $\frac{3}{23} \simeq 0.13$. Therefore, this code satisfies the minimum bit error correction requirement.

## 6.14 SUMMARY AND CONCLUSIONS

Digital transmission over mobile radio channels is greatly impaired by fading effects. Many techniques, if used either individually or in concert, can substantially improve the performance of bit error rate. The improvement is also dependent on the digital modulation scheme, with the coherent detection giving approximately 2 dB gain on the SNR with respect to noncoherent detection.

Branch diversity by itself is a powerful technique able to improve the SNR substantially (a 12-dB improvement in the SNR for a bit error probability of $10^{-3}$). Error detecting and correcting codes are only effective if combined with interleaving. Multiple transmission is also effective but usually requires larger bandwidth.

Data and control signalling involve digital transmission but with different requirements. Data transmission, generally, can tolerate delays and bit errors, whereas signalling usually requires quick response and a very low error rate. Generally speaking, data are transmitted as long message blocks and signalling uses short messages.

Accordingly, coding is an appropriate technique to be used with data and signalling transmission. Multiple transmission is more suitable for signalling whereas ARQ is more convenient for data.

## APPENDIX 6A   CHANNEL CODING: A BASIC INTRODUCTION

The aim of this appendix is to give a quick look at a special branch of digital signal processing—channel coding—in order to supply the basic concepts to the reader who is not familiar with this subject. It should be emphasized that the approach used here is the simplest possible and a more rigorous treatment will be left for the appropriate literature (e.g., References 2, 4, 5, and 6).

### 6A.1   Coding Jargon

As far as error control is concerned coding can be used in two different approaches, namely, *error detection* and *error correction*. The codes used in the former case are very simple and usually combined with an *automatic repeat request* (ARQ) scheme. The latter approach requires more elaborate codes so that up to a certain number of bit errors can be corrected at the reception. This scheme is known as *forward error correction* (FEC).

The redundancies are added to the message by the *channel encoder* at the transmission end, to produce the encoded data at *higher bit rate*. At the reception end, the *channel decoder* conveniently treats the whole set of received data in order to restore the initial message.

Codes have been classified into two basic groups, namely, *block codes* and *convolutional codes*. They are distinguished from one another by the absence or presence of memory, respectively. Codes can also be classified as *linear* or *nonlinear*. In linear codes two code words can be added in modulo-2 arithmetic to generate a third code word.

## 6A.2   Linear Block Codes

In an $(n, k)$ linear block code a $k$-bit message is encoded into an $n$-bit code word. Accordingly, the number of redundant bits is $n - k$. Given that an uncoded message is transmitted at a rate $z$, the transmission rate of the encoded message is increased to $(n/k)z$ so that the total transmission time can be kept unaltered. The dimensionless ratio $r = k/n$ is known as code rate and ranges from 0 to 1.

Let $\mathbf{u}$ be a $k$-dimensional row vector containing the uncoded message and let $\mathbf{c}$ be an $n$-dimensional row vector containing the encoded message. Define $\mathbf{G}$ as a $k \times n$ *generator matrix*, such that

$$\mathbf{c} = \mathbf{u}\mathbf{G} \tag{6A.1}$$

Because $\mathbf{c}$ contains the uncoded message plus the redundant bits, the generator matrix should be composed of a $k \times k$ identity matrix and a $k \times (n - k)$ coefficient matrix. We denote the identity matrix by $\mathbf{I}_k$ and the coefficient matrix by $\mathbf{P}$. Assuming that the $k$ rightmost bits of the code word correspond to the uncoded message, we have

$$\mathbf{G} = \begin{bmatrix} \mathbf{P} & | & \mathbf{I}_k \end{bmatrix} \tag{6A.2}$$

A block code with this representation is called a *linear systematic block code*.

Another important definition is the parity-check matrix, with dimension $(n - k) \times n$, written as

$$\mathbf{H} = \begin{bmatrix} \mathbf{I}_{n-k} & | & \mathbf{P}^T \end{bmatrix} \tag{6A.3}$$

where $\mathbf{I}_{n-k}$ is an $(n - k) \times (n - k)$ identity matrix and $\mathbf{P}^T$ is an $(n - k) \times k$ matrix, the transpose of $\mathbf{P}$. It can be seen that, with modulo-2 arithmetic,

$$\mathbf{H}\mathbf{G}^T = \begin{bmatrix} \mathbf{I}_{n-k} & | & \mathbf{P}^T \end{bmatrix} \begin{bmatrix} \mathbf{P}^T \\ --- \\ \mathbf{I}_k \end{bmatrix} = \mathbf{P}^T + \mathbf{P}^T = 0$$

The matrices $\mathbf{G}$ and $\mathbf{H}$ are used in the coding and decoding operations, respectively.

### 6A.2.1 *Error Detection: Syndrome*

When transmitting a code word $\mathbf{c}$, it arrives at the reception end as $\mathbf{r} = \mathbf{c} + \mathbf{e}$, where $\mathbf{e}$ is an error vector containing 1s at the error positions and 0s elsewhere. At the reception end the decoder performs the following operation:

$$\mathbf{s} = \mathbf{r}\mathbf{H}^T \tag{6A.4}$$

known as *syndrome* of $\mathbf{r}$. It is easy to show that if $\mathbf{s} \neq \mathbf{0}$ then errors have been detected. If $\mathbf{s} = \mathbf{0}$ then errors have not been detected (although they could still have occurred). Because $\mathbf{r} = \mathbf{c} + \mathbf{e}$ and $\mathbf{c} = \mathbf{u}\mathbf{G}$, then

$$\mathbf{s} = \overset{0}{\cancel{\mathbf{u}\mathbf{G}\mathbf{H}^T}} + \mathbf{e}\mathbf{H}^T$$

$$\mathbf{s} = \mathbf{e}\mathbf{H}^T \tag{6A.5}$$

This equation shows that the syndrome depends only on the error pattern $\mathbf{e}$ and not on the transmitted code word. Note that because $\mathbf{e}$ is an $n$-dimensional row vector and $\mathbf{H}^T$ is an $n \times (n - k)$ matrix, then $\mathbf{s}$ is an $(n - k)$-dimensional row vector. In other words there are $2^{n-k}$ possible syndromes.

## 6A.2.2 *Minimum Distance*

The Hamming distance $d(\mathbf{x}, \mathbf{y})$ between two code words $\mathbf{x}$ and $\mathbf{y}$ is defined as the number of 1s obtained after the bit-by-bit exclusive-or operation between the two code words. Hence the distance between 1001 and 0111 is 3. The Hamming weight $\omega(\mathbf{x})$ of a code word $\mathbf{x}$ is defined as the number of nonzero elements of such code. In the examples just given, the respective weights are 2 and 3. From the definition of Hamming distance it follows that

$$d(\mathbf{x}, \mathbf{y}) = \omega(\mathbf{x} + \mathbf{y}) \tag{6A.6}$$

The minimum distance $d_{min}$ of a linear block code is the smallest Hamming distance between any pair of code vectors in the code, i.e.,

$$d_{min} = \min\{d(\mathbf{x}, \mathbf{y}): \mathbf{x}, \mathbf{y} \in \mathbf{c}, \mathbf{x} \neq \mathbf{y}\}$$

Therefore

$$d_{min} = \min\{\omega(\mathbf{x} + \mathbf{y}): \mathbf{x}, \mathbf{y} \in \mathbf{c}, \mathbf{x} \neq \mathbf{y}\}$$

Because $\mathbf{z} = \mathbf{x} + \mathbf{y}$ is another code vector in $\mathbf{c}$ it follows that

$$d_{min} = \min\{\omega(\mathbf{z}): \mathbf{z} \in \mathbf{c}, \mathbf{z} \neq \mathbf{0}\}$$

or

$$d_{min} = \omega_{min} \tag{6A.7}$$

In other words, the minimum distance is the smallest Hamming weight of the nonzero code vectors in the code.

## 6A.2.3 *Error Correction*

An $(n, k)$ code admits $2^n$ code vectors containing $2^k$ code words. The best decoding strategy is such that, upon receiving any code vector, the decoder associates it with the nearest code word. Define nonoverlapping spheres with

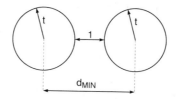

**Figure 6A.1.**   Distance between two neighboring code words.

radius $t$ around each code word so that within each sphere all the code vectors are decoded correctly. Because (1) the spheres do no overlap and do not touch each other and (2) the distance between two neighboring code words is the minimum distance, it is straightforward to show that (see Figure 6A.1)

$$d_{\min} \geq 2t + 1 \qquad (6A.8a)$$

Consequently, we can say that an $(n, k)$ linear block code can correct up to $t$ errors such that

$$t \leq \left\lfloor \frac{d_{\min} - 1}{2} \right\rfloor \qquad (6A.8b)^*$$

### 6A.2.4 Hamming Bound

Let $i$ be the number of error locations in the $n$-bit word. Accordingly, there are $\binom{n}{i}$ possible error patterns. If $t$ is the maximum number of error locations, the total number of all possible error patterns is $\sum_{i=0}^{t}\binom{n}{i}$. Consequently, an $(n, k)$ linear block code, able to correct up to $t$ errors, admits $2^{n-k}$ syndromes (as seen before). It is obvious that the number of syndromes cannot be less than the total number of possible error patterns, i.e.,

$$2^{n-k} \geq \sum_{i=0}^{t} \binom{n}{i} \qquad (6A.9)$$

Equation 6A.9 is known as the Hamming bound. Any binary code satisfying the equality condition

$$2^{n-k} = \sum_{i=0}^{t} \binom{n}{i} \qquad (6A.10)$$

is known as perfect code.

---

$^*\lfloor x \rfloor$ denotes the largest integer less than or equal to $x$.

### 6A.2.5 *An Example: Hamming Codes*

The Hamming codes have the following parameters, for any positive integer $m \geq 3$:

Block length                $n = 2^m - 1$
Number of message bits       $k = 2^m - m - 1$
Number of redundant bits   $n - k = m$
Minimum distance             $d = 3$

For $m = 3$ we have the (7, 4) Hamming code, correcting up to $(3 - 1)/2 = 1$ bit in a 7-bit word. The corresponding generator and parity check matrices are

$$\mathbf{G} = \begin{bmatrix} 1 & 1 & 0 & \vdots & 1 & 0 & 0 & 0 \\ 0 & 1 & 1 & \vdots & 0 & 1 & 0 & 0 \\ 1 & 1 & 1 & \vdots & 0 & 0 & 1 & 0 \\ 1 & 0 & 1 & \vdots & 0 & 0 & 0 & 1 \end{bmatrix} = \begin{bmatrix} \mathbf{P} & \vdots & \mathbf{I}_k \end{bmatrix}$$

$$\mathbf{H} = \begin{bmatrix} 1 & 0 & 0 & \vdots & 1 & 0 & 1 & 1 \\ 0 & 1 & 0 & \vdots & 1 & 1 & 1 & 0 \\ 0 & 0 & 1 & \vdots & 0 & 1 & 1 & 1 \end{bmatrix} = \begin{bmatrix} \mathbf{I}_{n-k} & \vdots & \mathbf{P}^T \end{bmatrix}$$

As an example, using Equation 6A.1, the code word corresponding to the message 1011 is 1001011. Similarly, the message 1110 has 0101110 as its code word.

The syndrome is obtained by multiplying the error pattern by $\mathbf{H}^T$ (see Equation 6A.5). There are seven possible single patterns that can be corrected. As an illustration, for an error at the leftmost bit, the error pattern is 1000000 and the corresponding syndrome is 100. For the error pattern 0000010 the syndrome is 111. For no error (0000000) the syndrome is 000, and so on.

## 6A.3    Cyclic Codes

Cyclic codes comprise a special class of linear block codes. Their algebraic structure is such that encoding, syndrome computation, and decoding are very simple to implement and can be done by means of shift registers with some feedback connections. One fundamental property of this code is that any cyclic shift of a code vector in **c** is also a code vector in **c**.

This suggests that the elements of a code word of length $n$ can be treated as the coefficients of a polynomial of degree $(n - 1)$. Let $c_i, i = 0, 1, \ldots, n - 1$, be the elements of a code word **c**. The corresponding code polynomial can be written as

$$c(D) = c_0 + c_1 D + c_2 D^2 + \cdots + c_{n-1} D^{n-1} \qquad (6A.11)$$

where $D$ is an arbitrary real variable. If we impose the condition that $D^n = 1$, then the multiplication of $c(D)$ by $D$ corresponds to a shift (rotation) of the elements of $c(D)$ to the right. It is easy to show that if we multiply $c(D)$ by $D$ $i$ times we get

$$D^i c(D) = c_{n-i} + c_{n-i+1}D + c_{n-i+2}D^2 + \cdots$$

$$+ c_0 D^i + c_1 D^{i+1} + c_2 D^{i+2} + \cdots + c_{n-i-1}D^{n-1} \quad (6A.12)$$

The main theorems of the cyclic codes are presented next.

**Theorem 6A.3.1.** *An $(n, k)$ cyclic code admits one and only one nonzero code polynomial $g(D)$ of minimum degree $(n - k)$ given by*

$$g(D) = 1 + g_1 D + g_2 D^2 + \cdots + g_{n-k-1}D^{n-k-1} + D^{n-k} \quad (6A.13)$$

This is easily shown as follows.

Assume that $g'(D)$ is another code polynomial with degree $(n - k)$. Using the linearity property we can add $g(D)$ and $g'(D)$ to obtain another polynomial. The degree of the resultant polynomial is less than $n - k$, because $D^{n-k} + D^{n-k} = 0$.

Note that in Equation 6A.13, $g_0$ is necessarily equal to 1. In the case $g_0 = 0$ and $g(D)$ is right-shifted $n - 1$ times, a code polynomial of degree less than $n - k$ is obtained.

The polynomial $g(D)$ is called the *generator polynomial*.

**Theorem 6A.3.2.** *Let $g(D)$ be the code polynomial as described in Theorem 6A.3.1. A binary polynomial with degree of $n - 1$ or less is a code polynomial if and only if it is a multiple of $g(D)$.*

Let $a(D)$ be a polynomial such that

$$a(D) = a_0 + a_1 D + \cdots + a_{k-1}D^{k-1} \quad (6A.14)$$

Then $c(D) = a(D)g(D) = a_0 g(D) + a_1 Dg(D) + \cdots + a_{k-1}D^{k-1}g(D)$, corresponding to a linear combination of the code polynomial $g(D), Dg(D), \ldots, D^{k-1}g(D)$. This implies that $c(D)$ is a code polynomial, given that $a(D)$ is a multiple of $g(D)$.

Now suppose we divide $c(D)$ by $g(D)$. The result is

$$c(D) = a(D)g(D) + b(D) \quad (6A.15a)$$

where the remainder $b(D)$ is either zero or a polynomial with degree less than that of $g(D)$. We can rewrite Equation 6A.15a as

$$b(D) = c(D) + a(D)g(D) \quad (6A.15b)$$

because $b(D) = -b(D)$ in modulo-2 arithmetic. Given that both $c(D)$ and $a(D)g(D)$ are code polynomials, $b(D)$ should also be a code polynomial. However, as $b(D)$ has a degree less than $g(D)$, it follows that $b(D)$ should be equal to zero, because $g(D)$ is the nonzero code polynomial of minimum degree.

**Theorem 6A3.3.** *The generator polynomial $g(D)$ is a factor of $1 + D^n$.*

It can be seen that

$$D^k g(D) = 1 + D^n + g'(D) \qquad (6A.16a)$$

where $g'(D)$ is $g(D)$ right-shifted $k$ times. This implies that $g'(D)$ is a multiple of $g(D)$. Hence, consider that $g'(D) = a(D)g(D)$. Rearranging Equation 6A.16a in modulo-2 arithmetic, we have

$$1 + D^n = g(D)h(D) \qquad (6A.16b)$$

where $h(D) = D^k + a(D)$. It can also be shown[6] that $h(D)$ is the parity check polynomial of the $(n, k)$ code.

### 6A.3.1 *Encoding a Message*

Let $\mathbf{u} = (u_0, u_1, \ldots, u_{k-1})$ be an uncoded message. Consider that $g(D)$ is a generator polynomial such that an $(n, k)$ systematic cyclic code is obtained. The message polynomial can be written as

$$u(D) = u_0 + u_1 D + \cdots + u_{k-1} D^{k-1} \qquad (6A.17)$$

In the systematic representation we want the $k$ bits of the message to occupy the $k$ rightmost bits of the $n$-bit code word. This is achieved by multiplying $u(D)$ by $D^{n-k}$,

$$D^{n-k} u(D) = u_0 D^{n-k} + u_1 D^{n-k+1} + \cdots + u_{k-1} D^{n-1} \qquad (6A.18)$$

Dividing $D^{n-k} u(D)$ by the generator polynomial $g(D)$, we obtain a quotient $a(D)$ and a remainder $b(D)$, i.e.,

$$D^{n-k} u(D) = a(D)g(D) + b(D) \qquad (6A.19a)$$

In modulo-2 arithmetic Equation 6A.19a can be rewritten as

$$a(D)g(D) = b(D) + D^{n-k} u(D) \qquad (6A.19b)$$

Because, according to Theorem 6A.3.2, $a(D)g(D) = c(D)$ is a code word, we can use this in Equation 6A.19b and obtain

$$c(D) = b(D) + D^{n-k} u(D) \qquad (6A.20)$$

Equation 6A.20 represents a code word polynomial of the $(n, k)$ cyclic code generated by $g(D)$. Therefore, the steps involved in the encoding process for an $(n, k)$ cyclic code according to a systematic representation are as follows:

1. Multiply the message polynomial $u(D)$ by $D^{n-k}$.
2. Divide $D^{n-k}u(D)$ by the generator polynomial $g(D)$ in order to obtain the remainder $b(D)$.
3. Add $b(D)$ to $D^{n-k}u(D)$ to obtain the code word polynomial $c(D)$.

### 6A.3.2 *Error Detection*

Upon receiving a code word, the first step in the decoding process is to calculate the syndrome. If the syndrome is zero, no errors are detected. If the syndrome is nonzero, errors are detected.

As far as cyclic codes are concerned, the syndrome calculation is carried out by dividing the received encoded message by the generator polynomial. Because $c(D) = a(D)g(D) + b(D) + b(D) = a(D)g(D)$, when dividing $c(D)$ by $g(D)$ no remainder is obtained, in case no errors are introduced in the transmitted code word. Consequently, the remainder of the division of the received code word by the generator polynomial is the syndrome polynomial itself. Let $r(D)$ be the received code polynomial. Then

$$r(D) = a(D)g(D) + s(D) \tag{6A.21}$$

where $s(D)$ is the syndrome polynomial.

### 6A.3.3 *An Example: Hamming Codes*

Let us illustrate the coding and decoding process of a cyclic code using the $(7, 4)$ Hamming code. Because $n = 7$, according to Theorem 6A.3.3, the generator polynomial is a factor of $1 + D^7$. Therefore, if $1 + D^7$ is factored into irreducible polynomials* we obtain

$$1 + D^7 = (1 + D)(1 + D^2 + D^3)(1 + D + D^3)$$

Knowing that the generating polynomial must have a minimum degree of $(n - k) = 3$, we have two options for $g(D)$ in this equation. Suppose we take the generating polynomial as being

$$g(D) = 1 + D + D^3$$

---

*Irreducible polynomials are polynomials that cannot be factored out using only coefficients from the binary field.

From Equation 6A.16b we have

$$h(D) = (1 + D)(1 + D^2 + D^3)$$

$$= 1 + D + D^2 + D^4$$

Let (0101) be the uncoded message. Its corresponding polynomial is $u(D) = D + D^3$. In order to obtain the encoded message we follow the steps as previously described.

1. Multiply the message by $D^{n-k}$:

$$D^{n-k}u(D) = D^{7-4}(D + D^3) = D^4 + D^6$$

2. Divide $D^{n-k}u(D)$ by $g(D)$:

$$
\begin{array}{ll}
D^6 + D^4 & \underline{D^3 + D + 1} \\
\underline{D^6 + D^4 + D^3} & D^3 + 1 \text{ (quotient)} \\
\quad\quad D^3 & \\
\quad\quad D^3 + D + 1 & \\
\quad\quad \underline{\phantom{D^3 + }} & \\
\quad\quad\quad D + 1 \text{ (remainder)} &
\end{array}
$$

3. Obtain $c(D)$:

$$c(D) = \underbrace{1 + D}_{b(D)} + \underbrace{D^4 + D^6}_{D^{n-k}u(D)}$$

The polynomial $c(D)$ corresponds to the code word (1100101). Note that the four rightmost bits constitute the uncoded message and the three leftmost bits, the redundant bits. It can be verified that the remainder of the division $c(D)/g(D)$ is zero. Any code word different from that produces a nonzero syndrome. As an example consider that the received code vector is (1100111) (an error on the 6th bit), represented by the polynomial $1 + D + D^4 + D^5 + D^6$. The division of this polynomial by $g(D)$ results in a remainder equal to $1 + D + D^2$, corresponding to the syndrome of this particular error (111). The respective error pattern* is (0000010), corresponding to an error on the 6th bit, as expected.

## 6A.4  Convolutional Codes

In block coding the code words are generated on a block-by-block basis. Moreover, the redundant bits at any time instant are a function only of the

---

*Refer to Section 6A.2.5.

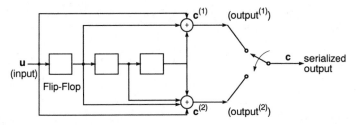

**Figure 6A.2.**    A $(2, 1, 3)$ binary convolutional encoder with constraint length $= 8$ and rate $= \frac{1}{2}$.

message at that instant. In convolutional coding the encoder is a finite-state machine accepting the message bits in a serial manner. The generated code word at a certain time instant is a function of both the input at that instant and the state of the machine.

An $(n, k, m)$ convolutional encoder accepts $k$ serial inputs, yields $n$ outputs, and has a state machine with $2^m$ states. It is composed of an $m$-stage shift register with the output of each stage conveniently combined with the $k$ inputs through $n$ adders. The adder outputs are then multiplexed so as to have one serialized encoded output. Consequently, if an information sequence of length $kL$ is convolutionally encoded, the corresponding code word has a length $n(L + m)$ and the code rate is given by

$$r = \frac{kL}{n(L + m)} \qquad (6A.22a)$$

This reduces to

$$r \simeq \frac{k}{n}, \quad \text{if } L \gg m \qquad (6A.22b)$$

The constraint length is defined as the maximum number of encoder outputs that can be influenced by a single information bit. Hence the constraint length is given by $n(m + 1)$.

As an example Figure 6A.2 shows a $(2, 1, 3)$ convolutional encoder.

### 6A.4.1 Encoding a Message

Let $\mathbf{u} = (u_0, u_1, u_2, \ldots)$ be the message to be encoded using the encoder of Figure 6A.2. Because there are two adders, two corresponding sequences $\mathbf{c}^{(1)} = (c_0^{(1)}, c_1^{(1)}, c_2^{(1)}, \ldots)$ and $\mathbf{c}^{(2)} = c_0^{(2)}, c_1^{(2)}, c_2^{(2)}, \ldots)$ are generated and finally combined to yield the encoded word $\mathbf{c} = (c_0^{(1)}c_0^{(2)}, c_1^{(1)}c_1^{(2)}, c_2^{(1)}c_2^{(2)}, \ldots)$. The sequences $\mathbf{c}^{(1)}$ and $\mathbf{c}^{(2)}$ can be easily obtained by following the connections of the encoder circuitry.* The corresponding expressions are shown in

---

*The initial state of the machine should be zero $(0, 0, \ldots)$ before any message feeds the input.

Equations 6A.23a and 6A.23b, where we have considered $\mathbf{c}^{(1)}$ and $\mathbf{c}^{(2)}$ to be the outputs of top adder and the bottom adder, respectively:

$$c_l^{(1)} = u_l^{(1)} + u_{l-1}^{(1)} + u_{l-3}^{(1)} \tag{6A.23a}$$

$$c_l^{(2)} = u_l^{(2)} + u_{l-1}^{(2)} + u_{l-2}^{(2)} + u_{l-3}^{(2)} \tag{6A.23b}$$

We assume $u_{l-j} \triangleq 0$ for all $l < j$. Note that these equations perform the convolution of the input sequence $\mathbf{u}$ with the respective response of each adder to the sequence $(1, 0, 0, \ldots)$. The sequence $(1, 0, 0, \ldots)$ is known as *impulse* and the output of the adder after this sequence is its *impulse response*. Let $\mathbf{g}^{(i)} = (g_0^{(i)}, g_1^{(i)}, g_2^{(i)}, \ldots)$ be the impulse response of adder $i$. For the example of Figure 6A.2 we have

$$\mathbf{g}^{(1)} = (1101)$$

$$\mathbf{g}^{(2)} = (1111)$$

Accordingly, the response of the adder $i$ to the sequence $\mathbf{u} = (u_0, u_1, \ldots)$ is the following discrete convolution:

$$c_l^{(i)} = \sum_{j=0}^{m} g_j^{(i)} u_{l-j}, \qquad i = 1, 2, \ldots, n; \, l = 0, 1 \ldots \tag{6A.24}$$

and $u_{l-j} = 0$ for all $l < j$.

Suppose, as an example, that the information sequence $\mathbf{u} = (11001)$ is fed into the encoder of Figure 6A.2. Then

$$\mathbf{c}^{(1)} = (10110101)$$

$$\mathbf{c}^{(2)} = (10000111)$$

and

$$\mathbf{c} = (11, 00, 10, 10, 00, 11, 01, 11)$$

The computations of Equation 6A.24 can be greatly simplified if we use an appropriate transform property by which the convolution in time domain is equivalent to multiplication in frequency domain. The frequency-domain representation of a sequence is its polynomial form. Accordingly, if the impulse response $\mathbf{g}^{(i)}$ and the message sequence $\mathbf{u}$ are expressed in their respective polynomial forms $g^{(i)}(D)$ and $u(D)$, the resultant convolution is the product of $g^{(i)}(D)$ by $u(D)$.

Hence,

$$g^{(i)}(D) = g_0^{(i)} + g_1^{(i)}D + \cdots + g_m^{(i)}D^m, \qquad i = 1, 2, \ldots, n$$

$$u(D) = u_0 + u_1 D + \cdots + u_{l-1}D^{l-1}$$

Therefore, the convolution represented by Equation 6A.24 can be written as

$$c^{(i)}(D) = g^{(i)}(D)u(D) \qquad (6A.25)$$

For our particular example,

$$g^{(1)}(D) = 1 + D + D^3$$

$$g^{(2)}(D) = 1 + D + D^2 + D^3$$

$$u(D) = 1 + D + D^4$$

Then

$$\mathbf{c}^{(1)}(D) = g^{(1)}(D)u(D) = 1 + D^2 + D^3 + D^5 + D^7$$

$$\mathbf{c}^{(2)}(D) = g^{(2)}(D)u(D) = 1 + D^5 + D^6 + D^7$$

which agrees with our previous calculations.

### 6A.4.2 *Decoding a Message*

Decoding algorithms play an essential role in the performance of a convolutional code. In this section we shall introduce the Viterbi algorithm initially implemented in a software form using a computer and now, more recently, implemented in VLSI chips. This algorithm is better understood if we make use of the trellis diagram. This is the expanded form of the encoder state diagram represented in time.

*State and Trellis Diagrams.* Let us first use the state diagram (SD) representation and then the trellis diagram (TD) representation. For the encoder of Figure 6A.2, the SD is shown in Figure 6A.3.

In this SD each branch is labelled with input/output$^{(1)}$output$^{(2)}$, where the outputs correspond to the respective outputs represented in Figure 6A.2. Any information (message) sequence can be encoded using the SD. We

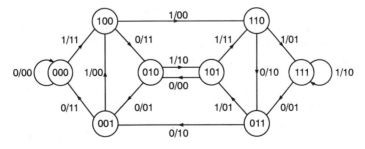

**Figure 6A.3.**   Encoder state diagram for a $(2, 1, 3)$ convolutional code.

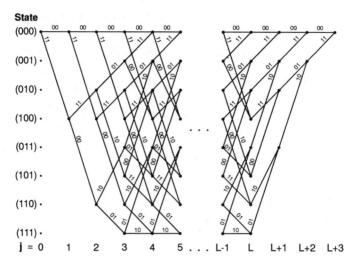

**Figure 6A.4.**    Trellis diagram for a $(2, 1, 3)$ convolutional encoder.

assume that the encoder is initially at state (000) and that after the last nonzero message block, it returns to the state (000) by means of an $m$-all-zero block appended to the message. Hence, the information sequence $\mathbf{u} = (11001)$ is encoded as $(11, 00, 10, 10, 00, 11, 01, 11)$.

Now, let us transform the SD of Figure 6A.3 into its corresponding trellis diagram. We start at a time instant, say, $j = 0$, with the decoder at state (000) and at each discrete increment of time, $j = 1, 2, \ldots$, we trace the evolution of the encoder state. For instance, at $j = 0$ we are at state (000). At $j = 1$ we may either move to state (100) or remain at (000), depending on whether the input is 1 or 0, respectively. The resultant trellis diagram is shown in Figure 6A.4, where the dots represent the states and each branch is labelled with the corresponding outputs. This diagram is drawn so as to have (1) the "downward transitions" (e.g., branch-connecting states (000) and (100)) occurring for an input equal to 1, and (2) the "upward transitions" (e.g., branch-connecting states (100) and (010)) occurring for an input equal to 0. The "horizontal transitions," corresponding to remaining at the same state, occur for an input equal to 0 for the state (000) and 1 for the state (111). The trellis diagram of Figure 6A.4 describes all the possible transitions for a word of length $L$. As an illustration for the sequence $\mathbf{u} = (11001)$, the corresponding code word is obtained straight from the trellis diagram, departing from state (000) and appending the $m$-all-zero sequence (000) to $\mathbf{u}$.

*Maximum-Likelihood Decoder.* Let $\mathbf{u}$ be a message and $\mathbf{c}$ its corresponding $N$-element code word, transmitted through a noisy channel. Let $\mathbf{c}'$ be the received $N$-element code vector. Consider a binary symmetric (BSC) channel where each element $c_i'$ of $\mathbf{c}'$ differs from the respective element $c_i$ of $\mathbf{c}$ with probability $p$. Then, the probability of receiving $\mathbf{c}'$, given that $\mathbf{c}$ has been transmitted, with $\mathbf{c}'$ differing from $\mathbf{c}$ by exactly $d$ elements is

$$p(\mathbf{c}'|\mathbf{c}) = p^d(1 - p)^{N-d} \qquad (6A.26)$$

The corresponding loglikelihood function is given by

$$\ln[p(\mathbf{c}'|\mathbf{c})] = d \ln\left(\frac{p}{1-p}\right) + N \ln(1 - p) \qquad (6A.27)$$

The probability of an error occurrence is usually low, so that it is reasonable to assume $p < \frac{1}{2}$. In this case $\ln[p/(1 - p)] < 0$. Because $N \ln(1 - p)$ is a constant for all $\mathbf{c}$, the decoder (maximum likelihood decoder) must choose $\hat{\mathbf{c}}$ as an estimate of $\mathbf{c}$ so that the Hamming distance $d$ between $\mathbf{c}'$ and $\mathbf{c}$ is minimized.

For a $Q$-ary discrete memoryless channel the loglikelihood function is, obviously,

$$\ln[p(\mathbf{c}'|\mathbf{c})] = \ln \prod_{i=0}^{N-1} p(c_i'|c_i) = \sum_{i=0}^{N-1} \ln[p(c_i'|c_i)] \qquad (6A.28)$$

The maximum likelihood decoder chooses the estimate $\hat{\mathbf{c}}$ if $\ln p(\mathbf{c}'|\mathbf{c})$ is maximum.

*The Viterbi Algorithm.* The Viterbi algorithm operates on a maximum-likelihood-rule basis. The general principle is to choose the path in the trellis diagram yielding the minimum accumulated Hamming distance between the received code vector and the output of the decoder given at each branch of the trellis. The Hamming distance between the received code vector and the sequence obtained by following a certain path in the trellis is defined as a "metric" for that particular path. The Viterbi algorithm comprises the following sequential steps:

1. Starting at time $j = m$, compute the metric for each single path entering each state. Store the path (survivor) and its metric for each state.
2. Increment $j$ by 1. Compute the metric for all the paths entering each state by accumulating the branch metric entering that state and the metric of the connecting survivor from the previous time unit. For each state, store the path with the lowest metric and eliminate all the other paths.
3. If $j < L + m$, repeat step 2. Otherwise, stop.

As an example, consider the message $\mathbf{u} = (110101)$. The corresponding transmitted encoded word for the encoder of Figure 6A.2 is $\mathbf{c} = (11, 00, 10, 01, 00, 10, 00, 01, 11)$. Suppose that the received code vector is $\mathbf{c}' = (11, 01, 10, 01, 00, 10, 00, 01, 01)$. After applying the Viterbi, algorithm the final survivor $(11, 00, 10, 01, 00, 10, 00, 01, 11)$ is shown highlighted in Figure 6A.5, where the circled numbers are the metrics for each state and the

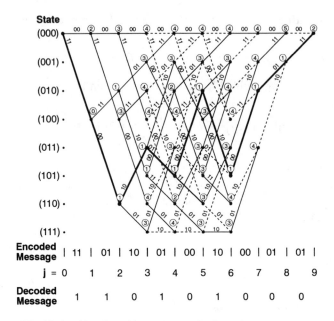

**Figure 6A.5.**  Viterbi algorithm for a binary symmetric channel.

dotted lines are the eliminated paths. The decoded message is then (110101) as wanted.

### 6A.4.3 Free Distance

In connection with the decoding algorithm, the distance properties also have a great influence in the performance of a convolutional code. As we have already mentioned, the distance of a code is directly related to the capacity of computing errors in a code word. As far as convolutional code is concerned, the most important distance measure is the minimum free distance, $d_{\text{free}}$. The free distance is defined as the minimum Hamming distance between any two code words in the code.

Equivalently, we may say that $d_{\text{free}}$ is the minimum weight code word of any length produced by a nonzero information sequence. Likewise, "it is the minimum weight of all paths in the state diagram that diverge from and remerge with the all-zero state."[6] Hence the free distance of the encoder shown in Figure 6A.2 is 6, obtained as we determine the weight of the code word (11, 00, 10, 10, 11) produced by the sequence (11000) (see state diagram of Figure 6A.3). The capacity of error correction is given by $t \leq (d_{\text{free}} - 1)/2$ in a total of bits equal to the number of branches necessary to obtain the $d_{\text{free}}$. Therefore, in our example, $t \leq (6 - 1)/2 = 2$ in a total of 5 bits (the code is capable of correcting 2 out of 5 bits).

## 6A.5   Trellis Code

Trellis codes for band-limited channels combine convolutional coding and modulation. Accordingly, channel coding and modulation are no longer performed separately as they are in the other coding schemes. This appendix does not aim at covering this subject, which is fully explored in References 8 through 15. An introduction to trellis coded modulation (TCM) can be found in Chapter 9.

## 6A.6   Some Important Codes

In this section we introduce some well-known codes and their main characteristics.

### 6A.6.1 *Hamming Codes*

These have been mainly used for error detection or correction in digital communications and data storage systems:

*Parameters*

| | |
|---|---|
| Code length | $n = 2^m - 1$ |
| Number of message bits | $k = 2^m - m - 1$ |
| Number of redundancy bits | $m = n - k \geq 3$ |
| Minimum distance | $d = 3$ |
| Error correction capability | $t = 1$ |

Some generator polynomials for Hamming codes can be found in Table 6A.1

### 6A.6.2 *Cyclic Redundancy Check Codes*

The cyclic redundancy check (CRC) codes are mainly used for error detection. The CRC can detect the following error patterns:

1. error bursts of length $\leq n - k$;
2. error bursts of length $= [1 - 2^{-(n-k-1)}](n - k + 1)$;

TABLE 6A.1.   Generator Polynomials for Hamming Codes

| m | g(D) | m | g(D) |
|---|---|---|---|
| 3 | $1 + D + D^3$ | 9 | $1 + D^4 + D^9$ |
| 4 | $1 + D + D^4$ | 10 | $1 + D^3 + D^{10}$ |
| 5 | $1 + D^2 + D^5$ | 11 | $1 + D^2 + D^{11}$ |
| 6 | $1 + D + D^6$ | 12 | $1 + D + D^4 + D^6 + D^{12}$ |
| 7 | $1 + D^3 + D^7$ | 13 | $1 + D + D^3 + D^4 + D^{13}$ |
| 8 | $1 + D^2 + D^3 + D^4 + D^8$ | 14 | $1 + D + D^6 + D^{10} + D^{14}$ |

**TABLE 6A.2.  Generator Polynomials for Some CRC Codes**

| Main CRC Codes | $g(D)$ | Character Length | Prime Factor |
|---|---|---|---|
| CRC-12 | $1 + D + D^2 + D^3 + D^{11} + D^{12}$ | 6 | $1 + D$ |
| CRC-16 | $1 + D^2 + D^{15} + D^{16}$ | 8 | $1 + D$ |
| CRC-CCITT | $1 + D^5 + D^{12} + D^{16}$ | 8 | $1 + D$ |

3. error bursts of length $\geq [1 - 2^{-(n-k)}](n - k + 1)$;
4. number of error bits $\leq d - 1$;
5. any error pattern, given that the number of errors is odd and the code generator polynomial presents as even number of nonzero coefficients.

Some generator polynomials for CRC codes can be found in Table 6A.2.

### 6A.6.3 *Golay Code*

The $(23, 12)$ Golay code is a multiple-correcting bindary code used in several communication systems:

*Parameters*

| | |
|---|---|
| Code length | $n = 23$ |
| Number of message bits | $k = 12$ |
| Number of redundancy bits | $m = 11$ |
| Minimum distance | $d = 7$ |
| Error correction capability | $t = 3$ |
| Generator polynomial | $g_1(D) = 1 + D^2 + D^4 + D^5 + D^6 + D^{10} + d^{11}$ |

or

$$g_2(D) = 1 + D + D^5 + D^6 + D^7 + D^9 + D^{11}$$

In fact,

$$1 + D^{23} = (1 + D)g_1(D)g_2(D)$$

### 6A.6.4 *Bose, Chaudhuri, and Hocquenghem Codes*

The Bose, Chaudhuri, and Hocquenghem (BCH) codes are a generalization of the Hamming codes for multiple-error correction.

*Parameters*

| | |
|---|---|
| Code length | $n = 2^m - 1$ |
| Number of message bits | $k \geq n - mt$ |
| Number of redundancy bits | $m \geq 3$ |
| Minimum distance | $d \geq 2t + 1$ |
| Error correction capability | $t < 2^{m-1}$ |

TABLE 6A.3.    Generator Polynomials for BCH Codes

| $n$ | $k$ | $t$ | $g(D)$ (octal) |
|---|---|---|---|
| 7 | 4 | 1 | 13 |
| 15 | 11 | 1 | 23 |
| 15 | 7 | 2 | 721 |
| 15 | 5 | 3 | 2467 |
| 31 | 26 | 1 | 45 |
| 31 | 21 | 2 | 3551 |
| 31 | 16 | 3 | 107657 |
| 31 | 11 | 5 | 5423325 |
| 31 | 6 | 7 | 313365047 |

Some generator polynomials for BCH codes can be found in Table 6A.3. As an example, the generator polynomial of the $(31, 21)$ BCH code is $D^{10} + D^9 + D^8 + D^6 + D^5 + D^3 + 1$ (from Table 6A.3, the corresponding octal representation is 3551, having the equivalent binary representation 011 101 101 001).

### 6A.6.5 *Reed – Solomon Codes*

The Reed–Solomon (RS) codes comprise a special and the most important subclass of $Q$-ary BCH codes. The encoder for an RS $(n, k)$ code on $m$-bit symbols groups the binary data stream into blocks, each block containing $k$ symbols ($km$ bits).

*Parameters*

Code length                                           $n = q - 1 = 2^m - 1$
Number of message symbols                   $k$
Number of redundancy symbols            $n - k = 2t$
Minimum distance                                   $d = 2t + 1$

### 6A.6.6 *Repetition Codes*

Repetition codes are the simplest form of linear block code. A single message bit ($k = 1$) can be encoded into a block of $n$ identical bits. Consequently, there are only two code words in the code, namely, an all-zero code word and an all-one code word, producing an $(n, 1)$ code. Accordingly, the minimum distance is equal to $n$. The decoding process makes use of majority voting. The number of correctable bits is $(n - 1)/2$.

### 6A.6.7 *Burst-Error-Correcting Codes*

The codes mentioned before are generally used for correcting random errors. As far as error bursts are concerned, special codes have been constructed both for block codes and for convolutional codes. Before introducing some of them, let us define the length of a burst. A burst of length $l$ is a vector having the nonzero components confined to $l$ consecutive digit positions, the first and the last of which are nonzero. For example, the error vector $\mathbf{e} = (001011000)$ is a burst of length 4.

### Block Codes

**Fire Codes**—These are a class of cyclic codes. The generator polynomial for the fire codes is constructed as follows:

$$g(D) = (D^{2l-1} + 1)p(D)$$

where $p(D)$ is an irreducible polynomial of degree $m$, $l$ is the burst length, such that $l \leq m$ and $2l - 1$ is not divisible by $\rho$ (see definition of $\rho$ that follows).

The code length $n$ is given by

$$n = \text{LCM}(2l - 1, \rho)$$

where $\text{LCM}(X, Y)$ is the least common multiple of $X$ and $Y$, and $\rho$ is the period of $p(D)$, given by $2^m - 1$.

The number of parity-check digits is given by $n - k = 2l - 1 + m$, constituting the higher-order exponent of the generator polynomial.

As an example, consider the irreducible primitive polynomial

$$p(D) = 1 + D + D^6$$

Then $\rho = 2^6 - 1 = 63$.

For $l = 6$ we have $2l - 1 = 11$ (note that 63 does not divide 11).

Hence $g(D) = (D^{11} + 1)(1 + D + D^6) = 1 + D + D^6 + D^{11} + D^{12} + D^{17}$ and $n = \text{LCM}(11, 63) = 693$.

Therefore, the $(693, 676)$ cyclic code is capable of correcting any burst error of length 6 or less.

**Some Other Codes**—Next we list a few computer-generated cyclic and shortened cyclic codes for correcting single bursts. In Table 6A.4 the generator polynomial appears in octal notation. Hence, as an example, for the code $(27, 17)$, $g(D) = 2671$ and the corresponding binary notation is $010\,110\,111\,001$, yielding

$$g(D) = D^{10} + D^8 + D^7 + D^5 + D^4 + D^3 + 1$$

**Convolutional Codes.** The burst-correcting convolutional codes are capable of correcting a given burst length, provided it is followed and preceded by an appropriate number of correct digits, known as guard space. This guard space is, in fact, a function of the decoding algorithm.[5]

Many convolutional codes are available for error-burst correction (e.g., Berlekamp–Preparata, Iwadare–Massey, etc.). The central idea of all of such codes is that the bits involved in the decoding process of a given bit are spread in time so as to allow the smallest possible number of bits to be

**TABLE 6A.4. Computer-Generated Cyclic and Shortened Cyclic Codes**

| $n - k - 2l$ | Code $(n, k)$ | Burst-Correcting Capability $l$ | Generator Polynomial $G(D)$ |
|---|---|---|---|
| 0 | $(7, 3)$ | 2 | 35 |
| 0 | $(15, 9)$ | 3 | 171 |
| 0 | $(27, 17)$ | 5 | 2671 |
| 0 | $(34, 22)$ | 6 | 15173 |
| 0 | $(50, 34)$ | 8 | 224531 |
| 1 | $(67, 54)$ | 6 | 36365 |
| 1 | $(103, 88)$ | 7 | 114361 |
| 2 | $(63, 55)$ | 3 | 711 |
| 2 | $(85, 75)$ | 4 | 2651 |
| 2 | $(131, 119)$ | 5 | 15163 |
| 2 | $(169, 155)$ | 6 | 55725 |

affected by an error. One well-known technique accomplishing this is interleaving: The data stream is broken into smaller independent streams according to the desired interleaving degree.

## REFERENCES

1. Arredondo, G. A., Feggler, J. C., and Smith, J. I., Advanced mobile phone service: voice and data transmission, *Bell Syst. Tech. J.*, 58(1), January 1979.

2. Haykin, S. S., *Digital Communications*, John Wiley & Sons, New York, 1988.

3. Schwartz, M., Bennet, W. R., and Stein, S., *Communications Systems and Techniques*, McGraw-Hill, New York, 1966.

4. Lin, S., *An Introduction to Error-Correction Codes*, Prentice-Hall, Englewood Cliffs, NJ, 1970.

5. Peterson, W. W., and Weldon, E. J., Jr., *Error-Correcting Codes*, MIT Press, Cambridge, MA, 1972.

6. Lin, S., and Costello, D. J., Jr., *Error Control Coding: Fundamentals and Applications*, Prentice-Hall, Englewood Cliffs, NJ, 1983.

7. Stein, S., Fading channel issues in system engineering, *IEEE J. Selected Areas Commun.*, SAC-5(2), February 1987.

8. Ungerboeck, G., Channel coding with multilevel/phase signals, *IEEE Trans. Inform. Theory*, IT-28, 55–67, January 1982.

9. Ungerboeck, G., Trellis-coded modulation with redundant signal sets, Part I: introduction, *IEEE Communications Magazine*, 25, 5–11, February 1987.

10. Ungerboeck, G., Trellis-coded modulation with redundant signal sets, Part II: state of the art, *IEEE Communication Magazine*, 25, 12–21, February 1987.

11. Anderson, J. B., and de Buda, R., Better phase-modulation error performance using trellis phase codes, *Electron. Lett.*, 12, 587–588, October 1976.

12. Anderson, J. B., and Taylor, D. P., A bandwidth efficient class of signal-space codes, *IEEE Inform. Theory*, IT-24, 703–712, November 1978.

13. Anderson, J. B., Aulin, T., and Sundberg, C. E., *Digital Phase Modulation*, Plenum, New York, 1986.

14. Calderbank, A. R., and Majo, J. E., A new description of trellis codes, *IEEE Trans. Inform. Theory*, IT-30, 784–791, November 1984.

15. Calderbank, A. R., and Sloame, N. J. A., New trellis codes based on lattices and cosets, *IEEE Inform. Theory*, IT-33, 177–195, March 1987.

16. Mota, J. C. M., Romano, J. M. T., and Souza, R. F., Blind equalization in data transmission systems (in Portuguese), RT-179, FEE, UNICAMP, Campinas, S.P., Brazil, 1989.

17. Foschini, G. V., Equalizing without altering or detecting data, *AT&T Tech. J.*, 64(8), October 1985.

18. Macchi, O., and Eweda, E., Convergence analysis of self-adaptive equalizers, *IEEE Trans. Inform. Theory*, IT-30(2), March 1984.

19. Haykin, S., *Adaptive Filter Theory*, Prentice-Hall, Englewood Cliffs, NJ, 1986.

20. Quareshi, S. V. H., Adaptive equalization, *Proc. IEEE*, 73(9), September 1985.

21. Oppenheim, A. V., and Schafer, R. W., *Digital Signal Processing*, Prentice-Hall, Englewood Cliffs, NJ, 1975.

22. Proakis, J. G., Adaptive equalization for TDMA digital mobile radio, *IEEE Trans. Vehicular Tech.*, 40(2), May 1991.

# PART IV

# Noise, Interference, and Modulation

# CHAPTER 7

# Noise and Interference

This chapter addresses the problems of noise and interference. Because noise is a well-known subject in any communication system, the treatment given here is rather brief and descriptive. The two types of noise, namely, additive and multiplicative, are studied but special attention is paid to the former because the latter is basically fading, which has already been extensively explored in previous chapters. The main objective of this chapter is, therefore, the interference problem, which may be classified into four main categories: intermodulation, intersymbol, adjacent channel, and cochannel. The intermodulation and intersymbol problems are examined, but the main focus will be on adjacent-channel and cochannel interference. The treatment of adjacent-channel interference is different from that usually given by other authors. Here we emphasize the influence of the traffic load on the probability of occurrence of adjacent-channel interference. More specifically, we study how this probability is affected by the presence of mobiles at the boundaries between cells. Cochannel interference problems are then tackled, with an initial objective of obtaining an analytical solution. This is achieved for the case of a pure Rayleigh environment, although more-complex cases are treated using numerical analysis. These cases include those with multiple interferers and with a fading having a lognormal or combined Rayleigh and lognormal distribution. An alternative approach is the use of Monte Carlo simulation, described at the end of the chapter. It is shown that this approach offers a great deal of flexibility and that the problem can be treated in a more realistic manner. The use of directional antennas as a means of counteracting cochannel interference is also discussed, and the discovery of novel cellular patterns is facilitated by using simulation.

## 7.1 INTRODUCTION

Noise and interference are two phenomena that limit the operating range of all radio equipment.[1] Noise is an unwanted disturbance within useful frequency band, arising from various sources and exhibiting different characteristics. Radio frequency interference originates from a communication system itself. This has increasingly become one of the biggest problems to be tackled, because such systems are growing very rapidly and in a chaotic manner. It is, therefore, essential to characterize the different types of noise and interference so that a system can be designed to ensure a signal level well above noise or interference. The characterization of these limiting factors is important because it helps in the development of prediction methods for communication systems performance. As a result, systems designed to operate adequately under unfavorable conditions may be achieved.

Different types and sources of noise and also their statistics will be described. An important performance measure is the carrier-to-noise ratio, which will be examined for the mobile radio system case. The performance measure concerned with interference is the carrier-to-interference ratio, calculated differently for the adjacent-channel and cochannel cases, because their characteristics are very distinct from each other. It will be seen that the performance with a given carrier-to-noise ratio is dependent on the modulation technique used, whereas the performance with a given carrier-to-interference ratio depends on the cell pattern. We will discuss the different methods of improving the system performance and introduce the novel cellular patterns that may arise when an appropriate use of directional antennas is accomplished.

## 7.2  NOISE

Noise disturbance is said to be of the additive or multiplicative type, depending on whether the modification introduced to the signal is of additive or multiplicative nature, respectively. The additive noise is just superimposed on the signal whereas the multiplicative noise can be viewed as an amplitude modulation of the signal by the noise.

There are many types of additive noise, such as ambient, background, static, radio, etc. We are mainly interested in the radio-type noise, of which the most important are atmospheric noise, galactic noise, artificial (i.e., man-made) noise, and receiver noise. Multiplicative noise, which causes fading, is a peculiarity of mobile radio systems. The two types of fading that are commonly observed are long-term and short-term fading, which will be briefly reviewed for convenience.

## 7.3  ADDITIVE NOISE

The objective of this section is to examine the frequency band within which there is significant noise power of some relevant radio noise sources. As mentioned before, the main sources to be considered are the radio receiver and the atmospheric, galactic, and man-made sources. We shall start by examining the radio receiver noise.

### 7.3.1  Types of Additive Noise

### 7.3.1  *Receiver Noise*

There are two type of noise in this category: thermal noise and shot noise.

*Thermal Noise.* The physical origin of this noise is the thermal agitation of the electric charge carriers in resistive materials. Thermal agitation is reduced if the resistive component is cooled down. Thermal noise is a rather common phenomenon that has already been noted by most of us—when listening to a steady hiss from a loudspeaker of a sensitive amplifier if turned on without an input signal.

J. B. Johnson[3] and H. Nyquist[4] investigated this phenomenon, the former experimentally and the latter theoretically. They showed that a resistor $R$ (in ohms) at a temperature $T$ (in kelvins) will deliver a power $kTB$ (in watts)* within a bandwidth $B$ (in hertz) into a matched load. Classical thermodynamic considerations were used to arrive at this result and will not be reproduced here, for brevity. It is possible to show that the power spectral density of the current, due to the random motion of the electrons in a resistor $R$, is[5] $S_i(\omega)$

$$S_i(\omega) = \frac{2kT/R}{1 + (\omega/\alpha)^2} \tag{7.1}$$

where $\alpha$ is the average number of collisions per second of an electron with the lattice structure. Usually $\alpha \simeq 10^{14}/\text{s}$, which leads us to conclude that, for angular frequencies up to $\omega = 10^{13}$ rad/s ($10^{12}$ Hz $= 1000$ GHz), the power spectral density is approximately constant ($(\omega/\alpha)^2 = 0.01 \ll 1$) and equal to

$$S_i(\omega) \simeq 2kT/R \tag{7.2}$$

The resistor $R$ can be represented by a noiseless resistance in series with a voltage source $v$ of uniform power density spectrum $2kTR$, because

$$v = Ri$$

and

$$S_v(\omega) = R^2 S_i(\omega) = 2kTR \tag{7.3}$$

The mean square noise voltage is given by (see Appendix 7A)

$$\overline{v^2} = 4kTRB \tag{7.4}$$

---

*$k = 1.38 \times 10^{-23}$ joule per kelvin (J/K) is Boltmann's constant.

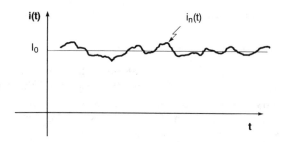

**Figure 7.1.** Random fluctuations of the current in a thermionic or semiconductor device.

Therefore, for $T = 20°C$ (293 K), $R = 10$ k$\Omega$, and $B = 1$ GHz, the mean square noise voltage is $\overline{v^2} = 1.6 \times 10^{-7}$ V$^2$, corresponding to an rms fluctuation of $\sqrt{\overline{v^2}} = 0.4$ mV on the voltage of any signal present at the resistor.

**Shot Noise.** Shot noise appears in both thermionic diodes (triodes) and semiconductor devices. In the former, shot noise is due to the anode current, consisting of the superposition of pulses of currents disposed randomly in time. Such pulses are electrons overcoming the surface potential barrier and reaching the anode. In semiconductor devices, a similar phenomenon occurs with the electrons and holes at the tail of the energy distribution overcoming a potential barrier. As shown in Figure 7.1, the net current $i(t)$ of the device fluctuates randomly about a mean value $I_0$ such that $i(t) = I_0 + i_n(t)$, where the fluctuation current $i_n(t)$ is named shot noise.

The mean value of $i_n(t)$ is zero. As shown in Appendix 7B, the power spectral density of the shot noise is

$$S_i(\omega) = \frac{4I_0 q}{(\omega\tau)^4}\left[(\omega\tau)^2 + 2(1 - \cos\omega\tau - \omega\tau\sin\omega\tau)\right] \quad (7.5)$$

where $q$ is the electronic charge $= 1.6 \times 10^{-19}$ C;

$\tau$ is the transit time of the electron between cathode and anode (second);

$I_0 = nq$, where $n$ is the mean number of electrons in motion.

This power density is plotted in Figure 7.2. Note that for $\omega\tau \leq 0.5$ the spectrum is approximately flat and equals $qI_0$. If we consider the typical value of $10^{-10}$ s for $\tau$ we obtain $\omega \simeq 5.10^9$ rad/s $\simeq 800$ MHz.

It is shown in Appendix 7B that the noise voltage $\overline{v^2}$ is

$$\overline{v^2} \simeq 2qI_0 R^2 B \quad (7.6)$$

Therefore, for $I_0 = 1$ mA, $R = 10$ k$\Omega$, and $B = 800$ MHz, we obtain $\overline{v^2} = 25.6$ $\mu$V, corresponding to an rms fluctuation of $\sqrt{\overline{v^2}} = 5$ mV (compare this result with that for thermal noise).

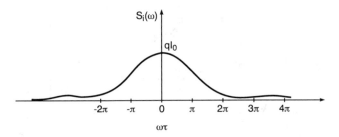

**Figure 7.2.** Power spectrum of the shot noise.

### 7.3.1.2 *Atmospheric Noise*

Lightning discharges in thunderstorms are the main source of this type of noise, and its level varies with the frequency, geographical region, season of the year, weather, time of the day, etc. It has been observed that the noise level decreases with the increasing latitude and that it has a significant value below 20 MHz, in quiet locations.

### 7.3.1.3 *Galactic Noise*

All sorts of noise that originate outside the Earth and its surrounding atmosphere are classified as galactic noise. The main sources are the Sun, the celestial radio–sky background radiation along the galactic plane and other cosmic sources. The galactic noise has significant values within the frequency range of 15 MHz–100 GHz. The ionosphere provides high-pass filtering, while the atmosphere provides low-pass filtering for this type of noise.

### 7.3.1.4 *Man-Made Noise*

Noise produced by human activities is mainly caused by electric motors, neon signs, power lines, and certain medical appliances. This type of noise is certainly more significant in urban areas than in suburban areas. The difference in power in these two areas may exceed 16 dB. In urban areas ignition noise predominates and is the dominant factor of performance degradation in mobile communications. Man-made noise is more noticeable when the sources are only a few kilometers or less away from the receiver. However, it can propagate through power lines and be noted at much greater distances.[2]

A comparison between the levels of different types of noise is shown in Figure 7.3, where the received noise power $F_a$ is given in decibels above the thermal noise $kTB$ ($T = 290$ K). Note that receiver noise has a more significant value than the atmospheric noise for frequencies above 30 MHz. In addition, the receiver noise has a more significant value than the galactic noise for frequencies above 250 MHz. On the other hand, galactic noise has a higher value than atmospheric noise for frequencies above 20 MHz. However, man-made noise usually predominates over all other sorts of noise,

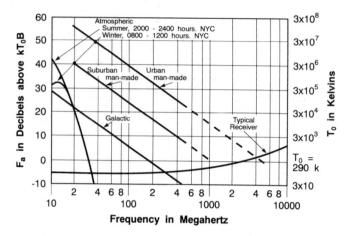

**Figure 7.3.** Noise power in decibels above thermal noise. (*Source: Reference Data for Radio Engineers*, 6th ed., Howard W. Sams, Indianapolis, IN, 1979. Reprinted with permission of Prentice-Hall Computer Publishing.)

being negligible (relative to receiver front-end noise) for frequencies above 4 GHz.

### 7.3.2    Characterization of Additive Noise

Noise is a random signal having amplitudes following a certain distribution. Both thermal noise and shot noise result from a relatively large ("infinite") number of statistically independent disturbances. From the central limit theorem of probability theory, the noise amplitude $r$ follows a normal (Gaussian) distribution $p(r)$, with a zero mean value and an rms value of $\sigma^2$, such that

$$p(r) = \frac{1}{\sqrt{2\pi}\,\sigma}\exp\left(-\frac{r^2}{2\sigma^2}\right) \tag{7.7}$$

The major difference between Gaussian noise and the other types of additive noise is the impulsive characteristics of the latter. Gaussian noise is simple to deal with and, usually, only knowledge of its mean power is relevant. The strong time and location dependence of impulsive noise makes its characterization difficult. Gaussian noise causes a "hiss" on voice channels whereas impulsive noise produces a "tick."

There is a variety of measurements that can be done to evaluate the impulsive noise characteristics, such as mean voltage, rms voltage, peak voltage, impulsiveness ratio, level crossing rate, and several distributions, such as amplitude probability, pulse height, pulse duration, etc. These

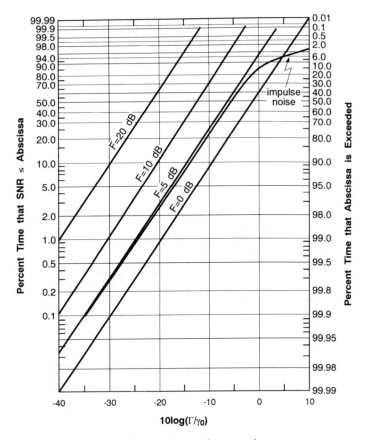

**Figure 7.4.** SNR at the output of an $F$-dB-noise-figure receiver.

measures may be useful in different situations, because the impulsive noise is greatly dependent on both time and geographical region.

Among the distributions, the amplitude probability distribution (APD) is of particular interest. It gives the proportion of time that the noise detected at the output of a receiver is above a certain level. This distribution is usually plotted on Rayleigh paper, where the probability distribution function of the signal-to-(Gaussian) noise ratio is a straight line (Section 5.6). When measured at the output of a receiver having a noise figure* $f$ ($f \geq 1$), the probability distribution function of the signal-to-noise ratio (SNR) $\Gamma$ is given by (see Appendix 7C)

$$P(\Gamma) = 1 - \exp\left(-f\frac{\Gamma}{\gamma_0}\right) \tag{7.8}$$

where $\gamma_0$ is the mean SNR. It is readily seen that any noise introduced by the

---

*In decibels the noise figure is $F = 10 \log f$.

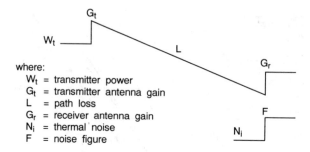

**Figure 7.5.** Diagram for calculation of $SNR_0$.

receiver ($f > 1$) will cause a deterioration of the output SNR. A plot of this distribution for different noise figures is shown in Figure 7.4. Also shown is a typical amplitude probability distribution for a noise of 5 dB. Note that there are two different regions for the APD curve: one at low signal levels, corresponding to the background noise, and the other at high signal levels, corresponding to the impulsive noise.

Finally, we conclude this section by determining the output signal-to-noise ratio $SNR_0$ for a given transmitter power $W_t$, transmitter antenna gain $G_t$, path loss $L$, receiver antenna gain $G_r$, thermal noise $N_i$, and receiver noise figure $F$. This can be accomplished with the help of the diagram of Figure 7.5.

If all the parameters are expressed in decibels, then

$$SNR_0 = W_t + G_t - L + G_r - N_i - F$$

## 7.4   MULTIPLICATIVE NOISE

A carrier frequency, when transmitted in a mobile radio environment, has its amplitude modulated by a random noise caused by both the obstructions encountered by the signal and the multipath propagation effect as illustrated in Figure 7.6. Such random amplitude modulation is known as fading, which has already been extensively studied in previous chapters, but will be briefly reviewed here. There are basically two types of fading in a mobile radio system: long-term fading and short-term fading.

### 7.4.1   Types of Multiplicative Noise

#### 7.4.11 *Long-Term Fading*

A radio wave propagating in a dispersive medium will be attenuated by several obstructions with varying thickness and electromagnetic characteristics (electric permittivity, magnetic permeability, conductivity). A vehicle

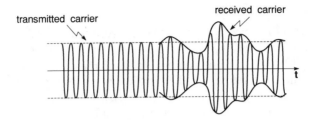

**Figure 7.6.** Multiplicative noise.

moving at a constant distance from the transmitter will receive a signal with an amplitude experiencing random attenuations due to the different properties of the attenuation constants of the random obstructions.

### 7.4.1.2 Short-Term Fading

This type of fading is a direct consequence of the multipath propagation effect. The resultant received signal is a sum of signals reaching the mobile through an "infinite" number of paths. These signals have random amplitudes and phases that may reinforce the signal at some instants and may destructively "reinforce" the signal at other instants.

## 7.4.2 Characterization of Multiplicative Noise

Due to their random behavior, both long-term and short-term fades cannot be treated in a deterministic way but, fortunately, they can be described on a statistical basis. The corresponding probability distributions are well known and will be presented here.

### 7.4.2.1 Long-Term Fading

The probability density function of the received fading signal envelope follows a normal distribution,

$$p(R) = \frac{1}{\sqrt{2\pi}\,\sigma} \exp\left[-\frac{1}{2}\left(\frac{R-M}{\sigma}\right)^2\right] \qquad (7.9)$$

where the envelope $R$ and its mean value $M$ and standard deviation $\sigma$ are given in decibels. The corresponding distribution function for an envelope $R_0$ is thus given by

$$p(R \leq R_0) = P(R_0) = \int_{-\infty}^{R_0} p(R)\, dR \qquad (7.10)$$

### 7.4.2.2 *Short-Term Fading*

Here we identify two situations: (1) the resultant signal is the sum of signals reaching the receiver through indirect paths only; (2) the resultant signal contains both the signal received through the direct path and also the signals from indirect paths. In the first case one has a Rayleigh density function, whereas the second case is associated with a Ricean density function. The Rayleigh density function is given by

$$p(r) = \begin{cases} \dfrac{r}{\sigma^2} \exp\left(-\dfrac{r^2}{2\sigma^2}\right), & r \geq 0 \\ 0, & \text{otherwise} \end{cases} \tag{7.11}$$

and the Ricean density function is given by

$$p(r) = \begin{cases} \dfrac{r}{\sigma^2} \exp\left(-\dfrac{r^2 + a^2}{2\sigma^2}\right) I_0\left(\dfrac{ar}{\sigma^2}\right), & r \geq 0 \\ 0, & \text{otherwise} \end{cases} \tag{7.12}$$

where $r$ and $\sigma^2$ are the envelope and variance, respectively; $a$ is the amplitude of the direct wave, and $I_0(\cdot)$ is the modified Bessel function of zeroth order. The distribution function for the Rayleigh case is given by

$$p(r \leq r_0) = \int_0^{r_0} p(r)\, dr = 1 - \exp\left(-\dfrac{r_0^2}{2\sigma^2}\right) \tag{7.13}$$

## 7.5  INTERFERENCE

Radio-frequency (RF) interference is one of the most important problems to be considered in the design, operation, and maintenance of mobile communication systems. Due to the fast growth of communication systems, it is almost impossible to maintain in operation an interference-free system, because the performance improvement of the equipment has not been able to keep pace with the increasing demand for services.

The two major interference problems in mobile radio systems are adjacent-channel interference and cochannel interference. Other types of interference include intermodulation and intersymbol. We shall address these problems but the main focus will be on the adjacent and cochannel problems in the cellular architecture. Intermodulation interference is generated in any nonlinear circuit when the product of two or more signals result into another signal, having a frequency that is equal or almost equal to the wanted signal. In the transmitter, the intermodulation interference usually occurs in the power amplifier, whereas in the receiver it is produced in the first converter (refer to Chapter 10).

Intersymbol interference is intrinsic to digital networks and it is a direct consequence of the limited bandwidth of the transmission medium. An ideal system should have an infinite bandwidth with a distortionless transmission, such that a sequence of pulses can be sent at a rate as high as desired. In other words, the pulse width could be made as short as desired. In practice, however, every system presents a limited bandwidth with unavoidable distortions, such that the shape of the transmitted pulses (symbols) is not kept the same. In fact, the symbols tend to spread out, with a consequent overlap among them. This is known as intersymbol interference (ISI). Equalization and coding can be used to minimize the effects of ISI. This has already been briefly studied in Chapter 6.

Intermodulation and intersymbol interferences are well-known phenomena that have already been extensively explored in the literature, and the treatment given is also applicable to a mobile radio environment. We shall then dedicate the rest of this chapter to the problem of adjacent-channel and cochannel interference and the techniques used to counteract their effects in the cellular mobile radio system.

One of the most difficult problems is to determine whether a particular type of interference is adjacent-channel or cochannel. Both of them present the same type of undesired output from a receiver. However, it may be possible to distinguish between them if specific identification tests are performed.

## 7.6 ADJACENT-CHANNEL INTERFERENCE

Adjacent-channel interference occurrence is basically due to equipment limitations such as frequency instability, receiver bandwidth, filtering, etc. Although equipment is designed for maximum interference performance of the system, a combination of factors, such as the cellular architecture and the random signal fluctuation, usually causes a deterioration of the received signal, primarily due to interference of adjacent channels. This is because channels are kept very near each other in the frequency spectrum for obtaining maximum spectrum efficiency.

### 7.6.1 Intracell Adjacent-Channel Interference

If adjacent channels are used in the same cell site, we may have a situation where a mobile station, transmitting from a short distance to the base station, will strongly interfere with the signal of another mobile, transmitting from a long distance to the same base station, on an adjacent channel. This is known as intracell adjacent-channel interference. One possible solution to this problem is to avoid the use of adjacent channels within the same cell.

## 7.6.2   Intercell Adjacent-Channel Interference

Even in the situation where adjacent channels are not used in the same cell but in adjacent cells, interference may still occur. For instance, consider two mobiles near the cell border, each one transmitting to its "own" base station through adjacent channels. Each base station receives the wanted signal plus a certain interference level of the unwanted signal. Adjacent-channel interference may occur because of two main factors: (1) both signals experience attenuation and fading; (2) their fadings are uncorrelated (the signals travel through different trajectories) so that the interfering signal may become larger than the wanted signal. This is known as intercell adjacent-channel interference or adjacent-channel near use. One "possible" solution to this problem is to avoid the use of adjacent channels in adjacent cells. However, if, for example, a seven-cell cluster (refer to Section 2.6) is to be implemented, adjacent channels are inevitably assigned to adjacent cells.

In this section we will investigate the statistics of the proportion of mobiles in a situation of having their adjacent-channel interference problems increased due to their geographical location within the cell. In other words, if receivers are considered to be ideal, we will be estimating the probability of intercell adjacent-channel interference. If receivers are not ideal, then this probability corresponds to the incremental percentage of the ratio of signal to adjacent-channel interference (in absolute values). This probability should be a function of the propagation parameters and traffic distribution of the system. The following investigation is based on the work by Sánchez[10] complemented and enhanced by some results obtained in Reference 11.

There are two situations to be considered: (1) interference at the mobile and (2) interference at the base station. Adjacent-channel interference is more likely to occur if the vehicles are near the boundaries of the cells, because communications can be provided by the channels of more than one cell. In order to investigate the possible cases of interference, let us define some parameters. Let channel $i$ be the reference channel, and let $p_1$, $p_2$, $\gamma$, $\delta$, and $\mu'$ be probability densities, such that

$p_1$ = probability of occurrence of only one adjacency, i.e., only one adjacent channel (say $i + 1$) is active and the other adjacent channel ($i - 1$) is not being used;

$p_2$ = probability of occurrence of two adjacencies, i.e., the two possible adjacent channels ($i - 1$ and $i + 1$) are both active;

$\gamma$ = proportion of mobiles with adequate communication with more than one base station, i.e., probability that a mobile can communicate through channels of two or more cells;

$\delta$ = proportion of mobiles with adequate communication with more than two base stations, i.e., probability that a mobile can communicate through channels of three or more cells;

$\mu'$ = proportion of mobiles within a cell with the signal level below a given threshold, i.e., the probability of a mobile receiving a signal level below a given value.

The probabilities $p_1$ and $p_2$ are dependent on both the channel assignment algorithm and the traffic load of the system. The probabilities $\gamma$, $\delta$, and $\mu'$, on the other hand, depend on the mean signal strength. Note that $\mu'$, and $\gamma$ and $\delta$ have been defined and extensively explored in Chapter 3 (Sections 3.5 and 3.6, respectively). Note also that $\mu'$ is a function of $\mu$ such that $\mu' = 1 - \mu$, where $\mu$ (the probability that the received signal is above a given threshold) is given by Equation 3.108 or 3.111, as required. In the examples to be given here we shall assume a path loss coefficient $\alpha = 3.5$ and a standard deviation $\sigma_w = 5$ dB. As we have seen in Chapter 3, for these parameters the probability $\mu$ calculated by either Equation 3.108 or 3.111 gives similar results. From the definitions given, we can say that $\gamma - \delta$ is the proportion of mobiles with access to two base stations only.

In order to simplify our reasoning, it is convenient to relate the probabilities $\mu'$, $(\gamma - \delta)$, and $\delta$ with the geographical distribution of the mobiles within a cell. In a hexagonal cell array, $\delta$ is the probability that a mobile is at the vicinity of the border of three mutually adjacent cells (the joint corner of three adjacent cells). Therefore, $\delta/6$ is the probability that this mobile is at one out of the six possible "corners". The difference $\gamma - \delta$ gives the probability that a mobile is at the vicinity of the borders of two adjacent cells. Hence, $(\gamma - \delta)/6$ gives the probability that this mobile is at one out of the six possible joint borders. The probability that the mobile lies within the weak signal area of the cell is $\mu'$. Now we analyze the two possible cases.

### 7.6.2.1 Adjacent-Channel Interference at the Mobile

The conditions causing intercell adjacent-channel interference at the mobile are depicted in Figure 7.7.

The mobile using channel $i$ may suffer interference from channel $i + 1$ and (or) channel $i - 1$ if both (either one) are (is) active in the adjacent cells. It is shown in Appendix 7D that the probability of adjacent-channel interfer-

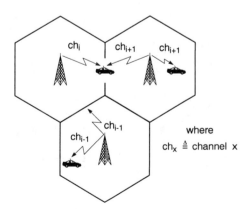

**Figure 7.7.**  Adjacent-channel interference at the mobile.

ence at the mobile ($p_M$) is

$$p_M = \frac{\gamma + \delta}{3} I_A \tag{7.14}$$

where

$$I_A = \frac{p_1 + 2p_2}{2} \tag{7.15}$$

The parameter $I_A$ gives the incidence of adjacency, i.e., the proportion of occurrence of adjacency.

### 7.6.2.2  Adjacent-Channel Interference at the Base Station

The conditions causing intercell adjacent-channel interference at the base station are depicted in Figure 7.8.

The base station $B_1$ receives the wanted signal from mobile unit $M_1$ and the interfering signal(s) from mobiles $M_2$ and/or $M_3$ transmitting in adjacent channels to base stations $B_2$ and $B_3$, respectively. Note that besides the strategic location of $M_2$ and/or $M_3$, near the cell borders, the mobile $M_1$ must be located in the weak signal area of cell 1. Appendix 7D shows that the probability of intercell adjacent-channel interference at the base station ($p_B$) is

$$p_B = \frac{\gamma + \delta}{3} \mu' I_A \tag{7.16}$$

where $I_A$ is given by Equation 7.15. We should note that $\gamma$ and $\delta$ are not independent variables. It has been shown (Section 3.6) that $\delta$ is a function of

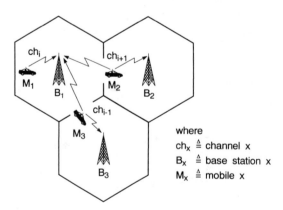

**Figure 7.8.**   Adjacent-channel interference at the base station.

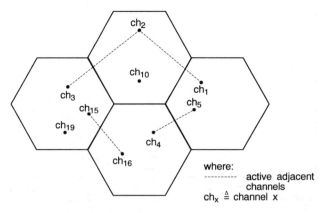

**Figure 7.9.** Adjacency of active channels.

$\gamma$ and the relation between them can be reasonably approximated by

$$\delta \simeq \begin{cases} 1.25\gamma^2, & 0 \le \gamma \le 0.8 \\ \gamma, & 0.8 \le \gamma \le 1.0 \end{cases} \qquad (7.17)$$

The probabilities $\gamma$ and $\delta$ are intimately related with the geographical locations from which the path losses are within some tolerance. They can, therefore, be written as functions of such tolerance (in decibels). The probability $\mu'$ also depends on the geographical location from which the signal is considered to be below a certain threshold. It can, therefore, be written as a function of such threshold (in decibels). The probability $\mu'$, however, is independent of $\gamma$ (and consequently of $\delta$). The probabilities $p_1$ and $p_2$ are not independent of each other, so that the parameter $I_A$ (incidence of adjacency) assumes values in the range from 0 to 1.

Consider, as an example, a seven-cell cluster system with all the cells having neighbors and all the channels having adjacent channels. If the traffic level is so high that all the channels are constantly active, then $p_2 = 1$ and obviously $p_1 = 0$, and $I_A = 1$. If there is no active channel then $p_1 = p_2 = 0$ and $I_A = 0$. As another example, consider the situation of Figure 7.9, where the dotted lines show adjacent channels in use. There are nine active channels, six of which (channels 1, 3, 4, 5, 15, and 16) having just one active adjacent channel and one (channel 2) having two active adjacent channels. Therefore $p_1 = \frac{6}{9}$ and $p_2 = \frac{1}{9}$, and

$$I_A = \frac{\frac{6}{9} + 2 \times \frac{1}{9}}{2} = \frac{4}{9}$$

***Adjacent-Channel Interference and Propagation Parameters.*** The probability $\mu'$ as a function of the threshold level (in decibels) may be extracted directly

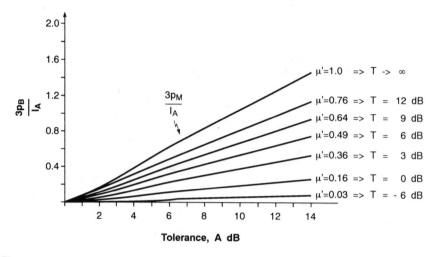

**Figure 7.10.** Probability of adjacent-channel interference versus tolerance.

from Figure 3.22 by $\mu' = 1 - \mu$. The probabilities $\gamma$ and $\delta$ as functions of the tolerance ($A$ dB) between the radio signal strength to two and three base stations are given by Figures 3.28 and 3.33, respectively.

We can now plot the probabilities of adjacent-channel interference $p_M$ and $p_B$ as functions of these tolerances and signal threshold (Figure 7.10). We chose as abscissa the path tolerance ($A$ dB) to at least two base stations and as ordinate the normalized interference probabilities $p_M/I_A$ and $p_B/I_A$. The probability of adjacent channel interference at the mobile is given by the upper curve only. The probability of adjacent-channel interference at the base station is given as a function of the tolerance ($A$ dB) having $\mu'$ (or, equivalently, the threshold $T$) as a parameter. Note that $p_M = p_B$ for $\mu' = 1$ (or $T \to \infty$).

As an example, suppose we consider that paths within $A = 6$ dB of each other are valid alternatives. Then, if the mobile is in a region where the received signal is below a threshold $T = 9$ dB (event with $\mu' = 64\%$ probability), then the normalized probability of adjacent-channel interference at the base station is $p_B/I_A \simeq 0.4/3 = 13.3\%$. Under the same condition, the normalized probability of adjacent-channel interference at the mobile is $p_M/I_A \simeq 0.6/3 = 20\%$ (in the latter case the value of $\mu'$ is irrelevant because the curve used is always that for $\mu' = 1$). Note that these are the maximum probabilities, because they are obtained from $p_B/I_A$ and $p_M/I_A$, where $I_A$ (the incidence of adjacency) is a parameter with positive values less than or equal to 1. The parameter $I_A$ depends on the traffic load and also on the channel assignment technique of the system.

Referring to Equations 7.14 and 7.16 for the values of $p_M$ and $p_B$, respectively, we can see that their maxima are $\frac{2}{3}$, obtained for $I_A = 1$, $\mu' = 1$, and $\gamma = 1$ (this latter implying $\delta = 1$). Consequently, in the extreme situation, the channels at the mobile, or at the base station, may have a 66.7% chance of adjacent channel interference. If we consider (1) tolerance $A = 8$ dB, (2) threshold $T = 12$ dB, and (3) incidence of adjacency $I_A = 70\%$ as acceptable (reasonable) figures, then $p_m \simeq 19.6\%$ and $p_B \simeq 15.4\%$.

*Adjacent-Channel Interference and Traffic Distribution.* Let us now estimate the distribution of the number of active adjacent channels. Let $p(n)$ be the probability of $n$ active adjacent channels $(n = 0, 1, 2)$. The state of the channels in use or not can be expressed by the Bernoulli distribution,

$$p(\omega) = p^{\omega}(1 - p)^{1-\omega}, \qquad \omega = 0, 1 \tag{7.18}$$

where $p$ is the probability of finding the channel active, and $\omega$ represents the channel state: 1 for active or 0 for idle. Assume that the channels are independent and uniformly distributed. Then

$$n = \omega_1 + \omega_2 \tag{7.19}$$

The probability $p(n)$ has a binomial probability density function,

$$p(n) = \binom{2}{n} p^n (1 - p)^{2-n}, \qquad n = 0, 1, 2 \tag{7.20}$$

Assuming that the cells have an equal capacity (i.e., same number $N$ of channels per cell) and that the traffic has an even distribution (i.e., the same blocking probability $B$ in each cell), then $p = B^{1/N}$ and

$$p(n) = \binom{2}{n} B^{n/N}(1 - B^{1/N})^{2-n} \tag{7.21}$$

Consequently, the probabilities $p_1$ and $p_2$ are given by

$$p_1 = p(1) = 2B^{1/N}(1 - B^{1/N}) \tag{7.22}$$

$$p_2 = p(2) = B^{2/N} \tag{7.23}$$

Then, using Equations 7.22 and 7.23 in Equation 7.15, the incidence of adjacency is

$$I_A = B^{1/N} = p \tag{7.24}$$

Note, as stated before, that $p_1$ and $p_2$ are dependent probabilities and can be written as

$$p_1 = 2\sqrt{p_2}\left(1 - \sqrt{p_2}\right) \tag{7.25}$$

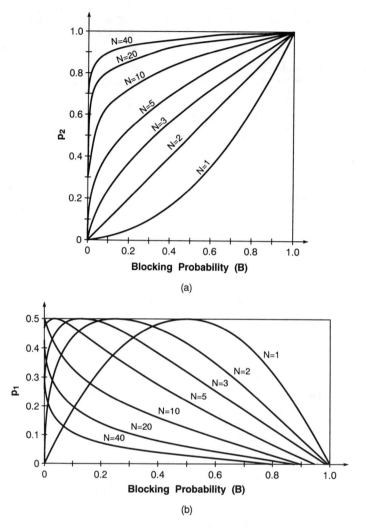

**Figure 7.11.** Adjacencies: (a) probability of two adjacencies; (b) probability of one adjacency; (c) incidence of adjacency.

The probability $p_1$ has its maximum at $B = 0.5^N$ with $p_1$ assuming the value 0.5. It can be noticed that the larger the number of channels per cell, the greater the incidence of adjacency (even for a low blocking probability). For instance, in a 40-channel-per-cell system the incidence of adjacency quickly reaches the figure of 90% even for a blocking probability near 0%. The curves of $p_1$, $p_2$, and $I_A$ as functions of $B$ (where $N$ is the parameter) are shown in Figure 7.11a–c.

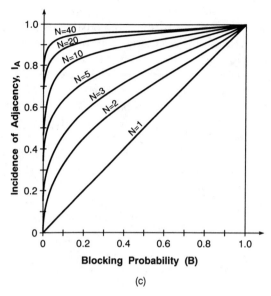

Figure 7.11. Continued

## 7.7 COCHANNEL INTERFERENCE

Cochannel interference is a complication that arises in mobile systems using cellular architecture, because channels are used simultaneously in as many cells as possible, with the minimum acceptable separation, in order to increase the reuse efficiency. A base station receiving the wanted signal from a mobile station within its cell may also receive unwanted signals (interferers) from mobiles within other cell clusters using the same channel. The determination of the frequency reuse distance, and hence the cell repeat pattern, has a direct influence on the cochannel interference levels: A bigger reuse distance implies a smaller cochannel interference. However, this also implies a bigger number of cells per cluster, resulting in a smaller reuse efficiency. Consequently, a trade-off between interference and efficiency must be found because they work in opposite directions.

The performance parameter of special interest is the carrier-to-cochannel interference ratio $C/I$. The ultimate objective of estimating $C/I$ is to determine the reuse distance and, thence, the repeat cell pattern. The $C/I$ ratio is a random variable, affected by the random phenomena such as (1) location of mobile, (2) Rayleigh fading, (3) lognormal shadowing, (4) antenna characteristics, and (5) cell site location.

In this section we will tackle the problem of determining $C/I$ by using two methods, analytical and simulation. In the analytical approach the effects of Rayleigh and lognormal fading will initially be considered independently.

Later they are taken into account simultaneously. The problem is first examined without any traffic distribution consideration, but later this is included for completeness of the investigation. In the simulation approach, a method using a Monte Carlo simulation will be described.

The $C/I$ ratio will be determined on a statistical basis. One measure of interest is the "outage probability," i.e., the probability of failing to achieve adequate reception of a signal. This probability will be indicated by $p(CI)$.

A mobile radio in a given cell receives a signal power $s$ from its base station, but it also receives interferers (unwanted signals) having powers $i_j$, $j = 1, 2, \ldots, n$ ($n$ is the number of active cochannels from its cocells). Cochannel interference will occur whenever the wanted signal does not simultaneously exceed the minimum required signal level $s_0$ and the interfering signals $i = \sum_{j=1}^{n} i_j$ by some protection ratio $r$. Consequently, the conditional outage probability, given $n$ as the number of interferers, is

$$p(CI|n) = 1 - \int_{s_0}^{\infty} p(s) \int_{0}^{s/r} p(i) \, di \, ds \qquad (7.26)$$

The difficulty of this method is to find the probability density function of the equivalent interfering signal $i$. An alternative to Equation 7.26 is to consider each interfering signal $i_j$ individually, as follows:

$$p(CI|n) = 1 - \int_{s_0}^{\infty} p(s) \int_{0}^{s/r} p(i_1) \int_{0}^{(s/r)-i_1} p(i_2) \cdots$$

$$\times \int_{0}^{(s/r)-i_1 \cdots - 1_{n-1}} p(i_n) \, di_n \cdots di_2 \, di_1 \, ds \qquad (7.27)$$

The difficulty now is to perform multiple integrations.

The total probability of cochannel interference can then be evaluated by

$$p(CI) = \sum_{n} p(CI|n) p(n) \qquad (7.28)$$

where $p(n)$ is the probability density function of the number of active interfering signals.

### 7.7.1  Cochannel Interference with Only One Type of Fading

In an interference-free system the outage probability given by Equation 7.26 reduces to $1 - \int_{s_0}^{\infty} p(s) \, ds$, which is the coverage requirement for an adequate reception. This has already been fully explored in the Section 3.5. In an interference environment the probability densities of both the signal and interferers must be known and the treatment is obviously more complicated. In a mobile radio environment Rayleigh and lognormal fadings occur simultaneously and again this is another complicating factor. However, it is possible to simplify the calculations by considering only one of these fadings and by allowing some tolerances due to the other fading as described by Hughes and Appleby.[12] The use of only one type of fading can be regarded as a valid approximation whenever either Rayleigh or lognormal fading is dominant when compared with the other. For example, shadowing effect is dominant in the case where an efficient diversity technique minimizes the Rayleigh fading. On the other hand, Rayleigh fading effect is dominant in the case where the standard deviation of the shadowing is very small.

The probability density function of SNR for the wanted signal and one interferer are the same, and in a Rayleigh environment this is given by Equation 7.29 (reproduced from Equation 5.17),

$$p(x) = \frac{1}{x_m} \exp\left(-\frac{x}{x_m}\right) \tag{7.29}$$

where $x = s$ for the wanted signal-to-noise ratio or $x = i$ for the interferer-to-noise ratio, $x_m = s_m$ for the local mean signal-to-noise ratio and $x_m = i_m$ for the local mean interference-to-noise ratio. In the case of shadowing only this probability is given by Equation 7.30, as demonstrated in Appendix 7E,

$$p(x) = \frac{1}{\sqrt{2\pi}\,\sigma x} \exp\left(-\frac{\ln^2\left(\frac{x}{x_m}\right)}{2\sigma^2}\right) \tag{7.30}$$

Using $p(s)$ and $p(i_j)$ (given by Equation 7.29) in Equation 7.27, the outage probability in a Rayleigh environment can be estimated. In the same way, with $p(s)$ and $p(i_j)$ (given by Equation 7.30) in Equation 7.27, the outage probability, in a lognormal environment, can be estimated. However, performing multiple integrals is a rather difficult task where the degree of difficulty increases with the increasing number of interferers. Let us analyze the two cases separately.

### 7.7.1.1 Multiple Interferers in a Rayleigh Fading Environment

The following description is based on a treatment given by Sowerby and Williamson,[14] where the outage probability for multiple interferers is calculated by following an analytical method using a recursive process.

We start by defining the following notation:

1. $p(CI|j) \triangleq p_j^{out} \triangleq$ outage probability given $j$ interferers
2. $\Delta p_j \triangleq p_j^{out} - p_{j-1}^{out}$

Using Equation 7.29 in Equation 7.27 it is straightforward to show that

$$p_0^{\text{out}} = 1 - \exp\left(-\frac{s_0}{s_m}\right) \tag{7.31}$$

and

$$p_1^{\text{out}} = 1 - \exp\left(-\frac{s}{s_m}\right) + \frac{1}{1 + \dfrac{\Lambda_1}{r}} \exp\left[-\frac{s}{s_m}\left(1 + \frac{\Lambda_1}{r}\right)\right] \tag{7.32}$$

where $\Lambda_1 = s_m/i_m$ is the ratio of the local wanted signal to the interfering signal.

It can be seen that

$$\Delta p_1 = p_1^{\text{out}} - p_0^{\text{out}} = \frac{1}{1 + \dfrac{\Lambda_1}{r}} \exp\left[-\frac{s_0}{s_m}\left(1 + \frac{\Lambda_1}{r}\right)\right] \tag{7.33}$$

By carrying out the calculations, for increasing number of Rayleigh interferers, a pattern emerges[14] and a generalization of Equation 7.33 is attainable. For $n$ interferers, we have

$$\Delta_n = p_n^{\text{out}} - p_{n-1}^{\text{out}} \tag{7.34}$$

where

$$\Delta p_n = \frac{1}{1 - \dfrac{\Lambda_n}{\Lambda_{n-1}}}\left(\frac{1}{1 - \dfrac{\Lambda_n}{\Lambda_{n-2}}}\left(\cdots\left(\frac{1}{1 - \dfrac{\Lambda_n}{\Lambda_1}}\left(\frac{1}{1 + \dfrac{\Lambda_n}{r}}\right.\right.\right.\right.$$

$$\left.\left.\left.\left.\times \exp\left(-\frac{s_0}{s_m}\left(1 + \frac{\Lambda_n}{r}\right)\right) - \Delta p_1\right) - \Delta p_2\right)\cdots\right) - \Delta p_{n-1}\right)$$

$$\tag{7.35}$$

With the recursive use of Equations 7.34 and 7.35 the outage probability, conditional on the number of interferers, can be obtained. Note that in Equation 7.35 an indeterminacy occurs if $\Lambda_n$ is equal to $\Lambda_1, \Lambda_2, \ldots, \Lambda_{n-2}$, or $\Lambda_{n-1}$. This can be avoided by using "slightly perturbed values of $\Lambda$ in place of those which have the same value."[14]

For an "interference-only" environment, i.e., if the minimum signal requirement is ignored ($s_0 = 0$), it is shown in Appendix 7F that

$$p_n^{\text{out}} = \sum_{j=1}^{n} \prod_{k=1}^{j} \frac{\Lambda_{k-1}/r}{1 + \Lambda_k/r} \tag{7.36}$$

where $\Lambda_0 \triangleq 1$.

If all the interferers are equal, then

$$p_n^{\text{out}} = 1 - \left( \frac{\Lambda/r}{1 + \Lambda/r} \right)^n \tag{7.37}$$

The numerical examples given here will consider this situation.

### 7.7.1.2 Multiple Interferers in a Lognormal Fading Environment

In the case where only lognormal fading is considered, the probability density function of the equivalent interfering signal $p(i)$ can be approximated by a lognormal distribution[15, 16] (see the next section) and is used in Equation 7.26.

### 7.7.2 Cochannel Interference with the Two Types of Fading

In this section we will consider the case where the received radio signal experiences both Rayleigh and lognormal fading simultaneously. The first step in this direction is to determine the resultant probability density function. It is shown in Section 3.4.4 that the combined probability of the received envelope is given by*

$$p(s) = \sqrt{\frac{\pi}{8\sigma^2}} \int_{-\infty}^{\infty} \frac{s}{10^{S/10}} \exp\left( - \frac{\pi s^2}{4 \times 10^{S/10}} \right) \exp\left( - \frac{(S - S_m)^2}{2\sigma^2} \right) dS,$$

$$s > 0, \tag{7.38}$$

where $S$  is the local mean in decibels;
    $S_m$ is the area mean in decibels (mean of the lognormal fading);
    $\sigma$  is the standard deviation in decibels (typically between 5 and 12 dB in urban locations).

---

*Note that, differently from Equations 7.29 and 7.30, this is the distribution of the envelope and not of the signal power.

The difficulty of dealing with multiple integrals in this case is really great due to the complexity of the joint density function itself. However, it can be shown[17] that the sum of these distributions is another distribution that can be approximated by a normal distribution, for $\sigma = 0$, and by a lognormal distribution, for $\sigma = 6$ and 12 dB. Consequently, the distribution of the equivalent interfering signal for $\sigma = 0$ is

$$p(i) = \frac{1}{\sqrt{2\pi}\,\sigma_i} \exp\left(-\frac{(i-\bar{i})^2}{2\sigma_i^2}\right), \quad \text{for } \sigma = 0 \qquad (7.39)$$

where

$$\bar{i} = n\exp\left(\frac{I_{m/2}}{10/\ln 10}\right) \triangleq \text{mean of } i \triangleq E[i]$$

$$\sigma_i^2 = \frac{\bar{i}^2}{n}\left(\frac{4}{\pi} - 1\right) \triangleq \text{variance of } i \triangleq E[i^2] - E^2[i]$$

$\bar{i}$ and $\sigma_i^2$ are obtained from Equation 7.38 with the application of the appropriate definition of moments.

For $\sigma > 0$, the distribution given in Equation 7.38 can be approximated by a lognormal distribution. Therefore the probability density of $i = \sum_n i_j$ (sum of lognormal distributions) is approximately another lognormal distribution[16] given by

$$p(i) = \frac{20/\ln 10}{i\sqrt{2\pi}\,\sigma_i} \exp\left(-\frac{1}{2\sigma_i}(20\log i - I_i)^2\right), \quad \text{for } \sigma > 0 \qquad (7.40)$$

where

$$\sigma_i^2 = \left(\frac{20}{\ln 10}\right)^2 \left\{\ln\left(\exp\left(\frac{\sigma^2 + (20/\ln 10)^2 \ln(4/\pi)}{(20/\ln 10)^2}\right) + n - 1\right) - \ln n\right\}$$

$$(7.41)$$

$$I_i = I_m - \left(\frac{10}{\ln 10}\right)\ln\left(\frac{4}{\pi}\right) + \frac{20}{\ln 10}\ln n + \frac{\sigma^2 + (20/\ln 10)^2 \ln(4/\pi) - \sigma_i^2}{40/\ln 10}$$

$$(7.42)$$

$\sigma_i^2$ and $I_i$ are, respectively, the variance and mean value of $i$, both given in decibels.

If $\sigma = 0$ (Rayleigh fading case) the density of $s$ in Equation 7.38 reduces to

$$p(s) = \sqrt{2\pi}\sqrt{\frac{\pi}{8\sigma^2}}\frac{s}{10^{S_m/10}}\exp\left(-\frac{\pi s^2}{4 \times 10^{S_m/10}}\right) \qquad (7.43)$$

because in Equation 7.38, $S = S_m$, and the integration in $dS$ equals $\sqrt{2\pi}$. Equation 7.43 is the Rayleigh distribution.

The conditional probability of cochannel interference can then be evaluated by using Equations 7.43 and 7.39 in Equation 7.26 for $\sigma = 0$ or Equations 7.38 and 7.40 in Equation 7.26 for $\sigma > 0$.

### 7.7.3  Some Outage Probability Results

The outage probabilities will be determined for the "interference-only" case (i.e., a minimum signal level is not required, $s_0 = 0$). It is convenient to analyze the outage probability in terms of a parameter $Z$ defined as the margin, in decibels, by which the mean value of the wanted signal $S_m$ exceeds the mean value of the intefering signal $I_m$ by a protection ratio $R_r$ (all in decibels), i.e.,

$$Z = S_m - (I_m + R_r) \qquad (7.44)$$

Note that $Z$ can be used directly in the equations where both types of fading are considered (Section 7.7.2). For the Rayleigh-fading-only case (Section 7.7.1) the following transformation must be accomplished in order to convert linear into logarithmic units

$$\frac{S_m}{ri_m} = \frac{\Lambda}{r} = 10^{Z/10} \quad \text{or} \quad \frac{S_m}{i_m} = 10^{(Z+R_r)/10} \qquad (7.45)$$

In order to estimate the probability of cochannel interference $p(CI)$, the probability density $p(n)$ of the number of active interfering signals must be determined. The status of the channels, in use or not in use, is described by the Bernoulli distribution,

$$p(\omega) = p^\omega(1-p)^{1-\omega}, \qquad \omega = 0,1 \qquad (7.46)$$

where $p$ is the probability of finding the channel active, and $\omega$ represents the channel state: 1 for active and 0 for idle. Assuming independent and uniformly distributed channels and considering only the six surrounding interferers,

$$n = \sum_{i=1}^{6} \omega_i \qquad (7.47)$$

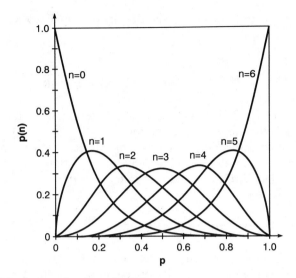

**Figure 7.12.**  Distribution of active cochannels.

Then $p(n)$ has a binomial probability density function

$$p(n) = \binom{6}{n} p^n (1 - p)^{6-n}, \qquad n = 0, 1, \ldots, 6 \tag{7.48}$$

For equal capacity cells and an evenly distributed traffic system,

$$p = B^{1/N} \tag{7.49}$$

where $B$ is the blocking probability of the cell with $N$ channels. The distribution $p(n)$ is shown in Figure 7.12.

Consequently, with $p(n)$ given by Equation 7.48 and $p(CI|n)$ calculated as explained in the previous section, we can evaluate $p(CI)$ as in Equation (7.28). Figure 7.13 shows the conditional and unconditional cochannel interference probabilities, $p(CI|n)$ and $p(CI)$, respectively. In this figure we have considered the cases of $n = 1$ and $n = 6$ interferers and $\sigma = 0$ dB (pure Rayleigh environment) and $\sigma = 6$ dB (Rayleigh and lognormal environment).

Note that the probability of interference increases both with the number of interferers and with the standard deviation. Note also that the influence of traffic load is not significant in a mature system where the number of cells is relatively large. This can be seen in Figure 7.13, where the unconditional probability $p(CI)$ is almost always reasonably close to the conditional probability $p(CI|6)$. We can explain this by reasoning that, because in a mature system the number $N$ of channels per cell tends to be large, the occupancy $p$ of the channels approaches 100% even for a low mean blocking probability ($\lim_{N \to \infty} p = \lim_{N \to \infty} B^{1/N} = 1$). Consequently, referring to Figure 7.12, we

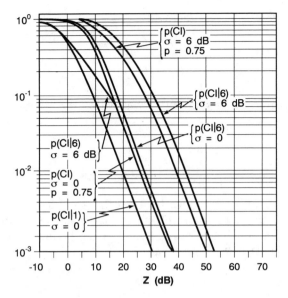

**Figure 7.13.**  Conditional and unconditional outage probability for $n = 1$ and $n = 6$ interferers with $\sigma = 0$ and $\sigma = 6$ dB.

can see that $p(n) \simeq 1$ for $n = 6$ and $p(n) \simeq 0$ for $n \neq 6$. Hence

$$p(CI) = \sum_n p(CI|n)p(n) \simeq p(CI|6)$$

In the case where the traffic is really low ($p \simeq 0$), the distribution of active cochannels will be predominantly given by $p(1)$.* Hence the final $P(CI)$ will be a small percentage of $p(CI|1)$. Consequently, we can assume that the upper bound for $p(CI)$ is $p(CI|6)$ and the lower bound is theoretically zero.

It is obvious that in a small system the traffic may have a great influence and we may expect that the curve for $p(CI)$ will lie between those for $p(CI|1)$ and $p(CI|6)$.

## 7.7.4  Reuse Distance and Cluster Size

Let $W_w$ and $W_i$ be the received power of the wanted and interfering signals at the mobile located at distances $d_w$ and $d_i$ from the wanted and interfering base stations, respectively. It has been demonstrated in Chapter 3 (Equation

---

*$p(0)$, although larger than $p(1)$, does not influence the result, because $p(CI|0) = 0$.

3.60a) that the following relation holds:

$$\frac{W_w}{W_i} = \left(\frac{d_w}{d_i}\right)^{-\alpha} \tag{7.50}$$

where $\alpha$ is an attenuation constant varying within the range from 2 to 4. It is obvious that

$$\frac{s_m}{i_m} = \frac{10^{S_m/10}}{10^{I_m/10}} = \frac{W_w}{W_i} = \left(\frac{d_w}{d_i}\right)^{-\alpha} \tag{7.51}$$

Now, (1) let $D$ be the distance between the wanted and interfering base stations, and (2) let $R$ be the cell radius. The cochannel interference worst case occurs when the mobile is positioned at the boundary of the cell, i.e., $d_w = R$ and $d_i = D - R$. In this case

$$\frac{d_i}{d_w} = \frac{D - R}{R} = \frac{D}{R} - 1 \tag{7.52}$$

where $D/R$ is the well-known *reuse distance* and is intimately related to the cell pattern of the system. This relation is given by (refer to Chapter 2, Equation 2.8)

$$\frac{D}{R} = \sqrt{3N} \tag{7.53}$$

where $N$ is the number of cells per cluster.

Using Equation 7.45 or 7.44, and Equations 7.51 through 7.53, we obtain

$$\frac{D}{R} = 1 + 10^{(Z+R_r)/10\alpha} = \sqrt{3N} \tag{7.54}$$

or

$$Z + R_r = 10\alpha \log(\sqrt{3N} - 1)$$

It has also been shown in Chapter 2 (Equation 2.7) that $N$ will only assume specific values, such as $1, 3, 4, 7, 9, 12, 13, \ldots$, given by the relation

$$N = i^2 + ij + j^2 \tag{7.55}$$

where $i$ and $j$ are integers.

Now let us compare the outage probability for some different cluster sizes. This can be done by means of Equation 7.54 and the curves of Figure 7.13. The results are shown in Table 7.1, where we assume a protection ratio $R_r = 0$ dB. The protection ratio depends on the modulation scheme and

TABLE 7.1.   Probability of Cochannel Interference in Different Cell Clusters

| N | Z + R_r (dB) | Outage Probability (%) | | | |
| | | Channel Occupancy p = 75% | | Channel Occupancy p = 100% | |
| | | σ = 0 dB | σ = 6 dB | σ = 0 dB | σ = 6 dB |
|---|---|---|---|---|---|
| 1 | − 4.74 | 100 | 100 | 100 | 100 |
| 3 | 10.54 | 31 | 70 | 40 | 86 |
| 4 | 13.71 | 19 | 58 | 26 | 74 |
| 7 | 19.40 | 4.7 | 29 | 7 | 42 |
| 12 | 24.46 | 1 | 11 | 2.1 | 24 |
| 13 | 25.19 | 0.9 | 9 | 1.9 | 22 |

varies typically from 8 dB (FM with 25-kHz spacing) to 20 dB (SSB modulation). Consequently, the results of Table 7.1 are in fact an underestimate of the true values of the outage probabilities. However, these true values can be easily determined, again with the use of both Equation 7.54, for a given protection ratio, and Figure 7.13.

Note that the influence of the traffic load in the outage probability is more significant for smaller clusters. Nevertheless, the smaller clusters are not feasible due to the unacceptable level of cochannel interference. For the larger clusters the traffic load is not as critical, but attention must be given to the effect of shadowing. Note that there is a remarkable difference between the outage probabilities when shadowing is not considered ($\sigma = 0$ dB) and when it is considered ($\sigma \neq 0$ dB). The increase of the standard deviation increases the cochannel interference probability. This shows that some kind of macroscopic diversity technique may be necessary to minimize the effect of shadowing.

### 7.7.5   Cochannel Interference and Simulation

This section is based on the work of Heeralall.[18]

As we have seen, the outage probability calculations are not simple at all, and they become even more complex when multiple interferers are involved. However, an approximate estimation can be accomplished with the use of an extended version of Equation 7.50. Because the effect of multiple interference is additive, it follows that, for $n$ interferers,

$$\frac{s_m}{i_m} = \frac{W_w}{W_i} = \frac{d_w^{-\alpha}}{\sum_{j=1}^{n} d_i^{-\alpha}} \tag{7.56}$$

The ratio $s_m/i_m$ is calculated for fixed mobiles, usually positioned for worst-case performance (i.e., $d_w = R$ and $d_i = D - R$) and ignoring fading

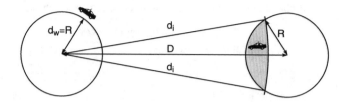

**Figure 7.14.** Illustration of the arrangement to obtain the distribution of the ratio of the signal to the cochannel interference for a single interferer.

effects. In order to get an insight into how the distribution of the carrier-to-cochannel interference ratio can be obtained by means of simulation, consider an elementary system with a single interferer as shown in Figure 7.14. Here only one degree of freedom, namely, the mobility of the interferer, is allowed.

The wanted mobile is fixed at the perimeter of the cell for minimum wanted signal. The probability that the interfering mobile is at a distance less than or equal to $d_i$ is equal to the proportion of the shaded area $S(d_i)$, i.e.,

$$p\left[CI \le \left(\frac{R}{d_i}\right)^{-\alpha}\right] = \frac{1}{4\pi R^2} \int_{D-R}^{d_i} dS(d_I) = \frac{S(d_i)}{4\pi R^2} \qquad (7.57)$$

because $S(D - R) = 0$, with $D - R \le d_i \le D + R$. The area $S(d_i)$ can be easily calculated as a function of $d_i$, $D$, and $R$. This cumulative distribution is obtained from a set of values of $s_m/i_m = (R/d_i)^{-\alpha}$ corresponding to different positions of the interfering mobile and the probability that the mobile is at that position. If another degree of freedom is allowed, say, the mobility of the wanted mobile, then a double integration is needed, i.e.,

$$p\left[CI \le \left(\frac{d_w}{d_i}\right)^{-\alpha}\right] = \frac{1}{\left(4\pi R^2\right)^2} \int_0^{d_w} dS(d_w) \int_0^{d_i} dS(d_I)$$

$$= \frac{S(d_w)S(d_i)}{\left(4\pi R^2\right)^2} = \frac{4\pi d_w^2 S(d_i)}{\left(4\pi R^2\right)^2} \qquad (7.58)$$

with $0 \le d_w \le R$ and $D - R \le d_i \le D + R$.

If more degrees of freedom—namely, Rayleigh fading, shadowing, and channel occupation—are allowed, the computations as well as the analysis of the results become extremely difficult. The difficulty is obviously increased with increasing number of interferers. The analysis in this case may be eased with the use of Monte Carlo simulation in a computer. The simulation consists of a large number of snapshots of the system layout where both the position of the mobiles (consequently the respective mean signal strengths) and the status of the corresponding signal (active or inactive) vary according to a given distribution. The basic steps used in Reference 18 are as follows:

1. The wanted signal is always active. Each interfering signal is set as active or inactive by sampling from a probability distribution that is derived from the channel occupation.
2. Step 2 is the determination of area mean signal power in decibels. It is different for the two directions of transmission between mobile and base. The two scenarios are described separately:

    a. *Mobile-to-base case*—For each active signal, the mobile must first be positioned randomly within its cell. Then the area mean signal strength from each mobile to the wanted base station must be calculated.

    b. *Base-to-mobile case*—The wanted mobile is positioned. Then mean signal strength from each base station to the wanted mobile is calculated.
3. Lognormal shadowing is simulated by sampling from a normal distribution (of standard deviation as set at the start of the simulation) and mean equal to the area mean signal strength in decibels. This gives a local mean strength in decibels.
4. Rayleigh fading is simulated in a similar manner. The local mean signal strength is converted to linear units (watts) and then equated to the mean of a negative exponential distribution from which the actual signal strength (in watts) is sampled.
5. Carrier-to-cochannel interference ratios are calculated as power ratios and then converted to decibels.
6. Histograms are updated.

Special simulation programs (i.e., SIMSCRIPT, SIMULA, etc.) can be used to facilitate the task of dealing with samples of probability distributions.

## 7.8   NOISE AND INTERFERENCE COUNTERACTIONS

The problem of additive noise is not confined to mobile radio systems. It is present in any communication network, and the measures to counteract its effect are well known in the literature. Basically, the strategy is to improve the equipment performance by designing better filters and by adjusting signal power levels. The multiplicative noise, on the other hand, constitutes a big problem in cellular mobile radio systems and the way to combat its effect is to use macroscopic and microscopic diversity as described in Chapter 5.

Adjacent-channel interference (as well as the additive noise) is a well-known problem of communication systems. However, its importance in cellular systems has been accentuated due to the spectrum efficiency improvement "neurosis." Adjacent-channel interference can be minimized by (1) improving the modulation techniques, (2) improving the filtering quality of the equipment, and (3) allowing some guard channels between channels allocated to each cell. By far, the biggest problem of all is cochannel interference, an intrinsic issue of the cellular network. We will dedicate the rest of this section to describing the method of combatting this kind of interference.

There is a general consensus that the use of directional antennas substantially improves the signal-to-cochannel-interference ratio. Heeralall[18] sums up the fundamental operation of directional antennas as follows:

"By restricting the reception of signals to directions where they are really needed, directional antennas cut down the number of significant cochannel interferers without affecting the level of carrier received. By restricting radiation of signals to directions where they are really needed, directional antennas minimize the total level of cochannel interference reaching a receiver. The result is a higher carrier–to–cochannel interference ratio."

Consequently, if the signal–to–cochannel-interference ratio is improved, a lower $D/R$ ratio becomes feasible and the implementation of a system with fewer hexagons per cluster can be attainable.

The performance of a cellular system with directional antennas may be evaluated in the same way as before, but Equation 7.50 must be modified to include the gain of the antennas in each direction as follows:

$$\frac{W_w}{W_i} = \frac{G_{Tw}(\theta_{Tw})G_{Rw}(\theta_{Rw})}{G_{Ti}(\theta_{Ti})G_R(\theta_{Ri})}\left(\frac{d_w}{d_i}\right)^{-\alpha} \tag{7.59}$$

where   $\theta_{XY}$ is the direction of the ray at antenna $X$ with respect to the direction of its maximum gain;

$G_{XY}(\theta_{XY})$ is the gain of antenna $X$ at the angle $\theta_{XY}$;

and $X = T$ for transmitting antenna or $X = R$ for receiving antenna;
$Y = w$ for the wanted antenna or $Y = i$ for the interfering antenna.

A diagram illustrating the arrangement for the cochannel interference calculations in a two-sector cell system is shown in Figure 7.15.

In such a simplified model, the wanted mobile is positioned at the edge of the sector and at a distance $R$ from the base station; $\varphi_1$ and $\varphi_2$ are the (independent) orientations of the wanted antenna and interfering antenna, respectively. The interfering mobile can move within its sector and for each position, the ratio $s_m/i_m$ can be calculated. This can be done by varying the vector $d_i$ from $D - R$ to $D + R$ in steps of $\Delta d_i$ (defined as desired) and the angle $\theta_i$ from $-\tan^{-1}(R/D)$ to $\tan^{-1}(R/D)$ in steps of $\Delta\theta_i$ (defined as desired). We may suppose that, if the mobile is within the sector, then there

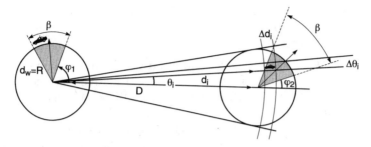

**Figure 7.15.** Illustration of the arrangement to obtain the distribution of the ratio of the signal to the cochannel interference in a two-sector cell system.

is interference; otherwise the interference is nil. The corresponding distribution can be obtained by averaging the respective area within the sector over $4\pi R^2$. This distribution is a function of $R$, $D$, $\varphi_1$, $\varphi_2$, and the antenna radiation pattern (beamwidth $\beta$).

When multiple interferers are involved, the problem becomes much more complex and a computer simulation is preferred.

Table 7.2 shows the results for some selected cellular patterns.[18, 19]

The sectorized patterns described in the literature are derived from the basic cell site lattice as $(N \times s)$, where $N$ is the number of cell sites per cluster and $s$ is the number of sectors per cell site. The most common ones are $(7 \times 3)$, $(4 \times 6)$, and $(3 \times 6)$ sectors, where the cluster is still composed of a contiguous group of hexagons. Some novel cell configurations based on noncontiguous clusters have been proposed by Heeralall[18, 19] and they are briefly described here. The general idea for generating these patterns is as follows:

"Place interfering sectors around a wanted sector such that there is a maximum spatial packing. Then each interferer is positioned and oriented to achieve the stated objective. Adjust that arrangement of cosectors until it looks like part of an infinite pattern."[19]

**TABLE 7.2. Performance of Some Selected Cellular Patterns**

| Cellular Pattern (cells / cluster) × (channel sets / cell) | Carrier-to-Cochannel Interference (dB) | |
|---|---|---|
| | First Decile | Median |
| 12 × 1 | 12.0 | 23.2 |
| 7 × 3 | 14.5 | 27.0 |
| 7 × 1 | 8.0 | 19.2 |
| 4 × 6 | 13.5 | 26.3 |
| 3 × 6 | 12.0 | 24.2 |

An example of a new cell configuration is given in Figure 7.16a. This pattern has been set up as follows:

- The first interferer was placed as near as possible to the wanted sector (for minimum spatial packing). It was positioned outside the major lobe, pointing into the opposite direction with respect to the wanted sector (for minimum interference).
- To build a regular pattern, a grid of 60° sectors was used and the sectors were placed only in positions that coincided with the grid and that allowed for an infinite regular structure.
- Another interferer was placed symmetrically with respect to the wanted sector. Then a row pattern with three channel sets per row emerged (i.e., $x$-direction expansion).
- To move in the $y$ direction and to obtain a two-dimensional pattern, a distance of four rows was tentatively found to meet the interference levels requirements. Consequently, a pattern of 12 channel sets (4 rows of 3 sets each) was set up.

Similarly, another pattern was tried, now allowing for coverage areas to overlap, (Figure 7.16b). The difference between these two patterns is that each set now serves four grid sectors instead of one, but each grid sector is also served by three other channel sets.

Note that in this new approach,

- reuse distance varies in different directions from any given sector;
- cosectors do not necessarily point into the same direction;
- clusters with cosectors in different directions have different shapes and may have different internal arrangements;
- each 60° sector is one-sixth of a hexagon, but the number of hexagons per cluster can vary from 1 and can have noninteger values as well, because different sets of cosectors are allowed to overlap;
- each cell site serves only two sectors that are 180° apart.

It is possible to obtain other patterns by visual inspection: in general, these patterns have fewer than 3 hexagons per cluster and fewer than 18 channel sets per cluster. It means that there is an improvement in both the spatial and time efficiencies (trunking efficiency is improved). In the example given in Figure 7.16a there are 2 hexagons per cluster and 12 channel groups per cluster. This obviously contrasts with the widely accepted formula $i^2 + ij + j^2 = N$, where this configuration is not allowed.

Note that each cell site serves only two 60° sectors that are 180° apart, so only one-third of the cell site angular range is being used. The reuse pattern is nonisotropic and, therefore, to maintain the cluster, all orientations are constrained to two possibilities only, leading to six cell sites per cluster.

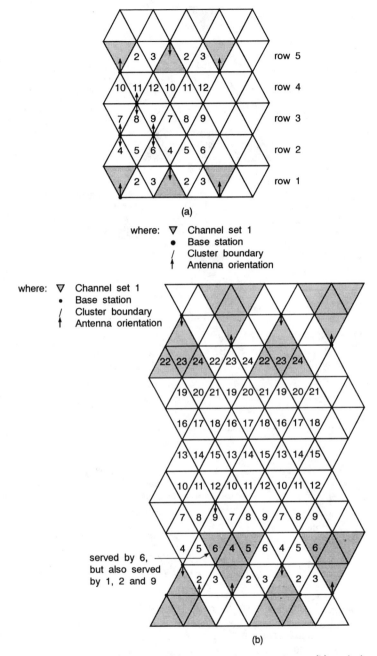

**Figure 7.16.** Novel sectorized patterns: (a) 12 channel sets per cluster; (b) equivalent to 6 channel sets per cluster. (*Source*: After S. Heerall and C. J. Hughes, High capacity cellular patterns land mobile radio systems using directional antennas, *IEE Proceedings*, 136(1), Part 1, 75–80, February 1989.)

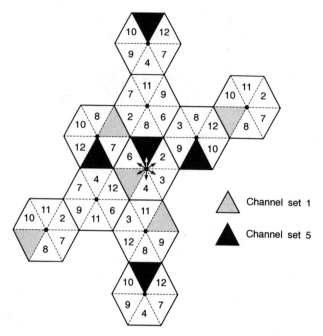

**Figure 7.17.** Modified version of Figure 7.16a. (*Source*: After S. Heeralall and C. J. Hughes, High capacity cellular patterns land mobile radio systems using directional antennas, *IEE Proceedings*, 136(1), Part 1, 75–80, February 1989.)

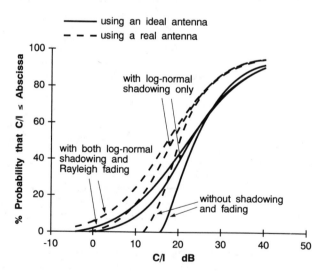

**Figure 7.18.** $C/I$ cumulative distribution functions for the cellular pattern of Figure 7.17 (transmission from base to mobile). (*Source*: After S. Heeralall and C. J. Hughes, High capacity cellular patterns land mobile radio systems using directional antennas, *IEE Proceedings*, 136(1), Part 1, 75–80, February 1989.)

This constraint can be removed if clusters are allowed to have noncontiguous cells. Rather than filling up a service area by building up around one set of cosectors, we can replicate the basic reuse pattern as many times as necessary and rotate it each time around one cell site. Then this procedure can be repeated around other cell sites with vacant positions. The resultant cellular pattern will have fewer cell sites, as illustrated in Figure 7.17.

These patterns have been developed heuristically and they lead to a substantial theoretical improvement in traffic-carrying capacity. The first decile of carrier–to–cochannel-interference has been determined to be 11 dB and the median 22.5 dB for a channel occupation of 68.8% (Figure 7.16a). The carrier-to-cochannel cumulative distribution of the pattern of Figure 7.17 is shown in Figure 7.18.

## 7.9  SUMMARY AND CONCLUSIONS

Noise and interference are limiting factors of good performance of any communication system. Their effects are usually very similar, but in mobile radio systems, noise and interference can be treated completely apart from each other.

Noise can modify the signal in an additive way or in a multiplicative way. The most significant type of additive noise is that caused by human activities, e.g., use of electric motors, neon lights, power lines, medical appliances, etc. It is possible to minimize its effect by designing better filters and adjusting the signal power levels. Fading can be considered as another type of noise that degrades the signal in a multiplicative manner. The ways of counteracting fading have already been extensively studied in Chapters 5 and 6.

Interference in a mobile radio system is a more complex problem to investigate. Adjacent-channel interference is usually minimized by avoiding the use of adjacent channels within the same cell. Adjacent channels in adjacent cells, however, are sometimes inevitably used because of the adopted cell pattern. In this case the presence of active mobiles near the cell borders substantially increases the probability of interference. Consequently, the probability of adjacent-channel interference depends not only on the geographical distribution of the mobiles but also on the traffic profile of the system. A heavily loaded system is more likely to have problems of adjacent-channel interference.

The biggest problem of all is cochannel interference, and this is an intrinsic issue of the cellular architecture. The determination of the cochannel interference must be carried out on a statistical basis due to the random factors involved in the process. Several studies have been made in this field, but they usually treat the problem in a simplified way, allowing very few degrees of freedom (mobiles positioned for the worst case, wanted and interfering mobiles fixed on a given position, etc.). Even in simplified form, the problem is still extremely complicated to deal with, because it involves either multiple

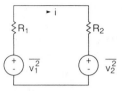

**Figure 7A.1.**   Two resistors exchanging power.

integrals (for multiple interferers) or the determination of the equivalent distribution of multiple interferers.

An alternative to the numerical analysis necessary to solve the multiple integrals is the use of Monte Carlo simulation on computer. Heeralall[18, 19] has used this approach extensively, with the flexibility of allowing all the degrees of freedom in an extremely easy manner, using the facility of an appropriate simulation program. The result was that not only was the carrier-to-cochannel interference problem investigated, but also novel cellular patterns were discovered with much better performance than that of the traditional patterns.

## APPENDIX 7A   THERMAL NOISE

Let $\overline{v_1^2}$ and $\overline{v_2^2}$ be the mean square noise voltages of two resistors $R_1$ and $R_2$, respectively, connected as shown in Figure 7A.1. If the resistors are at the same temperature, then in a given time interval each must, on average, receive as much energy as it delivers to its partner.[20] Hence

$$\frac{\overline{v_1^2}}{R_1} = \frac{\overline{v_2^2}}{R_2} \tag{7A.1}$$

From Equation 7A.1 we conclude that for thermal noise, $\overline{v^2} \propto R$. Moreover, in equilibrium, the balance of power must hold for any frequency band. Hence, the mean power exchange per unit bandwith is a function only of the frequency band $B$, resistance $R$, and temperature $T$. Nyquist[4] determined that the power delivered by the resistor to a matched load is equal to $kTB$, where $k$ is Boltzmann's constant ($k = 1.38 \times 10^{-23}$ J/K). Maximum noise power transfer will occur when $R_1 = R_2 = R$. In this case, the total power dissipated in $R$ is $\overline{R_i^2}$. From Figure 7A.1,

$$\overline{R_i^2} = R\overline{\left(\frac{v_1 - v_2}{2R}\right)^2} = \frac{\overline{v_1^2}}{4R} + \frac{\overline{v_2^2}}{4R} \tag{7A.2}$$

because $\overline{v_1 v_2} = \overline{v_1}\, \overline{v_2} = 0$.

The term $\overline{v_1^2}/4R$ corresponds to the mean power received by $R$ due to $v_1$. The same reasoning applies to the term $\overline{v_2^2}/4R$. Therefore, the power delivered by one resistor is $\overline{v^2}/4R$, where $\overline{v^2}$ is its mean square noise voltage. This power, obviously, equals that estimated by Nyquist. Hence

$$\frac{\overline{v^2}}{4R} = kTB$$

or, equivalently,

$$\overline{v^2} = 4kTBR \qquad (7A.3)$$

The power spectral density of $v$ is $S_v(\omega)$ such that

$$S_v(\omega) = \overline{v^2}/2B$$

Therefore,

$$S_v(\omega) = 2kTR \qquad (7A.4)$$

The power spectral density of the current $i$ is

$$S_i(\omega) = \frac{S_v(\omega)}{R^2} = \frac{2kT}{R} \qquad (7A.5)$$

## APPENDIX 7B   SHOT NOISE[6]

Let $d$ be the distance between the anode and cathode of a vacuum tube and let $V$ be the supply voltage. An electron $q$ with mass $m$ will travel with speed $v$ within a time $t$ such that

$$m\frac{v}{t} = qE = q\frac{V}{d}$$

where $E$ is the electric field intensity. Then

$$v = \frac{qV}{md}t \qquad (7B.1)$$

If the induced electric charge in the anode is $Q$, then using the balance of energy

$$\tfrac{1}{2}mv^2 = QV \qquad (7B.2)$$

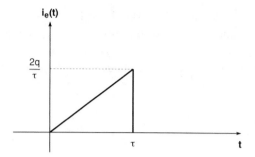

**Figure 7B.1.** Shape of the current from the cathode to the anode.

Therefore, with Equation 7B.1 in Equation 7B.2 we have

$$Q = \frac{1}{2} \frac{q^2 V}{md^2} t^2 \tag{7B.3}$$

The resultant current is

$$i_e(t) = \frac{dQ}{dt} = \frac{q^2 V}{md^2} t = \frac{2qt}{\tau^2} \tag{7B.4}$$

where $\tau = [2m/(qV)]^{1/2} d$ is the time taken by the electron to reach the anode.

Note that the current increases with $t$, but falls abruptly to zero when the electron reaches the anode, as shown in Figure 7B.1.

The triangular function of Figure 7B.1 can be written as

$$i_e = \frac{2q}{\tau^2} [tu(t) - \tau u(t - \tau) - (t - \tau)u(t - \tau)] \tag{7B.5}$$

where $u(t) = 1$ for $t > 0$ and $u(t) = 0$, otherwise.

The Fourier transform of $i_e(t)$ is $I_e(\omega)$ such that

$$I_e(\omega) = \frac{2q}{-(\omega\tau)^2} [1 - \exp(-j\omega\tau) - j\omega\tau \exp(-j\omega\tau)] \tag{7B.6}$$

Because the total current consists of the superposition of the $n$ individual* currents $i_e(t)$, then

$$S_i(\omega) = n|I_e(\omega)|^2$$

---

*$n$ is the mean number of electrons per second.

Accordingly,

$$S_i(\omega) = \frac{4I_0 q}{(\omega\tau)^4}\left[(\omega\tau)^2 + 2(1 - \cos\omega\tau - \omega\tau\sin\omega\tau)\right] \qquad (7B.7)$$

where $I_0 = nq$ is the mean current.

For $\omega\tau \leq 0.5$ the spectrum is approximately flat and equal to $I_0 q$. Therefore,

$$\overline{v^2} = 2BS_v(\omega) = 2BR^2 S_i(\omega) = 2qI_0 R^2 B \qquad (7B.8)$$

## APPENDIX 7C    SIGNAL-TO-NOISE RATIO AT THE OUTPUT OF A RECEIVER

The aim of this appendix is to determine the probability distribution of the SNR at the output of a receiver having a noise figure equal to $f$. The same distribution at the input of the receiver has already been determined in Chapter 5 (Section 5.6) and is given by

$$p(\gamma_i \leq \Gamma_i) = P(\Gamma_i) = 1 - \exp\left(-\frac{\Gamma_i}{\gamma_m}\right) \qquad (7C.1)$$

where $\gamma_m$ is the mean SNR, $\gamma_i$ is the SNR, and $\Gamma_i$ is a given SNR level at the receiver input.

At the output of the receiver, however, there is an additional noise introduced by the amplifier. The grade of deterioration is given by the noise figure $f$, defined as

$$f = \frac{\text{SNR at the input}}{\text{SNR at the output}} = \frac{S_i/N_i}{S_0/N_0} \triangleq \frac{\gamma_i}{\gamma_0} \qquad (7C.2)$$

where $S_i$ and $N_i$ are, respectively, the signal and noise powers at the input of the amplifier. Similarly, $S_0$ and $N_0$ are, respectively, the signal and noise powers at the output of the amplifier. If the power gain of the amplifier is $G$, then $S_i/S_0 = 1/G$ and

$$f = N_0/GN_i \qquad (7C.3)$$

But $N_0 = G(N_i + N_a)$, where $N_a$ is the noise power introduced by the amplifier. Thus,

$$f = 1 + N_a/N_i \qquad (7C.4)$$

From Equation 7C.2,

$$\gamma_0 = \gamma_i/f \qquad (7C.5)$$

The probability density of $\gamma_i$ is (refer to Equation 5.17)

$$p(\gamma_i) = \frac{1}{\gamma_m} \exp\left(-\frac{\gamma_i}{\gamma_m}\right) \tag{7C.6}$$

Then, by changing variables,

$$p(\gamma_0)|d\gamma_0| = p(\gamma_i)|d\gamma_i| \tag{7C.7}$$

Using Equations 7C.5 and 7C.6 in Equation 7C.7, we obtain

$$p(\gamma_0) = \frac{f}{\gamma_m} \exp\left(-f\frac{\gamma_0}{\gamma_m}\right) \tag{7C.8}$$

The corresponding probability distribution is

$$\text{prob}(\gamma_0 \leq \Gamma_0) = P(\Gamma_0) = \int_0^{\Gamma_0} p(\gamma_0) \, d\gamma_0 = 1 - \exp\left(-f\frac{\Gamma_0}{\gamma_m}\right) \tag{7C.9}$$

Note from Equation 7C.9 that there is a deterioration of the SNR for any value of the noise figure different from unity (i.e., $f \geq 1$).

## APPENDIX 7D    PROBABILITY OF ADJACENT-CHANNEL INTERFERENCE

The aim of this appendix is to determine the probability of occurrence of adjacent-channel interference as a function of the following parameters:

1. Proportion of active channels with only one adjacent channel in use ($p_1$)
2. Proportion of active channels with two adjacent channels in use ($p_2$)
3. Proportion of mobiles within the weak signal area of the cell ($\mu'$)
4. Proportion of mobiles with access to two or more base stations ($\gamma$)
5. Proportion of mobiles with access to three or more base stations ($\delta$)

Note that the parameters $p_1$, $p_2$, $\mu'$, $\gamma$, and $\delta$ constitute probability distributions. The probabilities $p_1$ and $p_2$ are functions of the traffic load and the channel assignment algorithms, whereas $\mu'$, $\gamma$, and $\delta$ are strictly related with the mean signal strength (the reader is referred to Section 7.6 for a better understanding of these probabilities). In our analysis we assume that adjacent channels are assigned to adjacent cells.

Adjacent-channel interference is more likely to occur when the mobile is near the cell borders. Consider three neighboring cells as shown in Figures 7D.1 and 7D.2, where the dotted circular lines delimit the base station service areas. Note that regions 2 and 4 represent overlapping areas where

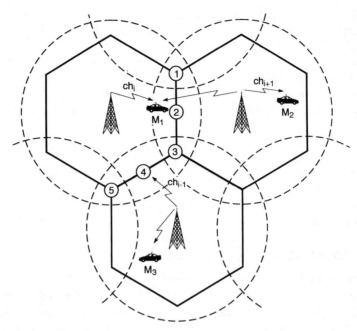

**Figure 7D.1.**   Adjacent-channel interference at the mobile.

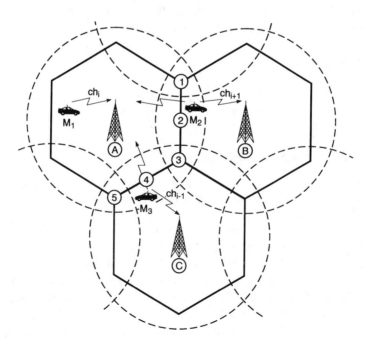

**Figure 7D.2.**   Adjacent-channel interference at the base station.

the mobiles are served by two base stations. In the same way, regions 1, 3, and 5 represent overlapping areas where the mobiles are served by three base stations.

The probability that a mobile is served by two base stations only is $\gamma - \delta$ for all of the six borders of the cell. If we consider only one border then this probability is $(\gamma - \delta)/6$. Similarly, the probability that a mobile is served by three base stations is $\delta$, whereas this is $\delta/6$ if we consider only one border of the cell.

## 7D.1 Adjacent-Channel Interference at the Mobile Station

Consider the situation as depicted in Figure 7D.1. We shall examine two cases as follows.

### 7D.1.1 *One Active Adjacent Channel*

Let channel $i$ be an active channel in a given cell. Let one of its adjacent channels ($i + 1$ or $i - 1$) be active in one of the adjacent cells (event with probability $p_1$). For ease of understanding, assume that the mobile $M_1$ uses channel $i$ whereas either $M_2$ uses channel $i + 1$ or $M_3$ used channel $i - 1$. Interference is likely to occur if

1. $M_1$ lies in region 1 (event with probability $\delta/6$), or
2. $M_1$ lies in region 2 (event with probability $(\gamma - \delta)/6$), or
3. $M_1$ lies in region 3 (event with probability $\delta/6$)

Hence the probability of event 1 is

$$p_1\left(\frac{\gamma - \delta}{6} + 2\frac{\delta}{6}\right)$$

### 7D.1.2 *Two Active Adjacent Channels*

We assume the same channel assignment as before, with the difference that all of the three channels are active (event with probability $p_2$). Interference is likely to occur if

1. $M_1$ lies in region 1 (event with probability $\delta/6$), or
2. $M_1$ lies in region 2 (event with probability $(\gamma - \delta)/6$), or
3. $M_1$ lies in region 3 (event with probability $\delta/6$—however, because the two adjacent channels may simultaneously interfere, this event contributes with $2\delta/6$ to the total probability), or
4. $M_1$ lies in region 4 (event with probability $(\gamma - \delta)/6$), or
5. $M_1$ lies in region 5 (event with probability $\delta/6$)

Then the probability of event 2 is

$$p_2\left(2\frac{\gamma - \delta}{6} + 4\frac{\delta}{6}\right)$$

Adjacent-channel interference at the mobile will occur with probability $p_M$ such that

$$p_M = p_1\left(\frac{\gamma - \delta}{6} + 2\frac{\delta}{6}\right) + p_2\left(2\frac{\gamma - \delta}{6} + 4\frac{\delta}{6}\right)$$

or, equivalently,

$$p_M = \left(\frac{\gamma + \delta}{3}\right)\left(\frac{p_1 + 2p_2}{2}\right) \qquad (7D.1)$$

## 7D.2   Adjacent-Channel Interference at the Base Station

Adjacent-channel interference at the base station is likely to occur when the home mobile is transmitting from a weak signal area and one or two "away" mobiles are transmitting in adjacent channels from the overlapping areas as shown in Figure 7D.2. As in the previous case, we shall consider one and two active adjacent channels. Moreover, the home mobile is always considered to be in the weak signal area (event with probability $\mu'$) and uses channel $i$.

### 7D.2.1 *One Active Adjacent Channel*

The analysis in this case is very similar to that given in Section 7D.1, item 1. Hence, the probability of this event is

$$\mu' p_1\left(\frac{\gamma - \delta}{6} + 2\frac{\delta}{6}\right)$$

### 7D.2.2 *Two Active Adjacent Channels*

There are various situations to be analyzed in this case. The analysis is a bit more intricate and the possible situations are summarized in Table 7D.1. The probability of this event is

$$\mu' p_2 \sum (\text{probabilities given in Table 7D.1})$$

$$= \mu' p_2\left(\frac{2\gamma + 2\delta}{6}\right) \quad (\text{believe or not!})$$

**TABLE 7D.1.   Possible Situations with Two Adjacent Channels in Use**

| $M_2$ is in Region | $M_3$ is in Region | Number of Interferers | Probability of the Event |
|---|---|---|---|
| 1 | 3 | 2 | $2(\delta/6)^2$ |
| 1 | 4 | 2 | $2(\delta/6)(\gamma-\delta)/6$ |
| 1 | 5 | 2 | $2(\delta/6)^2$ |
| 1 | Elsewhere | 1 | $\left(\dfrac{\delta}{6}\right)\left(1-\dfrac{\delta+\gamma}{6}\right)$ |
| 2 | 3 | 2 | $2\left(\dfrac{\delta}{6}\right)\left(\dfrac{\gamma-\delta}{6}\right)$ |
| 2 | 4 | 2 | $2\left(\dfrac{\gamma-\delta}{6}\right)^2$ |
| 2 | 5 | 2 | $2\left(\dfrac{\delta}{6}\right)\left(\dfrac{\gamma-\delta}{6}\right)$ |
| 2 | Elsewhere | 1 | $\left(\dfrac{\gamma-\delta}{6}\right)\left(1-\dfrac{\delta+\gamma}{6}\right)$ |
| 3 | 3 | 2 | $2(\delta/6)^2$ |
| 3 | 4 | 2 | $2\left(\dfrac{\delta}{6}\right)\left(\dfrac{\gamma-\delta}{6}\right)$ |
| 3 | 5 | 2 | $2(\delta/6)^2$ |
| 3 | Elsewhere | 1 | $\left(\dfrac{\delta}{6}\right)\left(1-\dfrac{\delta+\gamma}{6}\right)$ |
| Elsewhere | 3 | 1 | $\left(\dfrac{\delta}{6}\right)\left(1-\dfrac{\delta+\gamma}{6}\right)$ |
| Elsewhere | 4 | 1 | $\left(\dfrac{\gamma-\delta}{6}\right)\left(1-\dfrac{\delta+\gamma}{6}\right)$ |
| Elsewhere | 5 | 1 | $\left(\dfrac{\delta}{6}\right)\left(1-\dfrac{\delta+\gamma}{6}\right)$ |

Adjacent-channel interference at the base station will occur with probability $p_B$ such that

$$p_B = \mu' p_1\left(\frac{\gamma-\delta}{6} + 2\frac{\delta}{6}\right) + \mu' p_2\left(\frac{2\gamma+2\delta}{6}\right)$$

or, equivalently,

$$p_B = \mu'\left(\frac{\gamma+\delta}{3}\right)\left(\frac{p_1+2p_2}{2}\right) \tag{7D.2}$$

## APPENDIX 7E    DISTRIBUTION OF THE SNR IN A LOGNORMAL FADING ENVIRONMENT

A lognormal fading with envelope $r$ has the following distribution, as given by Equation 3.78b,

$$p(r) = \frac{1}{\sqrt{2\pi}\, r\sigma_r} \exp\left[ -\frac{\ln^2(r/m)}{2\sigma_r^2} \right] \tag{7E.1}$$

Define $\gamma$ as the ratio of the local mean signal power and the mean noise power. In the presence of Gaussian noise having a mean power equal to $N$, the SNR is then

$$\gamma = \frac{r^2/2}{N} \tag{7E.2}$$

The distribution of $\gamma$ can be easily obtained by a simple change of variables as follows:

$$p(\gamma)|d\gamma| = p(r)|dr| \tag{7E.3}$$

Using Equations 7E.3, 7E.2, and 7E.1, we obtain

$$p(\gamma) = \frac{N}{\sqrt{2\pi}\, r^2\sigma_r} \exp\left[ -\frac{\ln^2\left(2\gamma N/m^2\right)^{1/2}}{2\sigma_r^2} \right] \tag{7E.4}$$

Defining

$$\gamma_m \triangleq \frac{m^2/2}{N} \tag{7E.5}$$

it is straightforward to show that with Equation 7E.5 in Equation 7E.4 we obtain

$$p(\gamma) = \frac{1}{\sqrt{2\pi}\, \sigma_\gamma \gamma} \exp\left[ -\frac{\ln^2(\gamma/\gamma_m)}{2\sigma_\gamma^2} \right] \tag{7E.6}$$

where $\sigma_\gamma = \gamma\sigma_r$.

## APPENDIX 7F    OUTAGE PROBABILITY FOR THE "INTERFERENCE-ONLY" CASE

If no minimum signal is required ($s_0 = 0$), then from Equation 7.34

$$\Delta p_1 = p_1 - p_0 = p_1$$

$$\Delta p_2 = p_2 - p_1$$

$$\Delta p_3 = p_3 - p_2$$

$$\vdots$$

$$\Delta p_n = p_n - p_{n-1}$$

By adding all of these equations we have

$$p_n = \sum_{j=1}^{n} \Delta p_j$$

Now from Equation 7.35

$$\Delta p_1 = \frac{1}{1 + \Lambda_1/r}$$

$$\Delta p_2 = \frac{\Lambda_1/r}{(1 + \Lambda_1/r)(1 + \Lambda_2/r)}$$

$$\vdots$$

$$\Delta p_n = \frac{(\Lambda_1/r)(\Lambda_2/r) \cdots (\Lambda_{n-1}/r)}{(1 + \Lambda_1/r)(1 + \Lambda_2/r) \cdots (1 + \Lambda_n/r)}$$

Then

$$p_n = \sum_{j=1}^{n} \prod_{k=1}^{j} \frac{\Lambda_{k-1}/r}{1 + \Lambda_k/r}, \quad \text{with } \Lambda_0 \triangleq 1$$

If all of the interferers are assumed to interfere equally with the wanted signal, then $\Lambda_1 = \Lambda_2 = \cdots = \Lambda_k = \Lambda$.

Therefore,

$$p_n = \frac{1}{\Lambda/r} \sum_{j=1}^{n} \left( \frac{\Lambda/r}{1 + \Lambda/r} \right)^j$$

This sum represents the sum of a geometric progression having a ratio equal to $(\Lambda/r)/(1 + \Lambda/r)$ that is less than 1. Then

$$p_n = 1 - \left( \frac{\Lambda/r}{1 + \Lambda/r} \right)^n$$

## REFERENCES

1. Watt, A. D., Coom, R. M., Maxwell, E. L., and Plush, R. W., Performance of some radio systems in the presence of thermal and atmospheric noise, *Proc. IRE*, 46, 1914–1923, December 1958.
2. *Reference Data for Radio Engineers*, 6th ed., Howard W. Sams & Co., Indianapolis, IN, 1979.
3. Johnson, J. B., Thermal agitation of electricity in conductors, *Phys. Rev.*, 32, 97–109, July 1928.
4. Nyquist, H., Thermal agitation of electric charge in conductors, *Phys. Rev.*, 32, 110–113, July 1928.
5. Lathi, B. P., *Random Signal and Communication Theory*, International Textbook Co., Scranton, PA, 1968.
6. Lathi, B. P., *Communication Systems*, John Wiley & Sons, New York, 1968.
7. Parsons, J. D., and Gardiner, J. G., *Mobile Communication Systems*, Blackie and Son Ltd., Glasgow, 1989.
8. Shepherd, R. A. et al., Measurement parameters for automobile ignition noise, Stanford Research Institute, NTIS PB 247766, June 1975.
9. Diney, R. T., and Spaulding, A. D., Amplitude and time-statistics of atmospheric and man-made ratio noise, Environment Science Services Administration Technical Report ERL ISO–ITS 98, February 1970.
10. Sánchez V., J. H., Traffic performance of cellular mobile radio systems, Ph.D. thesis, University of Essex, England, June 1988.
11. Yacoub, M. D., Mobile radio with fuzzy cell boundaries, Ph.D. thesis, University of Essex, England, May 1988.
12. Hughes, C. J., and Appleby, M. S., Definition of a cellular mobile radio system, *IEE Proc.* 132(5), Part F, August 1985.
13. Hata, M., Kunoshita, K., and Hirade, K., Radio link design of cellular mobile systems, *IEEE Trans. Vehicular Tech.*, VT-31, 1982.
14. Sowerby, K. W., and Williamson, A. G., Outage probability calculation for multiple cochannel interferers in cellular mobile radio, *IEE Proc.*, 135(3), Part F, June 1988.

15. Yeh, Y. S., and Schwartz, S. C., Outage probability in mobile telephony due to multiple log-normal interferers, *IEEE Trans.* COM-32, 380–388, 1984.

16. Fenton, L. F., The sum of log-normal probability distributions in scatter transmission systems, *IRE Trans. Commun. Syst.*, CS-8, 57–67, March 1960.

17. Muammar, R., and Gupta, S. C., Cochannel interference in high capacity mobile radio systems, *IEEE Trans. Commun.* COM-30(8), August 1982.

18. Heeralall, S., The applications of direction antennas in cellular mobile radio systems, Ph.D. thesis, University of Essex, July 1988.

19. Heeralall, S., and Hughes, C. J., High capacity cellular patterns land mobile radio systems using directional antennas, *IEE Proce.*, 136(1), Part 1, February 1989.

20. Goodyear, C. C., *Signals and Informations*, Butterworth & Co. Ltd., London, 1971.

# CHAPTER 8

# Analog Modulation for Mobile Radio

This chapter is concerned with the performance of some analog modulation techniques, namely, AM, SSB, and FM, in a mobile radio environment. We start the chapter by defining the performance measure parameters, consisting basically of the signal-to-noise ratio and its variations. Thereafter, we analyze each technique separately, reviewing basic principles, the required transmission bandwidth, and ways of generating and detecting the modulated signals. We then formulate a general model for the received signal and analyze the performance of the modulation technique in the presence of additive and multiplicative noises. We show that multiplicative noise (fading) has a disastrous effect on AM and SSB systems and that the capture effect of the FM systems is not so prominent in the presence of fading. We also consider some emerging alternatives that may enhance the performance of the SSB systems, making them suitable for mobile radio applications.

## 8.1 INTRODUCTION

Modulation is a process by which one or more parameters of a carrier (amplitude or angle) are varied in accordance with a message signal (modulating wave). As a result of the modulation process, the frequency band of the message signal is shifted to a suitable region within the spectrum so that transmission over a communication channel becomes feasible. At the reception side the receiver carries out the inverse of the modulation process, named demodulation or detection, so that the message signal can be recovered.

First-generation mobile radio systems used analog modulation for voice transmission. The early systems started with standard amplitude modulation (AM), because this was the only feasible technique available at that time, but moved quickly into frequency modulation (FM) as soon as it was demonstrated that its practical implementation was viable. The single-sideband (SSB) technique was also experimented with in some mobile radio services. However, FM ended up predominating and today all the analog cellular systems have adopted this scheme of modulation for voice transmission.

Some enhanced forms of SSB, such as transparent tone in band (TTIB), feedforward signal regeneration (FFSR), and amplitude companded single sideband (ACSSB), have been proposed and intensively investigated, with a good chance of being used by some mobile radio services.

In this chapter we will introduce the basics of the main analog modulation techniques and analyze their performance in the presence of both additive and multiplicative noise, having in mind that the multiplicative is what mainly characterizes the mobile radio channel.

## 8.2  PERFORMANCE MEASURES OF MODULATION TECHNIQUES

Any communication channel of electrical systems is subject to random fluctuations of power known as noise. Noise is a limiting factor on the power required to transport information-bearing signals and may affect the demodulation process in different ways, according to the modulation technique that is used.

The performance of the modulation techniques can be evaluated by several measures, including (1) output signal-to-noise ratio ($SNR_O$), (2) channel signal-to-noise ratio ($SNR_C$), (3) signal-suppression noise ratio (SSN), (4) carrier-to-noise ratio ($\gamma$), and (5) figure of merit ($F$), defined as follows:

$$SNR_O = \frac{\text{average power of message signal at receiver output}}{\text{average power of noise at receiver output}} \tag{8.1}$$

$$SNR_C = \frac{\text{average power of modulated message signal}}{\text{average power of noise measured in message bandwidth}} \tag{8.2}$$

$$SSN = \frac{\text{average power of message signal at receiver output}}{\text{average power of signal-suppression noise at the receiver output}}$$

$$\tag{8.3}$$

$$\gamma = \frac{\text{average carrier power}}{\substack{\text{average noise power in bandwidth} \\ \text{of modulated signal at receiver input}}} \tag{8.4}$$

$$F = \frac{SNR_O}{SNR_C} \tag{8.5}$$

The SSN is relevant in the case where the signal is affected by multiplicative noise, when it is not possible to distinguish signal from noise. Accordingly, the noise $n_s(t)$ (termed signal-suppression noise) is taken as the difference between the output signal $y(t)$ (signal affected by noise) and the mean signal $\bar{y}(t)$, i.e.,

$$n_s(t) = y(t) - \bar{y}(t) \tag{8.6}$$

$$\text{SSN} = \frac{\langle \bar{y}(t)^2 \rangle}{\langle (y(t) - \bar{y}(t))^2 \rangle} \tag{8.7}$$

where $\langle \cdot \rangle$ denotes the time average.

In the analysis carried out in this chapter we shall use the notation $\tilde{x}$ to represent the signal $x$ in its complex form, such that $x = \text{Re}[\tilde{x}]$.

## 8.3  AMPLITUDE MODULATION

In this section we shall review the basic principles of the amplitude modulation technique and then investigate its performance in the presence of noise, first considering only Gaussian noise and later Rayleigh fading.

### 8.3.1  Basic Principles

Amplitude modulation is defined as a process in which the amplitude of the carrier wave is varied linearly with the message signal.[1] The standard form of the AM signal $s(t)$ is given by

$$s(t) = A[1 + k_a m(t)]\cos(2\pi f_c t) \tag{8.8}$$

where $f_c$ is the carrier frequency, $A$ is the carrier amplitude, and $k_a$ is the amplitude sensitivity of the modulator.

Note that the modulated wave $s(t)$ is a modified sinusoid having a constant frequency $f_c$ and a time-varying amplitude $A[1 + k_a m(t)]$. The modulus of this amplitude is known as the *envelope* of the modulated wave. The factor $|k_a m(t)|_{max}$ is referred to as the modulation factor or the percentage modulation, if expressed as a percentage.

In the case where the modulation factor $|k_a m(t)|_{max}$ is less than or equal to 1, the term $1 + k_a m(t)$ is always nonnegative and the envelope of the wave equals $A[1 + k_a m(t)]$. If, however, the factor $|k_a m(t)|$ exceeds unity for any $t$, the envelope of the wave is $A|1 + k_a m(t)|$ and there may occur overmodulation. The two cases are depicted in Figure 8.1 for a general message signal $m(t)$.

It is noteworthy that when $|k_a m(t)|_{max} \leq 1$, the envelope of the modulated wave, apart from its DC component, corresponds to the message signal $m(t)$. If, however, overmodulation occurs ($|k_a m(t)| > 1$ for some $t$), this one-to-one correspondence between envelope and message does not hold any longer and the wave is said to have envelope distortion.

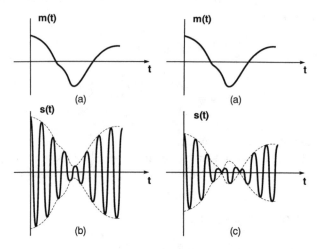

**Figure 8.1.**  Amplitude modulation: (a) message signal; (b) AM wave for $|k_a m(t)|_{max} < 1$; (c) AM wave for $|k_a m(t)| > 1$ for some $t$.

### 8.3.2  Transmission Bandwidth

Equation 8.8 is the time-domain description of the standard AM wave. In order to determine the transmission bandwidth of this modulation scheme, it is convenient to find the corresponding frequency-domain description of the modulated wave. This is accomplished by simply taking the Fourier transform of both sides of Equation 8.8. Let $S(f)$ and $M(f)$ be the Fourier transforms of $s(t)$ and $m(t)$, respectively. Therefore

$$S(f) = \frac{A}{2}[\delta(f - f_c) + \delta(f + f_c)] + \frac{k_a A}{2}[M(f - f_c) + M(f + f_c)]$$

(8.9)

where $\delta(\cdot)$ is the Dirac delta (impulse) function.

Consider that the message signal is band-limited to the interval $-W \leq f \leq W$ and that the carrier frequency $f_c$ is greater than the bandwidth $W$ to avoid spectral overlap (frequency distortion). For a generic message signal $m(t)$ and an AM wave $s(t)$, the corresponding frequency spectra are shown in Figure 8.2.

Taking only the positive portion of the spectrum for the modulated wave, it can be seen that the difference between its highest-frequency component $f_c + W$ and its lowest-frequency component $f_c - W$ defines the transmission bandwidth $B$. Therefore, for an AM wave this is exactly twice the message bandwidth, i.e.,

$$B = 2W$$

(8.10)

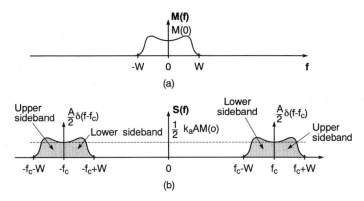

Figure 8.2.   Signal spectrum: (a) spectrum of a generic signal message; (b) spectrum of the AM wave.

### 8.3.3   Generation of AM Signals

Many efficient ways of generating AM waves are available in the literature. We will only describe some of the basic methods (without going into too much detail) just for the sake of a global overview of the subject.

High-power AM signals can be generated by means of a class C radio-frequency (RF) amplifier. The modulating signal is used to vary the supply voltage to the amplifier between zero and twice the nominal supply voltage. The modulator acts as a multiplier, and a DC voltage is added to the modulating signal to provide the carrier component. Low-power AM signals can be generated by means of modulators using nonlinear devices (such as diodes or transistors) conveniently biased to operate in a restricted portion of their characteristic curves. Let us describe two such circuits.

#### 8.3.3.1 *Switching Modulator*

The switching modulator can be implemented as shown in Figure 8.3. The nonlinear device can be a diode working as an ideal switch, i.e., having a zero impedance when forward-biased and an infinite impedance when reverse-biased. If the message signal $m(t)$ is such that $|m(t)| \ll A$, then the switching of the diode will be controlled by the carrier itself. In other words,

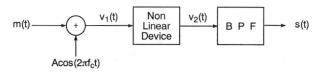

Figure 8.3.   Switching and square-law modulator.

$v_2(t)$ equals $v_1(t)$ when the carrier is positive and it equals zero when the carrier is negative. This corresponds to multiplying $v_1(t)$ by a nonnegative square wave having a period equal to $1/f_c$. Because such a wave comprises a DC voltage plus an infinite number of odd harmonic components, $v_2(t)$ will contain the AM signal plus unwanted terms. The unwanted terms can be filtered out by means of a band-pass filter (BPF). This is shown in the following equations:

$$v_1(t) = m(t) + A\cos(2\pi f_c t) \tag{8.11}$$

$$v_2(t) = v_1(t) \times (\text{square wave}) \tag{8.12}$$

where

$$\text{square wave} = a_1 + a_2\cos(2\pi f_c t)$$

$$+ \text{other odd harmonic components}$$

$$(a_1 \text{ and } a_2 \text{ are constants}) \tag{8.13}$$

Then

$$v_2(t) = a_1 A\left[1 + \frac{a_2}{a_1 A}m(t)\right]\cos(2\pi f_c t) + \text{unwanted terms} \tag{8.14}$$

and

$$s(t) = a_1 A\left[1 + k_a m(t)\right]\cos(2\pi f_c t) \tag{8.15}$$

where $k_a = a_2/a_1 A$.

### 8.3.3.2 *Square-Law Modulator*

The square-law modulator can also be implemented as shown in Figure 8.3, by using a convenient nonlinear device. Let the nonlinear device be modelled as follows:

$$v_2(t) = a_1 v_1(t) + a_2 v_1^2(t) \tag{8.16}$$

where $a_1$ and $a_2$ are constants.

Then, with $v_1(t)$ given by Equation 8.11, we have

$$v_2(t) = a_1 A\left[1 + \frac{2a_2}{a_1}m(t)\right]\cos(2\pi f_c t) + \text{unwanted terms} \tag{8.17}$$

and

$$s(t) = a_1 A\left[1 + k_a m(t)\right]\cos(2\pi f_c t) \tag{8.18}$$

where $k_a = 2a_2/a_1$.

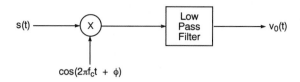

**Figure 8.4.**   Block diagram for coherent detection.

### 8.3.4   Detection of AM Signals

There are two basic ways of detecting AM signals: coherently (synchronously) and noncoherently (asynchronously). The coherent techniques use a replica of the carrier wave to reverse the modulation process. Noncoherent detection uses extra information about the carrier, transmitted in the modulated wave.

#### 8.3.4.1 *Coherent Detection*

The basic block diagram for coherent detection is shown in Figure 8.4. Consider that $s(t)$ is the AM wave as described by Equation 8.8 and that the local carrier differs from the transmitter carrier by a phase $\phi$. After performing the multiplication and filtering out the high-frequency components, it is straightforward to show that

$$v_0(t) = \tfrac{1}{2}A\left[1 + k_a m(t)\right]\cos\phi \qquad (8.19)$$

The DC component can be extracted by means of a capacitor, and the message will be left with an attenuation constant $\cos\phi$ depending on the locally generated carrier.

Another way of performing coherent detection is by using the square-law modulator of Figure 8.3. It can be seen that if $v_1(t)$ is the AM wave as described by Equation 8.8, then

$$v_2(t) = a_2 k_a A^2 m(t) + \tfrac{1}{2}a_2 k_a^2 m^2(t) + \text{other unwanted terms} \quad (8.20)$$

The "other unwanted terms" in Equation 8.20 include a DC component plus higher-frequency terms ($f_c, 2f_c$), eliminated by appropriate filters. Note, however, that $m^2(t)$ in Equation 8.20 also contains unwanted frequency components. They can be made harmless if we choose $|k_a m(t)|$ to be conveniently small.

#### 8.3.4.2 *Noncoherent Detection*

Asynchronous detection can be provided by an envelope detector as shown in Figure 8.5. The circuit is based on the fact that the signal envelope follows the same shape as that of the modulating waveform $m(t)$. Consequently, a

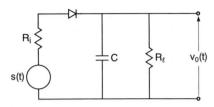

**Figure 8.5.** Envelope detector.

simple rectifier followed by a low-pass filter (LPF) can be used to recover the transmitted message signal. The time constant $R_\ell C$ should be chosen so that the discharging time of the capacitor through the load can be (1) slow enough to smooth out the positive peaks of the carrier wave and (2) fast enough to follow the maximum rate of change of the modulating wave. Therefore

$$1/f_c \ll R_\ell C \ll 1/W \tag{8.21}$$

As far as charging time is concerned, if we consider that the AM wave applied to the envelope detector is supplied by a voltage source having internal impedance $R_i$, then it is required that

$$R_i C \ll 1/f_c \tag{8.22}$$

### 8.3.5 A General Model For the Received AM Signal

Consider an AM signal transmitted from the base station in a mobile radio environment. Due to the multipath effect this signal is deteriorated by fading. Moreover, additive noise (such as receiver noise and others) deteriorate even more the already corrupted signal. The additive noise $n_c(t)$ is modelled as a sample function of a white noise process having mean value equal to zero and a constant power spectral density equal to $\eta/2$. The parameter $\eta$ is the average noise power per bandwidth (watts per hertz) measured at the front end of the receiver. The objective of our study is to determine the effects of the fading and Gaussian noise on the received wave.

#### 8.3.5.1 *Receiver Model*

In order to carry out this investigation, let us assume a simplified AM receiver model as shown in Figure 8.6a. The receiver signal consists of a fading AM signal component $e(t)$ plus the channel noise $n_c(t)$. The compound wave reaches the radio-frequency (RF) section (not shown in the figure) of the AM receiver, where the signal is down-converted to an intermediate frequency (IF). The converted signal then feeds the IF section, where most of the amplification and selectivity are provided.

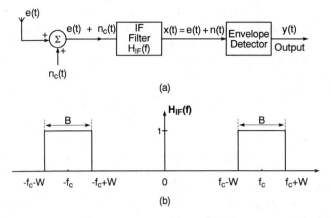

**Figure 8.6.** AM reception: (a) receiver model; (b) ideal IF filter characteristic.

The IF filter is tuned so that its midband frequency coincides with the carrier frequency. Its bandwidth $B$ is just wide enough to accommodate the AM signal, i.e., $B = 2W$. An ideal IF filter has band-pass characteristics as shown in Figure 8.6b. In our model we shall use asynchronous detection, as that provided by the envelope detector.

### 8.3.5.2 Received Signal

A carrier $\tilde{c}(t)$ having an amplitude $A$ and a frequency $\omega_c = 2\pi f_c$ is expressed in its exponential form as

$$\tilde{c}(t) = A \exp(j\omega_c t) \tag{8.23}$$

The carrier $\tilde{c}(t)$ transmitted by the base station arrives at the mobile's antenna through multiple paths. Accordingly, the received signal $\tilde{c}'(t)$ is composed of a number of "carriers" having amplitudes and phases varying randomly. In Section 4.5 we showed that $\tilde{c}'(t)$ is described by Equation 4.18a, reproduced here for convenience:

$$\tilde{c}'(t) = A \sum_{i=1}^{n} a_i \exp[j(\omega_c t + \omega_i t - \omega_c T_i)] \tag{8.24}$$

We may rewrite Equation 8.24 as

$$\tilde{c}'(t) = \tilde{R}(t) A \exp(j\omega_c t) = \tilde{R}(t)\tilde{c}(t) \tag{8.25}$$

where

$$\tilde{R}(t) = \sum_{i=1}^{n} a_i \exp[j(\omega_i t - \omega_c T_i)] = a_r(t)\exp[j\phi_r(t)] \tag{8.26}$$

Note that the transmitted carrier $\tilde{c}(t)$ is affected by a multiplicative noise $\tilde{R}(t)$ with Rayleigh-distributed amplitude $a_r(t)$ and with phase $\phi_r(t)$ uniformly distributed in the range $0–2\pi$ rad. Because the phase is not relevant for voice transmission in amplitude modulation, we may neglect $\phi_r(t)$ for the moment and take the real part of $\tilde{c}'(t)$ to be

$$c(t) = \mathrm{Re}[\tilde{c}'(t)] = a_r(t)A\cos(2\pi f_c t) \tag{8.27}$$

Now consider that the carrier $c(t)$ is amplitude-modulated by a message signal $m(t)$. The resultant signal is, as we have seen previously, $A[1 + k_a m(t)]\cos(2\pi f_c t)$. If we assume that the modulation bandwidth is much smaller than the coherence bandwidth, then fading will affect all the frequency components of the signal in the same manner ("flat fading") (see Section 4.5), and the resultant signal $e(t)$ at the antenna of the mobile is (refer to Figure 8.6a)

$$e(t) = a_r(t)A[1 + k_a m(t)]\cos(2\pi f_c t) \tag{8.28}$$

As shown in Figure 8.6a the signal $e(t)$ is added to the Gaussian white noise $n_c(t)$. The noise signal $n_c(t)$ becomes a narrowband noise signal $n(t)$ after the IF filter because the filter bandwidth is small compared to its midband frequency $f_c$. The band-limited noise can be expressed as

$$\tilde{n}(t) = [n_I(t) + n_Q(t)]\exp(j2\pi f_c t) \tag{8.29}$$

where $n(t) = \mathrm{Re}[\tilde{n}(t)]$
    $n_I(t)$ is the in-phase component
    $n_Q(t)$ is the quadrature component

Hence,

$$n(t) = n_I(t)\cos(2\pi f_c t) - n_Q(t)\sin(2\pi f_c t) \tag{8.30}$$

Its power spectral density follows the shape $H_{IF}(f)$ of the IF fil.er and is given by $(\eta/2)H_{IF}(f)$. Accordingly, the compound signal $x(t)$ after the IF filter is

$$x(t) = e(t) + n(t)$$

$$= \{a_r(t)A[1 + k_a m(t)] + n_I(t)\}\cos(2\pi f_c t) - n_Q(t)\sin(2\pi f_c t) \tag{8.31}$$

It is convenient to represent the components of $x(t)$ by means of phasors as in Figure 8.7. The output $y(t)$ of the envelope detector is the envelope of

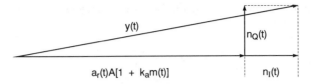

**Figure 8.7.** Phasor diagram of the AM fading signal plus Gaussian noise.

$x(t)$, easily extracted from the phasor diagram of Figure 8.7,

$$y(t) = \left\{ \left[ a_r(t) A[1 + k_a m(t)] + n_I(t) \right]^2 + n_Q^2(t) \right\}^{1/2} \quad (8.32)$$

### 8.3.6 The Effect of Additive Noise on AM Systems

Consider the case where the effect of Rayleigh fading is negligible (e.g., in an open area propagation case or when some diversity technique is used). Then $a_r(t) \simeq 1$ and this constitutes the standard analysis carried out in many textbooks (e.g., Reference 1). From Equation 8.28 we see that the average power of the modulated message signal is equal to $A^2(1 + P_m)/2$, where $P_m = k_a^2 \langle m^2(t) \rangle$ is the message power. The average noise power within the message bandwidth is $\eta W$, so that

$$\text{SNR}_C = \frac{A^2(1 + P_m)}{2\eta W} \quad (8.33)$$

If the average carrier power is large compared with the average noise power, then the signal term $A[1 + k_a m(t)]$ will be large compared with the noise terms $n_I(t)$ and $n_Q(t)$. In this case the receiver operates adequately. Accordingly, we may approximate Equation 8.32 by

$$y(t) \simeq A[1 + k_a m(t)] + n_I(t) \quad (8.34)$$

Filtering out the DC component $A$, the receiver output will be given by $Ak_a m(t) + n_I(t)$, where $Ak_a m(t)$ constitutes the signal itself and $n_I(t)$ is the unwanted noise. The average noise power in the message bandwidth can be obtained by multiplying the noise power spectral density ($\eta$) by the message bandwidth ($2W$), resulting in an average power of $2\eta W$ (see Appendix 8A). The signal power is $A^2 P_m$, where $P_m = k_a^2 \langle m^2(t) \rangle$ is the message power. Therefore

$$\text{SNR}_O = \frac{A^2 P_m}{2\eta W} \quad (8.35)$$

In a like manner, the carrier-to-noise ratio is given by

$$\gamma = \frac{A^2/2}{2\eta W} \tag{8.36}$$

The figure of merit is, therefore

$$F = \frac{\text{SNR}_O}{\text{SNR}_C} = \frac{P_m}{1 + P_m} \tag{8.37a}$$

In the special case of having a sinusoidal wave of frequency $W$ and amplitude $A_m$ as the modulating wave, we obtain $P_m = k_a^2 A_m^2/2 \triangleq \mu^2/2$, so that

$$F = \frac{\mu^2}{2 + \mu^2} \tag{8.37b}$$

If the carrier-to-noise ratio at the receiver input is very low, the message signal becomes unintelligible and the noise term dominates. The detected envelope $y(t)$ is mainly constituted by the narrowband noise with a Rayleigh distribution.* Therefore

$$p(y) = \frac{y}{\sigma^2} \exp\left(\frac{y^2}{2\sigma^2}\right) \tag{8.38}$$

where $\sigma^2$ is the variance of the noise $n(t)$, equal to $2\eta W$ in AM systems.

The probability that the envelope exceeds the carrier amplitude is

$$\text{prob}(y \geq A) = \int_A^\infty p(y)\, dy = \exp\left(-\frac{A^2}{4\eta W}\right) = \exp(-\gamma) \tag{8.39}$$

We may assume that, if this event occurs with 1% probability, i.e., $\text{prob}(y \geq A) = \exp(-\gamma) = 0.01$, then the envelope detector is expected to operate satisfactorily. For this case $\gamma = 4.6 = 6.6$ dB. If such an event occurs with a 50% probability, i.e., $\text{prob}(y \geq A) = \exp(-\gamma) = 0.5$, then the message at the output of detector can be assumed to be lost. For this case $\gamma = 0.69 = -1.6$ dB.

---

*Refer to Equation 8.29 for the exponential representation of the narrowband noise and compare this with Equation 3.82a of Section 3.4.2, where the Rayleigh distribution is obtained.

### 8.3.7 The Effect of Multipath Propagation on AM Systems

Now consider the case where only multiplicative noise corrupts the received signal. Using Equation 8.32, the envelope $y(t)$ is given by

$$y(t) = a_r(t) A[1 + k_a m(t)] \qquad (8.40)$$

Just for convenience let $r \triangleq a_r(t)$, so that

$$y(t) = rA[1 + k_a m(t)] \qquad (8.41)$$

Note that $r$ is a random variable with a Rayleigh distribution given by Equation 8.38. Because this represents a multiplicative noise, the true signal output is taken as being the average of the output. Then

$$\bar{y}(t) = A[1 + k_a m(t)] \int_0^\infty r p(r) \, dr$$

$$= A[1 + k_a m(t)] \sqrt{\pi/2} \, \sigma \qquad (8.42)$$

The message signal at the envelope detector output is* $Ak_a m(t) \sqrt{\pi/2} \, \sigma$ and the corresponding signal power is $(\pi/2)(A\sigma)^2 k_a^2 \langle m^2(t) \rangle = (\pi/2)(A\sigma)^2 P_m$. The signal-suppression noise is

$$n_s(t) = y(t) - \bar{y}(t) = A(r - \sqrt{\pi/2} \, \sigma)[1 + k_a m(t)] \qquad (8.43)$$

It is shown in Appendix 8B that the signal–to–signal-suppression-noise ratio SSN is

$$SSN = \frac{P_m}{1 + P_m} \frac{\pi/2}{(2 - \pi/2)} \qquad (8.44)$$

The message power $P_m$ is usually assumed to be on the order of 0.1 to avoid overmodulation. Therefore, the SSN is approximately 0.333, corresponding to a signal-suppression noise power three times as large as the signal power. This shows that Rayleigh fading by itself has a disastrous effect on AM systems.

## 8.4 SINGLE-SIDEBAND MODULATION

In this section single-sideband (SSB) modulation is briefly described and special attention is given to its advantages with respect to the standard AM system.

---

*The DC component does not contribute to the useful signal power.

### 8.4.1  Basic Principles

From the standard AM studies we learned that the upper and lower sidebands are symmetrical about the carrier frequency. This implies that, given either sideband, the other sideband can be determined. Therefore, it is understood that if both the carrier and one sideband are suppressed, we can recover the complete signal at the receiver end.

### 8.4.2  Transmission Bandwidth

One of the biggest advantages of SSB modulation over the standard AM system is that it requires half of the bandwidth of that required by the AM systems. In other words it uses exactly the message bandwidth, i.e.,

$$B = W \tag{8.45}$$

This is accomplished by band-pass filters used to suppress either the upper sideband or the lower sideband. The filter characteristic suppressing the upper sideband is shown in Figure 8.8a. The corresponding low-pass characteristic is depicted in Figure 8.8b.

If the message $m(t)$ has a Fourier transform $M(f)$, then the Fourier transform of the complex envelope of the SSB signal $S(f)$ is

$$S(f) = H'_L(f)M(f) \tag{8.46}$$

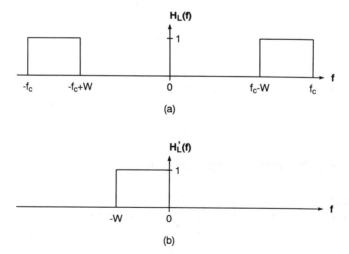

**Figure 8.8.**  SSB filtering: (a) SSB filter characteristic used to suppress the upper band; (b) equivalent low-pass filter characteristic.

The transfer function $H'_L(f)$ of the filter can be expressed as

$$H'_L(f) = \tfrac{1}{2}[1 - \operatorname{sgn}(f)] - u(f + W) \tag{8.47}$$

where $\operatorname{sgn}(\cdot)$ is the signum function and $u(\cdot)$ is the step function. Because $M(f)$ has its spectrum within the range $-W \le f \le W$, the product $u(f + W)M(f)$ is zero and

$$S'(f) = \tfrac{1}{2}[1 - \operatorname{sgn}(f)]M(f) \tag{8.48}$$

The inverse Fourier transform of Equation 8.48 is $\tilde{s}(t)$

$$\tilde{s}(t) = \tfrac{1}{2}[m(t) - j\hat{m}(t)] \tag{8.49}$$

where $\hat{m}(t)$ is the Hilbert transform* of $m(t)$. The SSB signal $s(t)$ is then

$$s(t) = \operatorname{Re}[\tilde{s}(t) A \exp(j2\pi f_c t)] \tag{8.50}$$

yielding

$$s(t) = \frac{A}{2}[m(t)\cos(2\pi f_c t) + \hat{m}(t)\sin(2\pi f_c t)] \tag{8.51}$$

It is easy to show that for lower sideband suppression, instead of the plus (+) sign we obtain a minus (−) sign in Equation 8.51.

### 8.4.3   Generation of SSB Signal

There are basically two methods of implementing SSB modulation, namely, the filter method and the outphasing method.

#### 8.4.3.1 *Filter Method*

This method uses a mixer (multiplier) to generate a double sideband suppressed carrier (as in Figure 8.2b without the carrier) and a filter to further suppress one of the sidebands. The difficulty in this case is to implement filters with very sharp cutoff characteristics. This becomes more critical with increasing frequency.

---

*The Hilbert transform of a function produces a phase shift of $-90°$ for all positive frequencies and a phase shift of $+90°$ for all negative frequencies. The amplitudes remain the same. For a function $x(t)$ the Hilbert transform $\hat{x}(t)$ is

$$\hat{x}(t) = \frac{1}{\pi}\int_{-\infty}^{\infty}\frac{x(\tau)}{t - \tau}\,d\tau$$

### 8.4.3.2 *Phase-Discrimination Method*

The modulator using this method is known as Hartley modulator. It implements the functions as expressed in Equation 8.51. The difficulty in this case is to design a phase shifter able to maintain the 90° phase shift over the full bandwidth of the baseband signal. This can be achieved by using two stages of quadrature modulation.[4]

### 8.4.4  Detection of SSB Signal

The baseband signal $m(t)$ can be recovered by means of coherent detection as mentioned before (refer to Figure 8.4). If the locally generated carrier has a phase shift of $\phi$, the detected wave $v_0(t)$ is

$$v_0(t) = \tfrac{1}{4}Am(t)\cos\phi - \tfrac{1}{4}A\hat{m}(t)\sin\phi \qquad (8.52)$$

Note that if the phase error is zero, the received signal contains the full message (scaled by an amplitude factor). For any $\phi \neq 0$, the recovered message suffers from phase distortion, having negligible effects as far as speech is concerned, but disastrous effects on data communication.

### 8.4.5  A General Model For the Received SSB Signal

Following the same reasoning as for the AM case, the resultant signal $e(t)$ at the antenna of the mobile, after the multipath propagation, is

$$e(t) = a_r(t)\frac{A}{2}\left[m(t)\cos(2\pi f_c t) - \hat{m}(t)\sin(2\pi f_c t)\right] \qquad (8.53)$$

where $a_r(t)$ is Rayleigh distributed.

The narrowband noise $n(t)$ has a midband frequency differing from the carrier frequency by $W/2$ and can be expressed as

$$n(t) = n_I(t)\cos\left[2\pi(f_c - W/2)t\right] - n_Q(t)\sin\left[2\pi(f_c - W/2)t\right] \qquad (8.54)$$

The compound signal after the IF filter is $x(t) = e(t) + n(t)$. The coherent detection carries out the operations $x(t)\cos(2\pi f_c t)$ in connection with low-pass filtering, so that

$$y(t) = a_r(t)\frac{A}{4}m(t) + \frac{1}{2}n_I(t)\cos(\pi Wt) + \frac{1}{2}n_Q(t)\sin(\pi Wt) \qquad (8.55)$$

provided that no phase error is encountered in the carrier.

### 8.4.6  The Effect of Additive Noise on SSB Systems

If fading is negligible, then $a_r(t) \simeq 1$. From Equation 8.53 we observe that the in-phase and quadrature components of the SSB wave contribute with an average power of $A^2P_m/8$ each, constituting a total of $A^2P_m/4$. The average noise power is $\eta W$. Thus

$$\text{SNR}_C = \frac{A^2P_m}{4\eta W} \tag{8.56}$$

The carrier-to-noise ratio is

$$\gamma = \frac{A^2/2}{\eta W} \tag{8.57}$$

From Equation 8.55 we note that the average power of the recovered message, $Am(t)/4$, is $A^2P_m/16$. The noise component is $[n_I(t)\cos(\pi Wt) + n_Q(t)\sin(\pi Wt)]/2$. The in-phase noise component has a power spectral density equal to $\eta$ in the message bandwidth $W$. Therefore, the corresponding average power is $\eta W$. In the case of the modulated noise $n_I(t)\cos(\pi Wt)$, the average power is $\eta W/2$, the same applying to $n_Q(t)\sin(\pi Wt)$. Accordingly, the total power of the noise component is $(\frac{1}{4}(2\eta W/2) = \eta W/4$. And the signal-to-noise ratio is

$$\text{SNR}_O = \frac{A^2P_m}{4\eta W} \tag{8.58}$$

Therefore the figure of merit is given by

$$F = \frac{\text{SNR}_O}{\text{SNR}_C} = 1 \tag{8.59}$$

### 8.4.7  The Effect of Multipath Propagation on SSB Systems

In the case where only multiplicative noise affects the received signal, the output of the receiver is (from Equation 8.55)

$$y(t) = a_r(t)\frac{A}{4}m(t) \tag{8.60}$$

As in the AM case, we rewrite Equation 8.60 as

$$y(t) = r\frac{A}{4}m(t) \tag{8.61}$$

Thus, following the same steps as before, we find that the mean signal $\bar{y}(t)$ is

$$\bar{y}(t) = \frac{A}{4}m(t)\sqrt{\frac{\pi}{2}}\,\sigma \tag{8.62}$$

The signal suppression noise is

$$y(t) - \bar{y}(t) = \frac{A}{4}\left(r - \sqrt{\frac{\pi}{2}}\,\sigma\right)m(t) \tag{8.63}$$

and the signal–to–signal-suppression noise ratio SSN is

$$\text{SSN} = \frac{\pi/2}{2 - \pi/2} \simeq 3.66 \tag{8.64}$$

If we compare this result with that given by the standard AM case, we can see that there is an improvement of 10.4 dB. The same improvement can be achieved for the case of the standard AM if an appropriate high-pass filter is used after the detector, as we explain next.

The effect of multipath propagation is to broaden each spectral line of the signal until it occupies a spectral width of $2f_m = 2v/\lambda$, where $f_m$ is the maximum Doppler shift, $v$ is the vehicle speed, and $\lambda$ is the wavelength (refer to Section 4.9). In Equation 8.43 we see that there is a component of noise $r$ due to the carrier fading. This can be suppressed by a high-pass filter if the spectrum of $r$ lies below the lowest baseband frequency, usually considered to be 150 Hz. In other words, we must have $f_m \leq 150$ Hz. Therefore, by filtering out the noise component due to the carrier fading, the improvement achieved is quite substantial as just calculated. The modulation of the sidebands, however, cannot be removed by filtering and remains as motion-induced noise.

## 8.4.8  Enhanced SSB Modulation Techniques

As far as spectrum efficiency is concerned, single-sideband modulation is indeed attractive. However, its performance in the presence of fading, when compared with FM, is very poor, perhaps disastrous. Accordingly, many efforts have been made toward enhancing the SSB performance for mobile radio purposes, and a number of techniques has emerged. These techniques individually or combined seem to yield very good (or at least promising) results.

McGeeham and Bateman[8,9] have proposed a scheme in which a gap within the speech band is created and a pilot tone is inserted in it prior to transmission. This technique named transparent tone in band (TTIB) has proved to be very effective, mainly if combined with a strategy termed feedforward signal regeneration (FFSR). The basic principle of TTIB plus FFSR is that, at the receiver end the pilot tone is extracted and used for three purposes:

1. To obtain a frequency reference for demodulation
2. To obtain a signal reference for the automatic gain control (AGC) circuit
3. To help reestablish the phase and amplitude of the transmitted signal, this constituting a powerful tool to combat the effects of Rayleigh (fast) fading. The basic idea is that the receiver, having the pilot tone as reference, corrects the phase and amplitude characteristics of the received sidebands so as to maintain constant the phase and amplitude of the pilot tone.

It is important to mention that this scheme will work well if there is high correlation between the pilot tone and the message frequencies. The envelope correlation producing an output signal–to–signal-suppression noise ratio of 20 dB is 0.9998 (page 206 of Reference 2). The correlation factor between two signals arriving at the same time instant is given by Equation 4.46 with $\tau = 0$ (time delay zero) and is written as

$$\rho_r = \frac{1}{1 + (\Delta\omega\,\overline{T})^2} \tag{8.65}$$

where $\Delta\omega$ is the frequency separation (rad/s) between the signals, and $\overline{T}$ is the delay spread. Let $\overline{T} = 1$ $\mu$s (in an urban area $\overline{T} \simeq 3$ $\mu$s, in a suburban area $\overline{T} \simeq 0.5$ $\mu$s). Therefore, the required frequency separation $\Delta f$ must be no greater than 2.25 kHz, which for a 3 kHz voice channel impels the pilot tone to be placed within the voice frequency band, preferably in the middle of the band.

Another enhanced SSB technique is the "amplitude-companded single sideband" (ACSSB) technique, using a process of (1) *comp*ressing the signal prior to modulation and transmission and (2) exp*anding* the compressed signal to restore its dynamic range at the reception end. This companding technique seems to strengthen the SSB against the effects of multipath propagation.

## 8.5  FREQUENCY MODULATION

Frequency modulation is well known for its outstanding performance as far as audio output signal-to-noise ratio is concerned. Analysis of the FM communication scheme is considerably more complicated than that of the amplitude modulation techniques but can be simplified if some approximations are carried out. In this section we shall review some of the FM fundamentals in connection with some detailed analysis based on the work by Rice.[5,6]

## 8.5.1  Basic Principles

Frequency modulation is a nonlinear process in which the frequency of the carrier is varied according to the message signal. Let $\theta(t)$ be the angular argument of a carrier $\tilde{s}(t)$ such that, expressed in its exponential form*

$$\tilde{s}(t) = A \exp[j\theta(t)] \tag{8.66}$$

Define the instantaneous frequency $f_i(t)$ as

$$f_i(t) = \frac{1}{2\pi} \frac{d\theta(t)}{dt} \tag{8.67}$$

In frequency modulation it is assumed that the instantaneous frequency is centered at $f_c$ but varies linearly with the message signal $m(t)$, i.e.,

$$f_i(t) = f_c + k_f m(t) \tag{8.68}$$

where $k_f$ is the frequency sensitivity. Integrating Equation 8.68 and multiplying by $2\pi$ we obtain the angular argument $\theta(t)$. Then, using $\theta(t)$ in Equation 8.66, we obtain the modulated wave $\tilde{s}(t)$

$$\tilde{s}(t) = A \exp[j(2\pi f_c t + \phi_s(t))] \tag{8.69}$$

where

$$\phi_s(t) = 2\pi k_f \int_0^t m(t)\, dt + \phi_s(0) \tag{8.70}$$

In our analysis we shall assume $\phi_s(0) = 0$. Note from Equation 8.69 that the envelope of the FM wave is constant. Moreover, its average power is independent of the message and is given by $A^2/2$.

### 8.5.1.1 Single-Tone FM

Let the message signal $m(t)$ be a sinusoidal wave given by

$$m(t) = A_m \cos(2\pi Wt) \tag{8.71}$$

In this case the FM wave is given by

$$\tilde{s}(t) = A \exp[j(2\pi f_c t + \beta \sin(2\pi Wt))] \tag{8.72}$$

---

*In FM we shall use the exponential representation of the sine wave because it is convenient. We note, however, that $\text{Re}[\tilde{s}(t)]$ is what really matters.

where $\beta \triangleq k_f A_m/W \triangleq \Delta f/W$ is the modulation index and $\Delta f$ is the frequency deviation.

The complex envelope $A \exp[j\beta \sin(2\pi Wt)]$ in Equation 8.72 is a periodic function having a fundamental frequency equal to $W$. Therefore, it can be expanded in the form of a complex Fourier series as

$$A \exp[j\beta \sin(2\pi Wt)] = A \sum_{n=-\infty}^{\infty} J_n(\beta)\exp(j2\pi nWt) \qquad (8.73)$$

where the complex Fourier coefficients $J_n(\beta)$ are given by

$$J_n(\beta) = \frac{1}{2\pi} \int_{-\pi}^{\pi} \exp[j(\beta \sin x - nx)] \, dx \qquad (8.74)$$

and $x = 2\pi Wt$.

The coefficients $J_n(\beta)$ are known as the $n$th-order Bessel functions of the first kind. Hence the FM wave in Equation 8.72 can be written in terms of the Bessel function as

$$\tilde{s}(t) = A \sum_{n=-\infty}^{\infty} J_n(\beta)\exp[j2\pi(f_c + nW)t] \qquad (8.75)$$

## 8.5.2 Transmission Bandwidth

Referring to Equation 8.75, we understand that the FM wave contains $2n$ side frequencies centered at $f_c$. In other words the transmission bandwidth is $B = 2nW$. Because $n$ varies from $-\infty$ to $+\infty$ the theoretical required bandwidth is infinite. However, in practice, it is possible to specify an effective bandwidth within which the distortion is considered to be under tolerable limits.

Consider, initially, the case of the single-tone modulation. If we examine the tables of the Bessel function, it can be seen that $J_n(\beta)$ diminishes very rapidly for $n > \beta$ and, more specifically, for large $\beta$. In fact, for large $\beta$ the ratio $n/\beta$ tends to unity. In this case, because $n = \beta$ and $B = 2nW$, we have $B = 2\beta W = 2 \Delta f$. For small $\beta$, only $J_0(\beta)$ and $J_1(\beta)$ are relevant. Therefore $B = 2W$. With these two limiting cases Carson[10] suggested that an approximate rule for transmission bandwidth of an FM wave generated by a single-tone modulating wave is

$$B = 2 \Delta f + 2W = 2W(1 + \beta) \qquad (8.76)$$

This is known as Carson's rule. In practice, this rule is applied even for a nonsinusoidal modulation where $W$ is assumed to be the highest frequency in the modulating waveform.

#### 8.5.2.1 *Narrowband FM*

Narrowband FM (NBFM) is of great interest in mobile radio communications where spectrum efficiency is critical. It is commonly used when the channel spacing is fairly small (12.5 or 25 kHz). This is achieved when the modulation index is small ($\beta \ll 1$). In this case $J_0(\beta) \simeq 1$, $J_1(\beta) \simeq \beta/2$, and $J_n(\beta) \simeq 0$ for $n > 1$. With these values we obtain from Equation 8.75 the expression for single-tone NBFM,

$$s(t) = \text{Re}[\tilde{s}(t)]$$

$$= A\cos[2\pi f_c(t)] + \frac{\beta A}{2}\cos[2\pi(f_c + W)t] - \frac{\beta A}{2}\cos[2\pi(f_c - W)t]$$

$$(8.77)$$

For a general case of modulating wave $m(t)$, when $\beta$ is very small then $\phi_s(t)$ is also very small and we may approximate $\cos[\phi_s(t)] \simeq 1$ and $\sin[\phi_s(t)] \simeq \phi_s(t)$. Therefore, from Equation 8.69 the NBFM equals

$$s(t) = \text{Re}[\tilde{s}(t)] \simeq A\cos(2\pi f_c t) - \phi_s(t)A\sin(2\pi f_c t) \qquad (8.78)$$

which closely resembles an AM wave. In this case, however, the sidebands are symmetrically placed about a line at 90° to the carrier and this generates a signal with a varying frequency and a quasiconstant amplitude. In other words, the NBFM contains a residual amplitude modulation. The bandwidth is $B = 2W$ as we determined previously.

#### 8.5.2.2 *Wideband FM*

The case where $\beta \gg 1$ is called wideband FM (WBFM). The FM wave contains an "infinite" number of side-frequency components symmetrically placed around the carrier.

### 8.5.3  Generation of FM Signals

Two basic methods are available to generate FM waves: indirect FM and direct FM.

#### 8.5.3.1 *Indirect FM*

In this method, first a narrowband FM wave is generated and then, by means of frequency multiplication, the increased frequency deviation is obtained, producing a wideband FM. A multiplier consists of a nonlinear device having an input–output relation given in a polynomial form. Let this

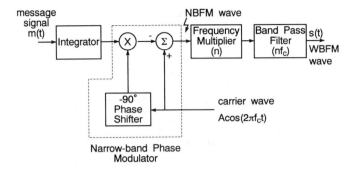

**Figure 8.9.** Block diagram of an FM modulator using the indirect method.

relation be given by

$$u = a_1v + a_2v^2 + \cdots + a_nv^n \tag{8.79}$$

where $a_1, a_2, \ldots, a_n$ are constants, $u$ is the output, and $v$ is the input. Suppose that $v$ is an NBFM wave (Equation 8.78) with frequency deviation $\Delta f$. The signal $u$ (Equation 8.79), therefore, contains a DC component plus $n$ frequency-modulated waves at $f_c, 2f_c, \ldots, nf_c$ with frequency deviation $\Delta f, 2\,\Delta f, \ldots, n\,\Delta f$, respectively. A band-pass filter can be used to select the appropriate WBFM. Figure 8.9 depicts a block diagram of an indirect FM modulator.

Instead of frequency multipliers, this function can be accomplished by phase-locked loops, which are now readily available in integrated-circuit form.

### 8.5.3.2 *Direct FM*

This can be accomplished by means of LC circuits where the capacitance is provided by a fixed capacitor shunted by a voltage-variable capacitor (varactor or varicap). The varicaps are diodes that, when reverse-biased, present a capacitance varying linearly with the voltage. As an example of this type of modulator we present the Hartley oscillator shown in Figure 8.10.

If the capacitance is such that $C = C_0 - km(t)$, then

$$f_i(t) = \frac{1}{2\pi\sqrt{(L_1 + L_2)C}} = f_0\left[1 - \frac{k}{C_0}m(t)\right]^{-1/2} \approx f_0 + k_f m(t) \tag{8.80}$$

where

$$f_0 = \frac{1}{2\pi\sqrt{C_0(L_1 + L_2)}} \quad \text{and} \quad k_f = f_0\frac{k}{2C_0}$$

for $km(t)/C_0 \ll 1$.

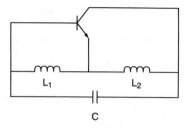

**Figure 8.10.**  Hartley modulator (principle of direct FM).

Equation 8.80 shows that the instantaneous frequency of the oscillator varies linearly with the message signal. Direct FM at microwave frequencies can be generated by Gunn semiconductor diodes, typically operating at 10 GHz. It is able to provide deviation of about 30 MHz/V.[4]

### 8.5.4   Detection of FM Signals

There are many ways of recovering the modulating signal from an FM wave. The basic idea of FM demodulators is frequency-to-voltage conversion. Let us describe some of the more popular techniques.

#### 8.5.4.1 *Balanced Frequency Discriminator*

Let an FM wave $\tilde{s}(t)$, as expressed by Equation 8.69, be the input of a differentiator with gain $a$. Its output $v_1(t)$ is therefore

$$v_1(t) = \mathrm{Re}\left[a\frac{d\tilde{s}(t)}{dt}\right] = 2\pi a A f_c\left[1 + \frac{k_f}{f_c}m(t)\right]\cos\left[2\pi f_c t + \phi_s(t) + \frac{\pi}{2}\right]$$

$$(8.81)$$

Provided that $|k_f m(t)/f_c| < 1$, an envelope detector may be used to recover the amplitude variations of this signal. The output of the detector is $|v_1(t)| = 2\pi a A f_c[1 + k_f m(t)/f_c]$. Note that the envelope of the wave contains the message $m(t)$ plus a DC component. The bias term (DC component) can be removed by means of a capacitor. Another way of removing this DC component is to use a symmetrical configuration of the envelope detector combined with the differentiator. If this symmetrical configuration is such that its output $|v_2(t)|$ equals $2\pi a A f_c[1 - k_f m(t)/f_c]$ then the baseband output $v_0(t)$ can be obtained by performing $v_0(t) = |v_1(t)| - |v_2(t)|$. A close realization of such a scheme is shown in Figure 8.11, where the upper and lower resonant filters are tuned to frequencies above and below the unmodulated carrier frequency $f_c$, respectively.

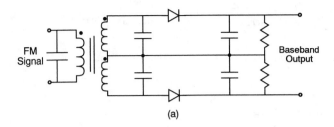

(a)

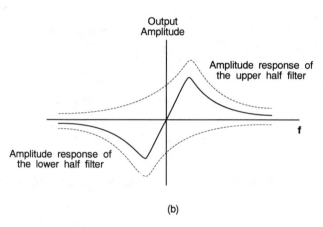

(b)

**Figure 8.11.** Balanced frequency discriminator: (a) circuit diagram; (b) frequency response.

### 8.5.4.2 *Instantaneous Frequency Detector*

Consider a time interval $T$ such that $f_c^{-1} \ll T \ll W^{-1}$. Hence, within $T$ (1) the message signal may be assumed as essentially constant and (2) a considerable number of zero crossings of the FM wave occurs. If the number of zero crossings of the FM wave is counted within $T$, then, effectively, an estimate of the instantaneous frequency of the wave is obtained. That is, let $n_0$ be the number of zero crossings within $T$. The instantaneous frequency $f_i$ is therefore approximately $n_0/2T$. Because the instantaneous frequency is proportional to the message signal (see Equation 8.68) we conclude that $m(t)$ can be recovered from a knowledge of $f_i$. A block diagram of this kind of demodulator is shown in Figure 8.12. The output of the limiter is a square-wave version of its FM wave. These pulses are shaped (shortened up) by the pulse generator and averaged over $T$ by the low-pass filter (integrator) to produce $m(t)$.

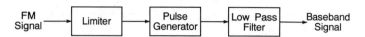

**Figure 8.12.**   Instantaneous frequency detector block diagram.

### 8.5.4.3 *Phase-Locked Loop*

A block diagram of a phase–locked loop (PLL) is depicted in Figure 8.13. The voltage-controlled oscillator (VCO) is a circuit that produces a sine wave whose frequency is varied according to the input voltage. In fact, the VCO is a frequency modulator (FM) with central frequency equal to the unmodulated carrier. Its output has a phase shift of 90° with respect to the unmodulated carrier when its input voltage is zero.

Let us analyze the PLL functioning. Refer to Figure 8.13. Consider that the FM signal $s_1(t)$ at the PLL input is

$$s_1(t) = A_1 \sin[2\pi f_c t + \phi_1(t)] \tag{8.82}$$

where

$$\phi_1(t) = 2\pi k_1 \int_0^t m(t)\, dt \tag{8.83}$$

Suppose that the VCO output $s_2(t)$ (output of an FM modulator) is expressed as

$$s_2(t) = A_2 \cos[2\pi f_c t + \phi_2(t)] \tag{8.84}$$

where

$$\phi_2(t) = 2\pi k_2 \int_0^t v_0(t)\, dt \tag{8.85}$$

The product wave $e(t) = s_1(t) \times s_2(t)$ would contain a high-frequency component $k_m A_1 A_2 \sin[4\pi f_c t + \phi_1(t) + \phi_2(t)]$ (eliminated by the filter and the VCO) and a low-frequency component $k_m A_1 A_2 \sin[\phi_1(t) - \phi_2(t)]$, where

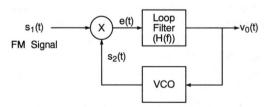

**Figure 8.13.**   Phase-locked loop block diagram.

$k_m$ is the multiplier gain. If the phase error $\phi_1(t) - \phi_2(t)$ is zero, the PLL is said to be in phase-lock. When this error is very small we may approximate the sine function by its argument, i.e.,

$$e(t) = k_m A_1 A_2 \sin[\phi_1(t) - \phi_2(t)] \simeq k_m A_1 A_2[\phi_1(t) - \phi_2(t)] \quad (8.86)$$

Differentiating both sides of Equation 8.86, we obtain

$$\frac{de(t)}{dt} \simeq k_m A_1 A_2 \left[ \frac{d\phi_1(t)}{dt} - 2\pi k_2 v_0(t) \right] \quad (8.87)$$

Let $V_0(f)$, $E(f)$, and $\Phi_1(f)$ be the frequency-domain representation of $v_0(t)$, $e(t)$, and $\phi_1(t)$, respectively. Taking the Fourier transform of Equation 8.87 and using the relation $V_0(f) = H(f)E(f)$, we have

$$V_0(f) = \frac{(jf/k_2)}{1 + 1/L(f)} \Phi_1(f) \quad (8.88)$$

where

$$L(f) = k_m k_2 A_1 A_2 \frac{H(f)}{jf} \quad (8.89)$$

is known as the open-loop transfer function of the PLL. When $L(f) \gg 1$ we obtain $V_0(f) \simeq (jf/k_2)\Phi_1(f)$. In this case the time-domain representation (inverse Fourier transform) is

$$v_0(t) \simeq \frac{1}{2\pi k_2} \frac{d\phi_1(t)}{dt} = \frac{k_1}{k_2} m(t) \quad (8.90)$$

In other words, the output $v_0(t)$ of the PLL is proportional to the original message signal $m(t)$.

## 8.5.5  A General Model For the Received FM Signal

Assume an FM receiver model as shown in Figure 8.14. The receiver FM signal $e(t)$ has a carrier frequency $f_c$ and a bandwidth $B$ such that its frequency band lies in the range $f_c - B/2 \leq |f| \leq f_c + B/2$. It is considered that only a negligible amount of power is encountered outside this band. The IF filter operates with a midband frequency $f_c$ and bandwidth $B$. An ideal IF filter has a band-pass characteristic as shown in Figure 8.6b. In practice, however, the filter may present a "bell shape" characteristic as depicted in Figure 8.15. Consider that this "bell-shaped filter" can be approximated by a

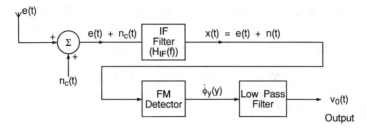

**Figure 8.14.**  FM receiver model.

Gaussian shape such that

$$H_{IF}(f) = \exp\left[-\pi(f - f_c)^2/B^2\right] + \exp\left[-\pi(f + f_c)^2/B^2\right] \quad (8.91)$$

Because the noise spectral density $S_n(f)$ follows the shape of the IF filter, we have

$$S_n(f) = \frac{\eta}{2} H_{IF}(f) \quad (8.92)$$

Following the same steps as those for the AM and SSB cases (refer to Sections 8.3.5 and 8.4.5), it is straightforward to determine the compound signal $x(t)$ at the FM detector input. It is appropriate, in the FM case, to represent the narrowband additive noise in exponential form, $[n_I(t) + jn_Q(t)]\exp(j2\pi f_c t)$. Accordingly,

$$x(t) = e(t) + n(t)$$

$$= \left\{a_r(t) A \exp\left[j(\phi_s(t) + \phi_r(t))\right] + n_I(t) + jn_Q(t)\right\}\exp(j2\pi f_c t)$$

$$(8.93)$$

where $\phi_s(t)$, given by Equation 8.70, is the phase due to the message itself and $\phi_r(t)$ is the random phase due to the multipath effect. Using the phasor

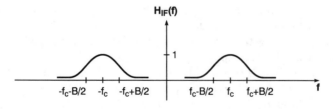

**Figure 8.15.**  Gaussian-shaped IF filter.

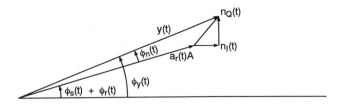

**Figure 8.16.**   Phasor diagram for FM wave plus noise.

representation as depicted in Figure 8.16, we obtain

$$x(t) = y(t)\exp\left[j\phi_y(t)\right]\exp(j2\pi f_c t) \qquad (8.94)$$

where

$$y^2(t) = \left\{a_r(t)A\cos\left[\phi_s(t) + \phi_r(t)\right] + n_I(t)\right\}^2$$

$$+ \left\{a_r(t)A\sin\left[\phi_s(t) + \phi_r(t)\right] + n_Q(t)\right\}^2 \qquad (8.95)$$

$$\phi_y(t) = \phi_s(t) + \phi_r(t) + \phi_n(t) \qquad (8.96)$$

### 8.5.6   Effect of Additive Noise on FM Systems

In a non-Rayleigh environment the multipath propagation effect is irrelevant, so multiplicative noise can be neglected, i.e., $a_r(t) = 1$ and $\phi_r(t) = 0$. The FM signal power is $A^2/2$ and the noise power is given by the integral of the noise power spectral density. Therefore, using Equation 8.91 in Equation 8.92, the IF signal-to-noise ratio $\gamma$ of $x(t)$ is

$$\gamma = \frac{A^2/2}{\int_{-\infty}^{\infty}S_n(f)\,df} = \frac{A^2}{2\eta B} \qquad (8.97)$$

The channel signal-to-noise ratio is easily obtained as

$$\mathrm{SNR}_C = \frac{A^2}{2\eta W} \qquad (8.98)$$

#### 8.5.6.1 *Output Signal Power*

Equating Equations 8.93 and 8.94 for the non-Rayleigh case, we obtain

$$y(t)\exp\left[j\phi_y(t)\right] = A\exp\left[j\phi_s(t)\right] + n_I(t) + jn_Q(t) \qquad (8.99)$$

As studied before, the output of the FM detector is proportional to the derivative of the phase $\phi_y(t)$ of the IF output signal. Hence, by first taking the natural logarithm of Equation 8.99 and then differentiating both sides with respect to time, we obtain

$$\frac{\dot{y}(t)}{y(t)} + j\dot{\phi}_y(t) = \frac{j\dot{\phi}_s(t)A\exp[j\phi_s(t)] + \dot{n}_I(t) + j\dot{n}_Q(t)}{A\exp[j\phi_s(t)] + n_I(t) + n_Q(t)}$$

$$\triangleq f\left(\phi, \dot{\phi}, n_I, \dot{n}_I, n_Q, \dot{n}_Q\right) \tag{8.100}$$

Because $\dot{\phi}_y(t)$ involves random variables, the mean signal output $v_0(t)$ must be estimated by the expectation (mean value) of the random variable $\dot{\phi}_y(t)$, i.e., $v_0(t) = E[\dot{\phi}_y(t)]$. The four random variables, namely, $n_I(t)$, $n_Q(t)$, $\dot{n}_I(t)$, and $\dot{n}_Q(t)$, are independent and Gaussian distributed. Rice[5] has shown that

$$E\left[\frac{\dot{y}(t)}{y(t)} + j\dot{\phi}_y(t)\right] = \int_{-\infty}^{\infty}\int_{-\infty}^{\infty}\int_{-\infty}^{\infty}\int_{-\infty}^{\infty} f\left(\phi, \dot{\phi}, n_I, \dot{n}_I, n_Q, \dot{n}_Q\right)$$

$$\times p\left(n_I, \dot{n}_I, n_Q, \dot{n}_Q\right) dn_I\, d\dot{n}_I\, dn_Q\, d\dot{n}_Q$$

$$= j\dot{\phi}(t)[1 - \exp(-\gamma)] \tag{8.101}$$

Given that $\dot{y}(t)/y(t)$ and $\dot{\phi}_y(t)$ are independent random variables,

$$E\left[\dot{y}(t)/y(t) + j\dot{\phi}_y(t)\right] = E[\dot{y}(t)/y(t)] + jE\left[\dot{\phi}_y(t)\right] \tag{8.102}$$

Comparing this expression with that of Equation 8.101, we notice that $E[\dot{y}(t)/y(t)] = 0$ and

$$E\left[\dot{\phi}_y(t)\right] = \dot{\phi}_s(t)[1 - \exp(-\gamma)] \tag{8.103}$$

Note that $\dot{\phi}_s(t)$ is proportional to the message $m(t)$.

Assuming a proportionality constant equal to 1, the output $v_0(t)$ of the demodulator is the message $m(t)$ attenuated by a factor $[1 - \exp(-\gamma)]$ due to the presence of noise. That is,

$$v_0(t) = E\left[\dot{\phi}_y(t)\right] = m(t)[1 - \exp(-\gamma)] \tag{8.104}$$

The output baseband signal $S_0$ is

$$S_0 = E[v_0^2(t)] = [1 - \exp(-\gamma)]^2 E\left[\dot{\phi}^2(t)\right]$$

$$= [1 - \exp(-\gamma)]^2 P_m \tag{8.105}$$

where $P_m$ is the input modulation signal power. For a single-tone modulation it is shown in Appendix 8C that

$$P_m = \alpha(2\pi\,\Delta f)^2 \qquad (8.106)$$

where $\alpha$ is a constant. With the use of Equation 8.76 in Equation 8.106, we obtain

$$P_m = \alpha\pi^2(B - 2W)^2 \qquad (8.107)$$

Using Equation 8.107 in Equation 8.105, we obtain

$$S_0 = \alpha\pi^2(B - 2W)^2[1 - \exp(-\gamma)]^2 \qquad (8.108)$$

The normalized baseband signal $10\log(S_0/W^2)$ is plotted versus the IF SNR $\gamma$ for different values of the baseband bandwidth ratios $B/2W$ in Figure 8.17. In these plots we have assumed $P_m = 10$ dB below $(2\pi\,\Delta f)^2$, i.e., $\alpha = 0.1$. This probably prevents the signal deviation peaks from exceeding the IF bandwidth.

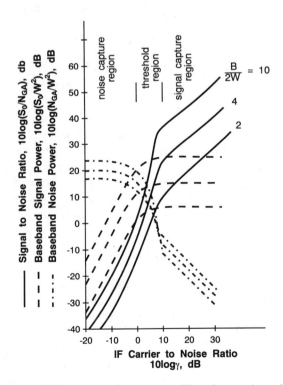

**Figure 8.17.** Baseband SNR, signal and noise versus IF carrier-to-noise ratio for some bandwidth ratios.

### 8.5.6.2 *Output Noise Power*

The output noise is given by $\dot{\phi}_n(t)$. The noise spectral density $S_{\dot{\phi}_n}(f)$ is the Fourier transform of its autocorrelation function $R_{\dot{\phi}_n}(\tau)$, i.e.,

$$S_{\dot{\phi}_n}(f) = \int_{-\infty}^{\infty} R_{\dot{\phi}_n}(\tau)\exp(j2\pi f\tau)\,d\tau \qquad (8.109)$$

with the autocorrelation function expressed as (refer to Appendix 9A)

$$R_{\dot{\phi}_n}(\tau) = E\left[\dot{\phi}_n(t)\dot{\phi}_n(t+\tau)\right] \qquad (8.110)$$

Rice[5] has shown that the baseband noise with modulation closely approximates that without modulation. Therefore, from Figure 8.16 with $\phi_s(t) = 0$ (note that $\phi_r(t) = 0$ and $a_r(t) = 1$ because no Rayleigh fading is being considered)

$$\phi_n(t) = \tan^{-1}\frac{n_Q(t)}{A + n_I(t)} \qquad (8.111)$$

Differentiating both sides of Equation 8.111,

$$\dot{\phi}_n(t) = \frac{\left[A + n_I(t)\right]\dot{n}_Q(t) - n_Q(t)\dot{n}_I(t)}{\sec^2\phi_n(t)\left[A + n_I(t)\right]^2}$$

$$= \frac{\left[A + n_I(t)\right]\dot{n}_Q(t) - n_Q(t)\dot{n}_I(t)}{\left[A + n_I(t)\right]^2 + n_Q^2(t)} \qquad (8.112)$$

With Equation 8.112 in Equation 8.110 the autocorrelation function can be evaluated, provided that the corresponding probability density function is known. Because there are four random variables involved, namely, $n_I(t)$, $n_Q(t)$, $\dot{n}_I(t)$, and $\dot{n}_Q(t)$, the autocorrelation turns out to be an eightfold integral that requires the evaluation of the joint distribution of these variables in $t$ and in $t + \tau$. This is obviously extremely difficult to carry out, but an approximation can be used. Rice[5] divided the noise spectrum $S_n(f)$ into three parts, such that

$$S_n(f) = S_1(f) + S_2(f) + S_3(f) \qquad (8.113)$$

$S_1(f)$ equals the output noise spectrum when the carrier is very large; $S_2(f)$ equals the output noise spectrum when the carrier is absent; $S_3(f)$ is a correction factor, vanishing for both a very small and a very large carrier. It predominates in the threshold region of $\gamma$.

Accordingly, $S_1(f)$ relates to $S_{\dot{\phi}_n}(f)$ in the same way as the output baseband signal power $S_0$ relates to $S_{\dot{\phi}_s}(f)$ (Equation 8.105),

$$S_1(f) = [1 - \exp(-\gamma)]^2 S_{\dot{\phi}_n}(f) \qquad (8.114)$$

Because differentiation in the time domain is equivalent to multiplication by $j2\pi$ in the frequency domain, the following relation holds:

$$S_{\dot{\phi}_n}(f) = (2\pi f)^2 S_{\phi_n}(f) \tag{8.115}$$

However, from Equation 8.111, if the carrier is very large,

$$\phi_n(t) \simeq \frac{n_Q(t)}{A} \tag{8.116}$$

and

$$\dot{\phi}_n(t) \simeq \frac{\dot{n}_Q(t)}{A} \tag{8.117}$$

Therefore, from Equation 8.115

$$S_{\dot{\phi}_n}(f) = \left(\frac{2\pi f}{A}\right)^2 S_{n_Q}(f) \tag{8.118}$$

It is shown in Appendix 8A that the spectrum of the in-phase component $n_I(t)$ and the quadrature component $n_Q(t)$ of a narrowband process $n(t)$ is

$$S_{n_I}(f) = S_{n_Q}(f) = S_n(f - f_c) + S_n(f + f_c) \tag{8.119}$$

Hence using Equations 8.91, 8.92, 8.97, 8.119, 8.118, and 8.114, we obtain

$$S_1(f) = \frac{\{2\pi f[1 - \exp(-\gamma)]\}^2 \exp(-\pi f^2 / B^2)}{2B\gamma} \tag{8.120}$$

As far as $S_2(f)$ and $S_3(f)$ are concerned, Rice[5] has shown that their sum equals $2\pi B\sqrt{2\pi}\,(1 - \mathrm{erf}\sqrt{\gamma})$ for each side of the spectrum. Davis[7] has found an empirical approximation given by

$$S_2(f) + S_3(f) \simeq \frac{4\pi B \exp(-\gamma)}{\sqrt{2(\gamma + 2.35)}} \tag{8.121}$$

The total output noise power $N_{GA}$ of the baseband filter due to a Gaussian noise is

$$N_{GA} = \int_{-W}^{W} [S_1(f) + S_2(f) + S_3(f)]\, df$$

$$= \frac{a[1 - \exp(-\gamma)]^2}{\gamma} + \frac{8\pi BW \exp(-\gamma)}{\sqrt{2(\gamma + 2.35)}} \tag{8.122}$$

where

$$a \triangleq \frac{(2\pi)}{2B} \int_{-W}^{W} f^2 \exp\left(-\frac{\pi f^2}{B^2}\right) df \tag{8.123}$$

Using the MacLaurin expansion,

$$a = \frac{(2\pi)^2}{2B} \int_{-W}^{W} f^2 \sum_{i=0}^{\infty} \frac{(-\pi f^2/B^2)^i}{i!} df$$

$$= \frac{4\pi W^3}{3B} \left\{ 1 - \frac{6\pi}{10} \left(\frac{W}{B}\right)^2 + \frac{12\pi^2}{56} \left(\frac{W}{B}\right)^4 + \cdots \right\} \tag{8.124}$$

The normalized baseband noise power $10\log(N_{GA}/W^2)$ is plotted versus $\gamma$ in Figure 8.17. The $SNR_O$ is therefore

$$SNR_O = \frac{S_0}{N_{GA}} = \frac{\alpha\pi^2(B-2W)^2[1-\exp(-\gamma)]^2}{\dfrac{a[1-\exp(-\gamma)]^2}{\gamma} + \dfrac{8\pi BW\exp(-\gamma)}{\sqrt{2(\gamma+2.35)}}} \tag{8.125}$$

which is plotted in Figure 8.17.

If $\gamma \gg 1$, then, from Equations 8.105 and 8.122, respectively,

$$S_0 \simeq P_m \tag{8.126}$$

and

$$N_{GA} \simeq \frac{a}{\gamma} \simeq \frac{4\pi W^3}{3B\gamma} = \frac{2}{3}\left(\frac{2\pi}{A}\right)^2 \eta W^3 \tag{8.127}$$

With $P_m = \alpha(2\pi \Delta f)^2$ the $SNR_O$ becomes

$$SNR_O = S_0/N_{GA} \simeq 3\alpha\beta^2(B/W)\gamma \tag{8.128}$$

where $\beta = \Delta f/W$. This shows that $SNR_O$ and bandwidth are interchangeable in FM systems. The figure of merit in this case is

$$F = \frac{SNR_O}{SNR_C} = 3\alpha\beta^2 \tag{8.129}$$

Note from Figure 8.17 that the threshold and capture effects are quite evident: When $\gamma$ is large the FM receiver captures on the signal and suppresses the noise. When $\gamma$ is small the opposite effect occurs.

### 8.5.7    The Effect of Multipath Propagation on FM Systems

The capture effect of FM receivers, so prominent in a nonfading environment, tends to vanish in multipath conditions due to rapid fading. When the signal fades in such a way that the receiver loses capture, the baseband output signal is suppressed. The rapid random suppression of the signal constitutes a noise component, known as signal-suppression noise. Additionally, deep fades are accompanied by rapid phase changes, producing another noise component: random FM noise (refer to Section 4.8). The total baseband output noise $N_0$ is, therefore, composed of the following elements: (1) signal-suppression noise $(N_{SS})$; (2) random FM noise $(N_{RF})$; and (3) Gaussian additive noise $(N_{GA})$. The statistical properties of the three noise components are different, so that an absolute measure of the quality of the mobile radio channel is not possible. In particular, the random FM noise is independent of the carrier-to-noise ratio $\gamma$. As for the measure of quality, subjective tests seemed to indicate that the ratio of average signal to average noise gives a good measure.[2] Consider initially an output noise $N_0$ constituted by the signal-suppression noise and the Gaussian noise, i.e., $N_0 \triangleq N_{SS} + N_{GA}$. The average-signal–to–average-noise ratio is, in this case $\bar{S}_0/\bar{N}_{SA}$. For the random FM situation this is $\bar{S}_0/\bar{N}_{RF} = \bar{S}_0/N_{RF}$, because $\bar{N}_{RF} = N_{RF}$.

If the fading rate is small compared to the IF bandwidth (this is usually the case for microwave carrier frequencies, where we have 50–1000 fades per second), then a quasistatic approximation can be used. In this approximation the signal and the noise can be expressed as functions only of the instantaneous IF SNR, $\gamma$. Accordingly, the averages can be taken over the statistics of $\gamma$ having a negative exponential distribution (refer to Section 5.6) written as

$$p(\gamma) = \frac{1}{\gamma_0} \exp\left(-\frac{\gamma}{\gamma_0}\right) \qquad (8.130)$$

where $\gamma_0$ is the average IF SNR.

Therefore, the mean signal output voltage averaged over $\gamma$ is

$$\bar{v}_0(t) = E[v_0(t)] = \int_0^\infty v_0(t) p(\gamma) \, d\gamma \qquad (8.131)$$

With Equations 8.104 and 8.130 in Equations 8.131, we obtain

$$\bar{v}_0(t) = m(t) \frac{\gamma_0}{1 + \gamma_0} \qquad (8.132)$$

Hence, the mean signal output power is

$$S_0 = \frac{\gamma_0^2}{(1 + \gamma_0)^2} P_m \tag{8.133}$$

where $P_m = \langle m^2(t) \rangle$ is the message power given by Equation 8.107.

The signal-suppression noise is $n_s(t) = v_0(t) - \bar{v}_0(t)$ and the corresponding power is

$$N_{SS} = \left\langle [v_0(t) - \bar{v}_0(t)]^2 \right\rangle = \left[ \frac{1}{1 + \gamma_0} - \exp(-\gamma) \right]^2 P_m \tag{8.134}$$

The average output noise power $N_{SA}$ is

$$\overline{N}_{SA} = \int_0^\infty (N_{SS} + N_{GA}) p(\gamma) \, d\gamma \tag{8.135}$$

With Equations 8.134, 8.122, and 8.130 in Equation 8.135, we get

$$\overline{N}_{SA} = \left[ \frac{1}{1 + 2\gamma_0} - \frac{1}{(1 + \gamma_0)^2} \right] P_m + \frac{a}{\gamma_0} \log \frac{(1 + \gamma_0)^2}{1 + 2\gamma_0}$$

$$+ 8BW \sqrt{\frac{\pi}{2\gamma_0(1 + \gamma_0)}} \exp\left[ 2.35 \frac{1 + \gamma_0}{\gamma_0} \right] \mathrm{erfc} \sqrt{2.35 \frac{1 + \gamma_0}{\gamma_0}} \tag{8.136}$$

where $a$ is given by Equation 8.123 and $\mathrm{erfc}(x) \triangleq 1 - \mathrm{erf}(x)$.

The ratio of average signal to average noise is $\overline{S}_0/\overline{N}_{SA}$, which can be obtained directly from the ratio between Equation 8.133 and Equation 8.136. As for the random FM, it is shown in Section 4.8 that the corresponding output noise $N_{RF}$ within the frequency range $\omega_1$ to $\omega_2$ is given by Equation 4.77, reproduced here for convenience,

$$N_{RF} = \frac{(\Gamma v)^2}{2} \ln\left( \frac{\omega_2}{\omega_1} \right) \tag{8.137}$$

where $\Gamma = 2\pi/\lambda$ and $v$ is the speed of the mobile. Assuming an audio band of $\omega_1 = 300$ Hz and $\omega_2 = 3000$ kHz, the average-signal–to–average-noise ratio is given by

$$\frac{S_0}{N_{RF}} = \frac{(B - 2W)^2}{20(v/\lambda)^2 \ln 10} \frac{\gamma_0^2}{(1 + \gamma_0)^2} \tag{8.138}$$

where $\alpha$ has been assumed to be 0.1.

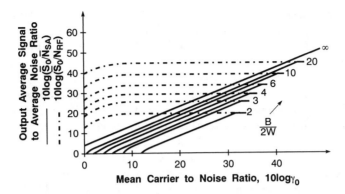

**Figure 8.18.** Output signal-to-noise ratio versus carrier-to-noise ratio in a Rayleigh environment.

Both ratios $\bar{S}_0/\bar{N}_{SA}$ and $\bar{S}_0/N_{RF}$ are plotted[7] in Figure 8.18 as functions of the average carrier-to-noise ratio $\gamma_0$, where we assume a mobile speed of 96 km/h (60 mi/h) and a carrier frequency of 900 MHz ($\lambda = \frac{1}{3}$ m), corresponding to a Doppler shift of $v/\lambda = 80$ Hz.

The interpretation of Figure 8.18 is eased if we carry out the following reasoning. The overall signal-to-noise ratio $SNR_O$ must include all the noise components as follows:

$$SNR_O = \frac{\bar{S}_0}{\bar{N}_{SA} + \bar{N}_{RF}} = \frac{(\bar{S}_0/\bar{N}_{SA})(\bar{S}_0/N_{RF})}{(\bar{S}_0/\bar{N}_{SA}) + (\bar{S}_0/N_{RF})} \qquad (8.139)$$

We can rewrite Equation 8.139 as $SNR_O = XY/(X + Y)$, where $X$ and $Y$ may assume the values of either ratio, indistinctly. Expressing $SNR_O$ in decibels, we have

$$10 \log SNR_O = 10 \log X + 10 \log Y - 10 \log(X + Y)$$

If $X \gg Y$, then $10 \log SNR_O \simeq 10 \log Y$. When $X \simeq Y$, then

$$10 \log SNR_O \simeq 10 \log X - 10 \log 2$$

$$\simeq 10 \log Y - 10 \log 2$$

Using the same reasoning, we may interpret the curves of Figure 8.18 as follows. For small values of $\gamma_0$, $\bar{S}_0/\bar{N}_{SA} \ll \bar{S}_0/N_{RF}$, so that $\bar{S}_0/\bar{N}_{SA}$ predominates. Therefore $SNR_O \simeq \bar{S}_0/\bar{N}_{SA}$. For large values of $\gamma_0$ the opposite situation occurs, so that $SNR_O \simeq \bar{S}_0/N_{RF}$. For intermediate values of $\gamma_0$, the maximum difference between $SNR_O$ and $\bar{S}_0/\bar{N}_{SA}$ or $\bar{S}_0/N_{RF}$ is $-3$ dB ($= -10 \log 2$), attained when $\bar{S}_0/\bar{N}_{SA} = \bar{S}_0/N_{RF}$. Finally, we may use the following interpretation for the overall $SNR_O$ curves. They coincide with those of $\bar{S}_0/N_{SA}$ for small $\gamma_0$ and with those of $\bar{S}_0/N_{RF}$ for large $\gamma_0$. For intermediate values of $\gamma_0$ the overall $SNR_O$ curves perform a smooth transition from the $\bar{S}_0/N_{SA}$ curves to the $\bar{S}_0/N_{RF}$ ones with a difference between them not exceeding 3 dB.

Note that the pronounced threshold due to capture observed in the non-Rayleigh case (Figure 8.17) is not present any longer. As the transmitter power is increased ($\gamma_0$ is increased) the output signal-to-noise ratio increases approximately linearly with $\gamma_0$. Threshold crossings are, therefore, less frequent but the random FM noise remains unchanged (it does not depend on $\gamma_0$) and becomes the dominant noise component. Hence, the limiting output signal-to-noise ratio is given by Equation 8.138 for large $\gamma_0$, i.e.,

$$SNR_{limiting} = \lim_{\gamma_0 \to \infty} \frac{S_0}{N_{RF}} = \frac{(B - 2W)^2}{20 f_m^2 \ln 10} \qquad (8.140)$$

where $f_m = v/\lambda$.

## 8.6 SUMMARY AND CONCLUSIONS

The standard AM scheme has in its favor ease of implementation and spectrum saving: Demodulation can be carried out by means of an envelope detector or a square-law detector, and the transmission bandwidth is only twice the message bandwidth. On the other hand, additional power is required to transmit the carrier. SSB modulation requires the minimum transmitter power and minimum transmission bandwidth but the receivers are more complex. The FM scheme usually needs more frequency spectrum but yields better performance in the presence of noise.

As far as additive noise is concerned the FM systems operating above a minimum carrier-to-noise ratio (FM threshold) usually give better performance than the SSB and AM systems. Because this performance depends on the transmission bandwidth, we may say that, for FM, bandwidth and signal-to-noise ratio are interchangeable. Comparative performance of these systems is shown in Figure 8.19.

When fading (multiplicative noise) is taken into account, standard AM and conventional SSB experience disastrous performance. The SSB scheme still gives better results than AM, with a signal–to–signal-suppression noise ratio 10.4 dB above that of the AM. The FM technique is also substantially affected by the effects of fading, but has noticeably better performance.

Generally speaking, the fading spectrum lies below the information spectrum. This suggests that the use of a convenient filter can improve the noise performance of AM systems. However, because fading depends on the speed of the vehicle, fading and information spectra may overlap with a consequent degradation of signal quality. This can be minimized by transmission of a

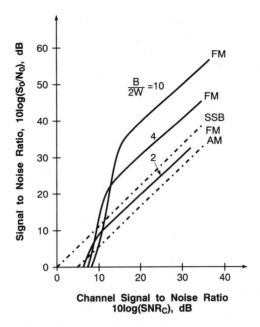

**Figure 8.19.** Comparative performance of analog modulation techniques.

pilot tone that can be used for frequency or gain control. In narrowband systems, both the pilot tone and the modulated signal are equally affected by fading (flat fading). For slow fading, the automatic gain control (AGC) circuit may be able to compensate for the variations of the amplitude of the signal, "cancelling out" the fading. However, this requires a minimum 30-dB carrier-to-noise ratio in order to have the AGC of the receiver working satisfactorily. For high fading rates, which occur in urban areas, the AGC is substantially less effective.

Some new SSB modulation schemes using an in-band pilot tone and companded amplitude have been proposed and are likely to be used in some mobile radio services such as satellite mobile communications.

## APPENDIX 8A   POWER SPECTRAL DENSITY
## OF A NARROWBAND NOISE

A narrowband noise $n(t)$ centered at a frequency $f_c$, can be written as a function of an in-phase component $n_I(t)$ and quadrature component $n_Q(t)$ as follows:

$$n(t) = n_I(t)\cos(2\pi f_c t) - n_Q(t)\sin(2\pi f_c t) \qquad (8A.1)$$

Given $n(t)$ we can obtain the in-phase component $n_I(t)$ by performing the multiplication $2n(t)\cos(2\pi f_c t)$ followed by a low-pass filtering of the resultant signal. In the same way, the quadrature component can be obtained by performing $2n(t)\sin(2\pi f_c t)$ and low-pass filtering the resultant signal. Therefore, we may write these components as

$$n_I(t) = 2n(t)\cos(2\pi f_c t) \tag{8A.2}$$

$$n_Q(t) = 2n(t)\sin(2\pi f_c t) \tag{8A.3}$$

The power spectral density $S_{n_I}(f)$ of the random process $n_I(t)$ is the Fourier transform of its autocorrelation function $R_{n_I}(\tau)$. This autocorrelation function is given by (see Appendix 9A)

$$R_{n_I}(\tau) = E\left[n_I(t + \tau)n_I(t)\right]$$

$$= E\left[4n(t + \tau)n(t)\cos(2\pi f_c t + 2\pi f_c \tau)\cos(2\pi f_c t)\right]$$

$$= 4E\left[n(t + \tau)n(t)\right]\tfrac{1}{2}E\left[\cos(2\pi f_c \tau) + \cos(4\pi f_c t + 2\pi f_c \tau)\right]$$

$$= 2R_n(\tau)\cos(2\pi f_c \tau) \tag{8A.4}$$

The Fourier transform of $R_n(\tau)$ is $S_n(f)$. The Fourier transform of the modulated process $R_n(\tau)\cos(2\pi f_c \tau)$ is $\tfrac{1}{2}S_n(f + f_c) + \tfrac{1}{2}S_n(f - f_c)$, such that

$$S_{nI}(f) = S_n(f - f_c) + S_n(f + f_c) \tag{8A.5}$$

The same reasoning can be applied to obtain an equal relation for $S_{n_Q}(f)$, i.e.,

$$S_{n_Q}(f) = S_n(f - f_c) + S_n(f + f_c) \tag{8A.6}$$

## APPENDIX 8B   SIGNAL – TO – SIGNAL-SUPPRESSION NOISE RATIO OF AM SYSTEM

Section 8.3.7 showed that the message signal power at the envelope detector output is $(\pi/2)(A\sigma)^2 P_m$. Using Equations 8.42 and 8.43, it follows that the signal-suppression noise power is

$$\left\langle \left[y(t) - \bar{y}(t)\right]^2 \right\rangle = A^2 \left\langle \left(r - \sqrt{\pi/2}\,\sigma\right)^2 \right\rangle \left\langle \left[1 + k_a m(t)\right]^2 \right\rangle \tag{8B.1}$$

Expanding the first term of the product we have

$$\left\langle \left(r - \sqrt{\pi/2}\,\sigma\right)^2 \right\rangle = \left\langle r^2 - 2r\sqrt{\pi/2}\,\sigma + \pi/2\sigma^2 \right\rangle$$

$$= \langle r^2 \rangle - 2\langle r \rangle\sqrt{\pi/2}\,\sigma + \pi/2\sigma^2 \tag{8B.2}$$

However,

$$\langle r^2 \rangle = \int_0^\infty r^2 p(r) \, dr = 2\sigma^2 \tag{8B.3}$$

where $p(r)$ is given by Equation 8.38. Also,

$$\langle r \rangle = \int_0^\infty r p(r) \, dr = \sqrt{\pi/2} \, \sigma \tag{8B.4}$$

Therefore, with Equations 8B.4 and 8B.3 in Equation 8B.2, we have

$$\left\langle \left( r - \sqrt{\pi/2} \, \sigma \right)^2 \right\rangle = (2 - \pi/2)\sigma^2 \tag{8B.5}$$

Expanding the second term of the product (Equation 8B.1) and assuming that the message $m(t)$ has a zero value, we obtain

$$\left\langle \left[ 1 + k_a m(t) \right]^2 \right\rangle = 1 + P_m \tag{8B.6}$$

where

$$P_m = k_a^2 \langle m^2(t) \rangle$$

Accordingly, with Equations 8B.6 and 8B.5 in Equation 8B.1, the signal–to–signal-suppression noise ratio SSN is

$$\text{SSN} = \frac{(\pi/2)\sigma^2 P_m}{\left\langle \left( y(t) - \overline{y(t)} \right)^2 \right\rangle} = \frac{P_m}{1 + P_m} \frac{\pi/2}{(2 - \pi/2)} \tag{8B.7}$$

## APPENDIX 8C   SINGLE-TONE MODULATION

Consider a sinusoidal wave of frequency $W$ as the modulating wave, and assume a frequency deviation of $\Delta f$. Hence the modulated signal is

$$\tilde{s}(t) = A \exp\left[ j2\pi f_c t + \frac{\Delta f}{W} \sin(2\pi W t) \right]$$

where we have used

$$2\pi k_f \int_0^t m(t) \, dt = \frac{\Delta f}{W} \sin(2\pi W t)$$

Differentiating both sides of this equation we obtain

$$m(t) = \frac{\Delta f}{k_f} \cos(2\pi W t)$$

Therefore, the average power $P_m$ is

$$P_m = \frac{(\Delta f / k_f)^2}{2}$$

$$= \alpha(2\pi \Delta f)^2$$

where $\alpha = 1/[2(2\pi k_f)^2]$, with $k_f$ in hertz per volt.

## REFERENCES

1. Haykin, S., *An Introduction to Analog and Digital Communications*, John Wiley & Sons, Singapore, 1989.
2. Jakes, W. C., Jr., *Microwave Communications Engineering*, John Wiley & Sons, New York, 1974.
3. Lee, W. C. Y., *Mobile Communications Engineering*, McGraw-Hill, New York, 1982.
4. Parsons, J. D., and Gardiner, J. G., *Mobile Communication Systems*, Blackie and Son Ltd., Glasgow, 1989.
5. Rice, S. O., Noise in FM receivers, in *Proc. Symp. Time Series Analysis*, M. Rowenblatt, Ed., Chapter 25, Wiley, New York, 1963.
6. Rice, S. O., Statistical properties of a sine wave plus random noise, *Bell Syst. Techn. J.*, 27, 109–157, January 1948.
7. Davis, B. R., FM noise with fading channels and diversity, *IEEE Trans. Commun.*, COM-19(6), Part II, 1189–1199, December 1971.
8. McGeeham, J. P., and Bateman, A. J., Phase-locked transparent tone-in-band (CTTIB): a new spectrum configuration particularly suited to the transmission of data over SSB mobile radio networks, *IEEE Trans. Commun.*, COM-32(1), 81–87, 1984.
9. McGeeham, J. P., and Bateman, A. J., Theoretical and experimental investigation of feed forward signal regeneration as a means of combating multipath propagation effects in pilot-base SSB mobile radio systems, *IEEE Trans. Vehicular Tech.*, VT-32, 106–120, February 1983.
10. Carson, J. R., and Fry, T. C., Variable frequency electric circuit theory with applications to the theory of frequency modulation, *Bell Syst. Tech. J.*, 16, 513–540, October 1937.

CHAPTER **9**

# Digital Techniques for Mobile Radio

This chapter examines two basic techniques used in digital mobile radio systems, namely, speech coding and modulation. Because there is a "countless" number of speech coding and modulation schemes, we chose to analyze those selected to be used in the European and American digital cellular radio systems, in connection with their closely related techniques. Initially, we introduce the basic principles of linear prediction coding (LPC) and vector quantization. The basics of linear prediction are briefly examined. Thence we describe some LPC techniques such as RELP, RPE-LPC, MPE-LPC, RPE-LTP, MPE-LTP, CELP, and VSELP. Because the speech-coding techniques RPE-LTP and VSELP have been selected for use in the European and the American systems, respectively, they are better explored. We then examine some higher-order modulation schemes such as QPSK, DQPSK, MSK, and GMSK, studying their basic principles and exploring the generation, detection, power spectra, and bit error rate performance of the respective signals. In the bit error rate studies the technique performance is examined taking into account a Gaussian channel and a Rayleigh fading channel. In the latter we consider the slow-fading case for all the modulation schemes and the fast-fading case for DQPSK and GMSK only, because these are the techniques selected for use in the European and American systems.

## 9.1 INTRODUCTION

An analog signal using a digital communication system experiences two processes prior to transmission: analog-to-digital (A/D) conversion, and modulation. The analog-to-digital conversion reduces the complex analog waveform to a convenient digital configuration. The modulation then processes the obtained digital waveform to make it suitable for transmission.

### 9.1.1 A / D Conversion

The A/D conversion comprises three basic steps, namely, sampling, quantizing, and coding. *Sampling* is a process of detecting the instantaneous value of a waveform, usually carried out at regular time intervals. The sampling theorem requires that a signal should be sampled at a frequency at least twice as high as the highest frequency present in the waveform. For telephony signals the speech is band-limited to 3.4 kHz. Allowing for the guard band, we approximate this to 4 kHz, leading to a sampling rate of 8 kHz (corresponding to an interval of 125 $\mu$s between samples). After sampling, a sequence of pulses is obtained. These pulses have amplitudes equal (or proportional) to the amplitude of the analog input signal at the sampling instant. The resultant waveform is known as a pulse amplitude modulation (PAM) signal.

The next step in the digitization process is *quantization*. Quantization consists in assigning a new value of amplitude to each pulse, corresponding to the nearest level available in the specially created finite discrete set of amplitude levels. This step introduces noise in the process, the quantization noise, due to the difference between the true amplitude and the assigned amplitude. The larger the number of quantization levels, the smaller the quantization noise. It has been found that a ratio of signal to quantization noise of 30–40 dB is required for telephony signals, achieved with a minimum of 256 quantization levels.[20]

*Coding* is the next step in the digitization process. The oldest and conceptually simplest coding technique used in speech and video signals is PCM,[1] in which the PAM signal is converted into a binary stream and is multiplexed for serial transmission. The 256 ($= 2^8$) quantization levels can be represented (encoded) by a minimum of 8 bits, resulting in a transmission rate of 8000 samples per second $\times$ 8 bits per sample = 64 kbit/s. Accordingly, a 3.4-kHz analog speech signal using PCM requires a 32-kHz bandwidth for baseband transmission, almost 10 times as large as that of the analog transmission. This constitutes one of the main constraints of using PCM-encoded signals for digital mobile radio, where the frequency spectrum must be carefully administered. There are some spectrally efficient modulation techniques that can be used to minimize this problem. The implications of using such techniques will be explored later in this section. Another solution to this problem is the use of modern speech-coding algorithms.

In conventional 64-kbit/s PCM, each sample is quantized and encoded independently, regardless of the redundancy present in the speech. It is reported that successive samples present a correlation coefficient of not less than 0.85.[20] If the correlation between samples is taken into account, fewer bits can be used to represent the speech. Modern speech-coding algorithms explore the redundancy of speech signals to reduce the transmission bit rate. There is a "countless" number of speech-coding techniques already developed as well as under investigation. A complete study of these techniques is well beyond the scope of this book. We shall restrict ourselves to the fundamentals of the speech-coding techniques, exploring both the algorithms chosen to be used in the European and American digital cellular systems and the related algorithms.

### 9.1.2  Modulation

Modulation can be considered as a second coding stage.[21] Many aspects must be considered in choosing a modulation technique, including (1) required bandwidth, (2) intersymbol interference, (3) adjacent-channel interference, and (4) bit error rate performance (BER).

Binary modulations, such as ASK, FSK, and PSK, are rather simple and robust but spectrally inefficient. In this sense, multilevel modulation techniques are preferred, despite their inferior BER performance. In order to gain insight into these problems, consider a $B$-kHz bandwidth channel. Roughly speaking, a binary modulator can transmit $B$ kbit/s through this channel. Suppose now that each pair of binary symbols is encoded as one symbol. For instance, in ASK this could be achieved by assigning the symbols 0, 1, 2, and 3 to the pairs 00, 01, 10, and 11, respectively. Accordingly, the symbol rate is divided by 2, with a corresponding decrease by a factor of 2 of the occupied bandwidth. On the other hand, the error rate performance is degraded due to the increase of the number of levels that must be distinguished from one another at the reception end.

In a way similar to speech coding, there is a myriad of digital modulation techniques. Therefore, a full investigation on these techniques is well beyond the scope of this book. We shall concentrate on the techniques chosen to be used in the European and American digital cellular systems and on those techniques more closely related to them.

## 9.2  SPEECH CODING FOR MOBILE RADIO

The ultimate objective of the speech-coding algorithms is to transmit, store, or synthesize speech at a given quality using fewer bits.[1] Bit-rate reduction is accomplished by taking advantage of both the redundancies in the speech signal and the perceptual limitations of the human ear. Speech redundancies are related to the following human tract characteristics:

- Speech spectrum changes relatively slowly compared to the sampling rate.
- Vibration rate of the vocal cords also changes relatively slowly.
- Speech energy is concentrated at the lower frequencies.
- Speech sounds can be modelled as periodic or noisy excitations passing through the vocal tract (filter), as shall be detailed later.

One limitation of human hearing that can be explored by speech-coding algorithm is its relatively low sensitivity to signal phase. Moreover, the hearing is more strongly sensitive to a small portion of the audible frequency spectrum.

Bit rate is directly related to transmission bandwidth. Accordingly, bit-rate reduction implies bandwidth reduction, accomplished at the expense of degrading speech quality. At a given bit rate, speech quality can be improved by increasing the complexity of the algorithm. Therefore, in assessing overall speech-coding algorithm performance, the three parameters, namely, *Rate*, *Quality*, and *Complexity*, must be taken into consideration.

The parameter Quality presents different requirements, varying according to the service. There are basically four Quality classes to be considered:

- *Broadcast quality*—referring to wideband transmission with high-quality speech
- *Toll quality*—referring to speech as heard over a switched telephone network
- *Communication quality*—referring to highly intelligible speech but with more distortion than toll-quality speech
- *Synthetic quality*—referring to "machinelike" speech

The speech quality can be evaluated by means of objective measures (e.g., SNR) or subjective measures (e.g., the mean opinion score [MOS]).

The Complexity parameter is related to the processing required to implement the coding algorithm. This is usually measured in mega-operations per second (MOPS).

Speech coders are usually grouped into two categories: waveform coders and voice coders (vocoders). Waveform coders are further divided into time-domain waveform (TW) and spectral-domain waveform (SW) coders. The former takes advantage of the periodicity and slowly varying intensity of the signal, whereas the latter explores the speech redundancies across frequencies. Vocoders assume a speech production model to reproduce the speech. Waveform coders usually yield superior speech quality but operate at higher bit rates. Examples of TW coders include PCM, DPCM and ADPCM. Among SW coders we have subband coding (SBC), adaptive transform coding (ATC), and others. The linear predictive coding (LPC) techniques, such as residual excited linear prediction (RELP) etc., are examples of vocoders. In this chapter we shall concentrate our attention on vocoders.

## 9.2.1  Vocoders

The basic speech production model of vocoders assumes a clear separation between excitation and vocal-tract filter information. The excitation is the sound-production mechanism, whereas the vocal tract is the "device" used to modulate the sound. Using this assumption, the information is encoded separately, with a substantial decrease in the bit rate.

The excitation may be either voiced or unvoiced. Voiced sounds are quasiperiodic, occurring in the larynx, where air flow can be periodically

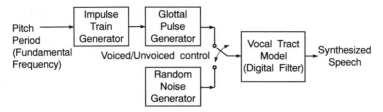

**Figure 9.1.**  Typical model of speech production.

interrupted by the vocal folds. Unvoiced sounds are noisy and aperiodic, generated at a narrow constriction of the vocal tract (usually toward the mouth end). The voiced sound production can be modelled as a response of the local-tract filter excited with a periodic sequence of impulses spaced by a fundamental period equal to the *pitch* period. Unvoiced sound corresponds to the response of the vocal tract when excited with a white-noise sequence.[11] The vocal tract is modelled as a filter having its parameters varying with time. Therefore, in theory, if (1) the filter coefficients, (2) a voiced or unvoiced parameter, and (3) the pitch period are available, the corresponding speech signal may be reproduced in its entirety, except for quantization error and noise. A block diagram illustrating a typical model of speech production is depicted in Figure 9.1.

A considerable number of low-bit-rate speech-coding techniques is available and analyzed in the literature.[1,12] The most popular technique is linear predictive coding, comprising a series of other algorithms using the LPC basic principles. In this chapter we shall examine the LPC fundamentals and the techniques chosen for use in the European and American digital cellular mobile radio systems. With the same objective, we shall describe vector quantization, a way of further reducing the bit rate.

## 9.3  LINEAR PREDICTIVE CODING

Linear predictive coding (LPC) owes its popularity to its simplicity, compactness, and precise representation of speech. Simplicity and compactness are accomplished at the expense of some approximations, sacrificing the nasal and unvoiced sounds. Before proceeding with our studies, we shall examine some of the fundamentals of linear prediction, aiming at speech-coding applications.

### 9.3.1  Linear Prediction

Linear prediction is a form of estimation using a linear combination of present and past samples of a stationary process to predict a sample of the process in the future. Let $S_{n-k}$, $k = 1, \ldots, M$, be random samples from a stationary process $S(t)$ and let $S_n$ be the sample to be predicted. The estimate of $S_n$ is $\hat{S}_n$ such that

$$\hat{S}_n = \sum_{k=1}^{M} h_k S_{n-k} \tag{9.1}$$

where $h_k$, $k = 1, \ldots, M$, are constants and $M$ is the number of delay elements. The filter designed to implement the estimate $\hat{S}_n$ is called a linear predictor. Accordingly, $h_k$ are the filter coefficients, and $M$ is referred to as the filter order. The difference between the true sample $S_n$ and its estimate $\hat{S}_n$ is called the prediction error $\varepsilon_n$, i.e.,

$$\varepsilon_n = S_n - \hat{S}_n$$

$$= S_n - \sum_{k=1}^{M} h_k S_{n-k} \tag{9.2}$$

The structure implementing Equation 9.2 is known as an analysis filter or inverse filter. We may invert Equation 9.2 to obtain $S_n$ given (1) the error, (2) the past samples, and (3) the filter coefficients.

Therefore,

$$S_n = \varepsilon_n + \sum_{k=1}^{M} h_k S_{n-k} \tag{9.3}$$

The structure implementing Equation 9.3 is called a synthesis filter. Three such filters are shown in Figure 9.2.

### 9.3.2 Linear Predictive Coding

The standard LPC vocoder provides two basic functions for speech production: analysis, carried out at the transmission end, and synthesis, carried out at the reception end. The analysis consists of three basic functions: (1) deciding whether the speech is voiced or unvoiced, (2) determining the pitch period in case of voiced sounds, and (3) calculating the filter coefficients. The synthesis consists in choosing either a periodic or white-noise waveform to excite its filter and modelling the vocal tract. The determination of the optimum filter coefficients at the transmission end may involve either the calculation and minimization of the prediction error over the relevant past samples or some specially designed algorithms.

The number of samples used in the analysis–synthesis process is such that the duration of the segment formed by these samples is 10–30 ms. During this period the speech production process may be considered as essentially

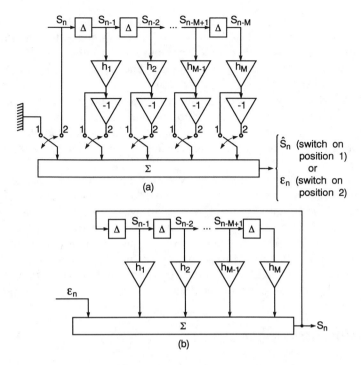

**Figure 9.2.** Linear prediction: (a) linear predictor (switch at position 1) or inverse filter (switch at position 2); (b) synthesis filter.

stationary. The resultant of this simplistic process is speech with a synthetic ("machinelike") quality and typical transmission rate of 2.4 kbit/s. Moreover, the intelligibility of the produced speech is poor for breathy or nasal sounds.

## 9.4 VECTOR QUANTIZATION

Vector quantization (VQ) is not a speech-coding technique on its own. On the contrary, it is a general coding principle used in connection with speech-coding techniques to reduce further the number of coding bits required. In general, speech coders assume each instantaneous speech sample $S_n$ is independent, having most of its redundancy removed in the analysis process. However, consecutive samples are far from random, keeping some correlation between them. Vector quantization considers consecutive samples of $S_n$ as a block (or vector) and encodes them appropriately to remove further redundancy. Consider, for example, 10 5-bit speech samples (a total of 50 bits). It may be possible to use 10 bits in a VQ scheme in order to encode the information contained in these 50 bits, because the number of distinct sounds in speech is not sufficiently large to warrant 50 bits.[12]

Let $k$ be the number of consecutive samples (the frame). After the normal analysis (without VQ) a set of $k$ scalar parameters is generated. A codebook containing $p$ $k$-dimensional vectors is created so that, instead of the $k$ scalar parameters, a $\log_2 p$-bit code word is sent. The index $\log_2 p$ identifies the vector most closely representing the initial set of $k$ scalar parameters. Accordingly, the complexity of the encoder is increased by the additional task of searching for the appropriate vector among the $p$ possible ones in the codebook to be encoded and sent. On the other hand, the complexity of the decoder remains unchanged, although more memory is required to accommodate a replica of the codebook.

The two key problems of VQ are creating and searching the codebook. Codebook creation may involve iterative procedures by using a large training sequence of speech containing a representative amount of phonemes. An initial codebook is assumed, but the optimum codewords are obtained by averaging all of the vectors of the training sequence that can be mapped onto the initial codeword. As for the searching problems, many suboptimal algorithms can be used in order to avoid the substantially time-consuming process represented by the full codebook search.

## 9.5   SOME LPC TECHNIQUES

In this section we shall examine some of the main techniques using LPC principles. The aim is to present the speech codec algorithms that have been adopted by the European as well as by the American digital cellular systems.

Waveform speech coders produce toll-quality speech at rates above 16 kbit/s, whereas standard LPC vocoders provide synthetic-quality speech at rates below 2.4 kbit/s. The higher-quality speech achieved by the former is due to the transmission of the residual error $\varepsilon_n$ (See Equation 9.2) in its entirely. Standard LPC vocoders, on the other hand, provide information about voicing, pitch period, and gain. Both of them yield the filter coefficients. It is possible to combine some of the characteristics of the waveform coders with those of the LPC vocoder in order to improve speech quality of the latter at the expense of a little increase in bit rate. These "hybrid" (vo)coders are still able to produce toll-quality speech at rate ranging from 4.8 to 13 kbit/s. Some of these techniques are described next.

### 9.5.1   Residual-Excited Linear Prediction

The residual-excited linear prediction (RELP) algorithm uses the same parameters as those of standard LPC in addition to the low-pass filtered residual error. The low-pass filtering ranges from 0 to typically 900 Hz, where the frequencies are supposed to carry the highest perceptual importance. This contributes to the naturalness of the speech as well as to a better reproduction of the voiced sounds (the weak points of the standard LPC vocoders). Because the filtered baseband retains the waveshape, pitch and gain parameters need not be sent. This greatly simplifies the RELP analyzer and improves the LPC synthesis.

### 9.5.2   Regular-Pulse Excited LPC

Instead of using spectral information to improve speech quality as in the RELP case, regular-pulse excited LPC (RPE-LPC) operates in the time domain. In the RPE-LPC the residual (error) signal is presented by a given number of impulses per frame of speech data. A reduction (decimation) factor of $8:1$ is quite common. For example, a frame containing 64 residual samples is reduced to 8 equally spaced samples. Like RELP, in RPE-LPC pitch and gain parameters are not sent.

### 9.5.3   Multipulse Excited LPC (MPE-LPC)

This technique operates in a way similar to RPE-LPC. The difference, however, is that the impulses chosen to represent the residual signal are not necessarily equally spaced. On the contrary, their positions and amplitudes are selected to yield the best representation of the error signal. In a sense, multipulse residual "resembles" a skeleton of the actual residual signal"[12].

Both RPE-LPC and MPE-LPC may include a long-term prediction (LTP) to provide information about the pitch period. In the former case we have the RPE-LTP and in the latter the MPE-LTP.

### 9.5.4   The GSM Codec

More than 20 different codec proposals were considered for evaluation by the European digital cellular radio, the GSM system. Out of them, 4 were selected for testing, namely, RPE-LPC, MPE-LTP, SBC-adaptive PCM in 14 subbands, and SBC-adaptive differential PCM in 6 subbands. The results are summarized in Table 9.1.[13]

Note that all of the codecs exceeded the speech quality of that given by a companded FM with carrier-to-noise ratios of 18 dB and 26 dB moving at 36 km/h. Note also that the RPE-LPC codec has the highest average quality score. Finally, the codec chosen was a combination of RPE-LPC and MPE-LTP resulting in RPE-LTP with a net bit rate of 13 kbit/s.

A simplified block diagram of RPE-LTP GSM codec is shown in Figure 9.3.[14] The speech signal, sampled at 8 kHz, is subdivided into 20-ms segments (160 samples per segment). During each segment interval the coefficients $r_n(k)$ of the LPC inverse filter are calculated by the LPC analysis block, so that the residual $\varepsilon'_n$ is minimized. The pitch is predicted by the LTP analysis filter (pitch predictor) and subtracted from $\varepsilon'_n$ to generate the "Gaussian-noise-like" error $\varepsilon_n$. The LTP filter is characterized by the pitch period $p_n$ and a gain factor $g_n$. The residual signal $\varepsilon_n$ is low-pass filtered (LPF) at

**TABLE 9.1.    Comparison between Codec Proposals for the GSM System**

| Codec | Speech Quality MOS (out-of-5) | Net Bit Rate (kbit /s) | Complexity (MOPS) |
|---|---|---|---|
| RPE-LPC | 3.54 | 14.77 | 1.5 |
| MPE-LTP | 3.27 | 13.20 | 4.9 |
| SBC-APCM | 3.14 | 13.00 | 1.5 |
| SBC-ADPCM | 2.92 | 15.00 | 1.9 |
| FM | 1.95 | | |

*Source*: P. Vary, K. Hellwig, R. Hofmann, R. J. Sluyter, C. Galand, and M. Rosso, Speech codecs for the European mobile radio system, in *Proc. Int. Conf. on Acoustics Speech and Signal Processing*, *ICASSP-1988*, pp. 227–230. Reprinted with permission.

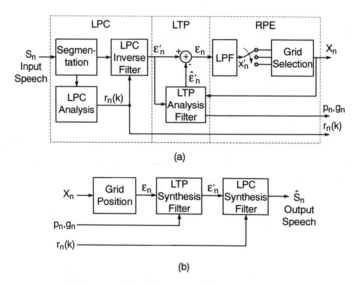

(a)

(b)

**Figure 9.3.**    RPE-LTP block diagram: (a) encoder; (b) decoder.

$\frac{4}{3}$ kHz and only every third sample of the resultant signal is selected for transmission. The coefficients $r_n(k)$, in number of 8, and the parameters $p_n$ and $g_n$ are encoded with 3.6 kbit/s, whereas the samples of the filtered residual are encoded with 9.6 kbit/s.

At the decoder the parameters $X_n$, $p_n$, $g_n$, and $r_n(k)$ are used in the grid position and synthesis filter to synthesize the speech.

## 9.5.5    Code-Excited Linear Prediction

The code-excited linear prediction (CELP) algorithm uses LPC techniques in connection with vector quantization through an analysis-by-synthesis pro-

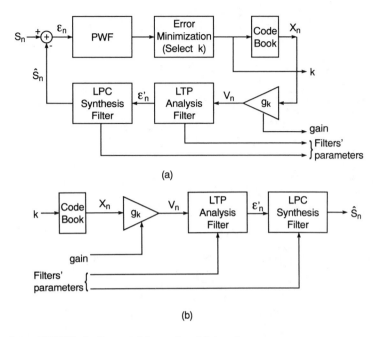

**Figure 9.4.** CELP block diagram: (a) encoder; (b) decoder.

cedure. Figure 9.4 shows the basic structure of the CELP codec. The basic analysis procedure consists in finding an optimum codeword (vector) $c_k$ in the codebook according to some subjective error criterion. Each codeword is scaled by a gain factor $g_k$ and processed through the LTP and LPC synthesis filters to synthesize the predicted speech term $\hat{S}_n$. The residual $\varepsilon_n = S_n - \hat{S}_n$ is processed through the perceptual weighting filter (PWF) and used in a full search procedure to find the best codeword that minimizes the energy of $\varepsilon_n$. The codebook index $k$, the gain $g_k$, and the filter parameters are sent and used at the decoder to synthesize the speech.

### 9.5.6 The American Codec[18]

The speech-coding algorithm used in the American digital mobile radio system is a variation on CELP, called vector-sum excited linear prediction (VSELP). It uses codebooks with predefined structure so that full search is avoided, significantly reducing the time required for the optimum codeword search. A block diagram of the VSELP codec is shown in Figure 9.5. The long-term predictor lag $l$ is determined from the past output of the LTP filter and the current input speech using a closed-loop approach. This VSELP utilizes two codebooks, each of which has its own gains. The long-term prediction lag is determined first, assuming no input from these two codebooks. Once the prediction lag is established, the first codebook is searched,

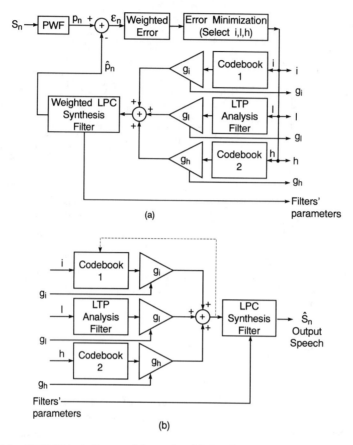

**Figure 9.5.** VSELP block diagram: (a) encoder; (b) decoder.

assuming no input from the other codebook. Having determined the optimum codeword, the other codebook is then searched in order to minimize the energy of the prediction error $\varepsilon_n$ even further. The signal $p_n$ is the weighted input speech for the subframe minus the zero input response of the weighted LPC synthesis filter. The sampling rate is 8 kHz and 160 samples are grouped to form a 20-ms frame. A subframe is composed of 40 samples (5 ms), and the order of the short-term predictor is 10. The basic data rate of the speech coder is 7950 bit/s.

## 9.6  DIGITAL MODULATION FOR MOBILE RADIO

The process of conversion of a digital baseband signal into an IF or an RF signal for transmission is called *digital modulation*. The inverse process, digital demodulation, is carried out at the reception end by means of coherent, differentially coherent, or noncoherent detection. In coherent detection both transmitter and receiver work synchronously. Consequently, a carrier and timing recovery circuitry are required at the receiver. Because this is not necessary in either differentially coherent or noncoherent detection, the corresponding receivers are less complex. On the other hand, coherent detection yields a better bit error rate performance than that of noncoherent and differentially coherent detection for a given carrier-to-noise ratio.

The choice of modulation technique must take into account two factors, namely, power and spectral efficiencies. Power-efficient techniques are those having low error probability ($10^{-4}$ or $10^{-8}$) in a environment with relatively low carrier-to-noise ratio (8.4 or 12 dB, respectively). Spectrally efficient techniques are those capable of transmitting high bit rate per bandwidth (2 bit/s$^{-1}$ Hz$^{-1}$ or more).

As far as mobile radio is concerned, due to the scarcity of bandwidth, spectral efficiency is what we must usually consider in a choice of modulation scheme. However, we must not forget that the mobile radio environment is rather "unkind", so that a spectrally efficient technique may give an untolerably poor performance. The 64-QAM modulation scheme, with 4.5 bit s$^{-1}$ Hz$^{-1}$, and the 256-QAM scheme with 6.6 bit s$^{-1}$ Hz$^{-1}$, are examples of such techniques.[1] On the other hand, QPSK with 2 bit s$^{-1}$ Hz$^{-1}$ and also MSK are good examples of power efficient techniques. Because DQPSK and GMSK are the techniques chosen for use in the American and European digital cellular radio systems, respectively, and QPSK and MSK are closely related to them, we shall concentrate our attention on them.

## 9.7 QPSK AND λ-SHIFTED DQPSK MODULATION SCHEMES

Quadriphase-shift keying (QPSK) and λ-shifted differentially encoded quadriphase-shift keying (λ-shifted DQPSK) are special cases of multiphase (*M*-ary PSK) modulation where the information is contained in the phase of the carrier. In particular, for a quadriphase modulation, the phase of the carrier takes on one of four equally spaced values such as $\lambda$, $\lambda + \pi/2$, $\lambda + \pi$, and $\lambda + 3\pi/2$, where $\lambda$ is the initial phase.

### QPSK Signal

According to the preceding definition, the general representation for a set of quadriphase signalling waveform is

$$s_i(t) = A \cos\left[2\pi f_c t + (i - 1)\frac{\pi}{2} + \lambda\right] \tag{9.4}$$

where $i = 1, 2, 3, 4$, $0 \le t \le T$, $T$ is the symbol duration, $A$ is the carrier amplitude, and $f_c$ is the carrier frequency.

Each of the four possible phases corresponds to a unique pair of information bits. The assignment of two information bits to the four possible phases is usually done so that the codewords representing the adjacent phases differ by one bit. This is known as Gray encoding. For example, we may assign the phases $\lambda$, $\pi/2 + \lambda$, $\pi + \lambda$, and $3\pi/2 + \lambda$ to the pairs 00, 01, 11, and 10, respectively.

Define $\phi_i$ as the instantaneous phase of the modulated signal (i.e., the signal phase at the current symbol interval)

$$\phi_i \triangleq (i - 1)\pi/2 + \lambda \tag{9.5}$$

Then, from Equation 9.4,

$$s_i(t) = A \cos(2\pi f_c t + \phi_i) \tag{9.6}$$

Expanding the cosine function in Equation 9.6, we have

$$s_i(t) = I_i A \cos(2\pi f_c t) - Q_i A \sin(2\pi f_c t) \tag{9.7}$$

where

$$I_i \triangleq \cos \phi_i \quad \text{and} \quad Q_i \triangleq \sin \phi_i \tag{9.8}$$

The signal described by Equation 9.7 can be viewed as two quadrature carriers with amplitudes $A \cos \phi_i$ and $A \sin \phi_i$ varying according to the transmitted phases in each signalling interval. In particular, when $\lambda = \pi/4$ we have $\phi_i = (2i - 1)\pi/4$. In this case the amplitudes of the quadrature carriers take on two possible values, $\pm A\sqrt{2}$, at each symbol interval.

### λ-Shifted DQPSK Signal

In DQPSK the information is differentially encoded. Accordingly, symbols are transmitted as changes in phase rather than as absolute phases. The DQPSK modulation can be viewed as the noncoherent version of the QPSK. In order to understand how it operates, we start by explaining the binary differentially encoded PSK (DPSK). In binary DPSK a bit 1 is sent by shifting the current phase of the carrier by $\pi$ rad. A bit 0 is transmitted by leaving the current phase of the carrier unchanged. In effect, in a more general binary DPSK the current carrier phase is shifted by $\lambda$ rad when a 0 is transmitted and by $\lambda + \pi$ rad when a 1 is transmitted. If $\lambda$ is chosen to be equal to 0, information containing a long string of zeros is transmitted with no phase shift in the carrier. Consequently, the spectrum of the transmitted signal can be very narrow in such interval. However, if $\lambda$ is chosen to be different from zero, the carrier phase is shifted in every signalling interval. In this case the width of the signal spectrum is approximately equal to $1/T$, where $T$ is symbol duration.

In a λ-shifted DQPSK the relative phase shifts between successive intervals are $\lambda$, $\lambda + \pi/2$, $\lambda + \pi$, and $\lambda + 3\pi/2$, where $\lambda$ takes on the usual values of 0 or $\pi/4$. Because in DQPSK modulation the information is transmitted as changes of phase, the current carrier phase ($\phi_i$) is written as the difference between the previous carrier phase ($\phi_{i-1}$) and the phase change ($\Delta\phi_i$), to be introduced. Note that $\Delta\phi_i$ is a function of the current symbol. Therefore

$$\phi_i = \phi_{i-1} - \Delta\phi_i \tag{9.9}$$

Using the definitions of Equation 9.8 applied to Equation 9.9, we have

$$I_i = I_{i-1}\cos\Delta\phi_i - Q_{i-1}\sin\Delta\phi_i$$
$$Q_i = I_{i-1}\sin\Delta\phi_i - Q_{i-1}\cos\Delta\phi_i \tag{9.10}$$

where $I_{i-1} = \cos\phi_{i-1}$ and $Q_{i-1} = \sin\phi_{i-1}$ are the amplitudes at the previous symbol interval. The DQPSK signal can be written in the same way as in Equation 9.7. If $\lambda$ is chosen to be equal to $\pi/4$, then the amplitudes of the quadrature carriers can take on one of five possible values, $0$, $\pm A$, $\pm A/\sqrt{2}$.

## 9.7.1  Generation and Detection of QPSK and DQPSK Signals

A quadriphase PSK signal can be viewed as two binary PSK signals impressed on the quadrature carriers. This approach can be used to simplify the generation and detection of quadriphase PSK signals. Without loss of generality we shall assume $\lambda = \pi/4$.

### 9.7.1.1  *Modulation*

Consider a binary data stream $b(t)$ entering the modulator. The first step in the modulation process is to split this binary sequence into two separate binary streams, namely **O** and **E**, by a serial-to-parallel converter. All of the odd-numbered bits form a stream **O** and all of the even numbered bits form a steam **E**. At a given symbol instant $i$ the pair $(O_i E_i)$ represents a symbol and the corresponding carrier phase change. Accordingly, the next step is to encode the two data streams so that the correspondence of binary data to carrier phase can be achieved.

Let the sequences **I** and **Q** be the encoded data. As far as the QPSK modulation scheme is concerned these sequences assume only two possible values, $\pm 1/\sqrt{2}$, as stated in the previous section. Moreover, these values correspond to absolute phases of the carrier and they are functions only of the current symbol. Hence, an absolute phase encoder can be implemented by means of a simple combinatorial logic. As for DQPSK, the phase change is a function of both the previous phase and the current symbol. Therefore, a sequential state machine may be used to implement a differential phase encoder.

**TABLE 9.2.    Data–Phase and Data–Phase-Change Correspondences for QPSK and DQPSK**

| | | | QPSK | | DQPSK | | |
|---|---|---|---|---|---|---|---|
| $i$ | $O_i$ | $E_i$ | $\phi_i$ | $I_i$ | $Q_i$ | $\Delta\phi_i$ | $I_i$    $Q_i$ |
| 1 | 0 | 0 | $\pi/4$ | $1/\sqrt{2}$ | $1/\sqrt{2}$ | $\pi/4$ | Depend on |
| 2 | 0 | 1 | $3\pi/4$ | $-1/\sqrt{2}$ | $1/\sqrt{2}$ | $3\pi/4$ | previous data |
| 3 | 1 | 1 | $-3\pi/4$ | $-1/\sqrt{2}$ | $-1/\sqrt{2}$ | $-3\pi/4$ | according to |
| 4 | 1 | 0 | $-\pi/4$ | $1/\sqrt{2}$ | $-1/\sqrt{2}$ | $-\pi/4$ | Equation 9.10 |

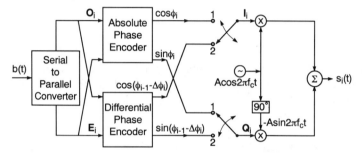

**Figure 9.6.**   QPSK modulator (switch at position 1) and DQPSK modulator (switch at position 2).

The data–phase correspondence for QPSK and the data–phase-change correspondence for DQPSK are shown in Table 9.2. Note that $I_i$ and $Q_i$ are equal to $\cos\phi_i$ and $\sin\phi_i$, respectively, at any signalling interval. Therefore, the final step in the modulation process is to implement the functions as described by Equation 9.7. The resultant block diagram is shown in Figure 9.6.

### 9.7.1.2 Demodulation

Demodulation of QPSK signals is carried out by means of a coherent detection with the carrier required to be recovered at the receiver. Multiplication of the received signal by the in-phase and quadrature carriers, followed by integration (low-pass filtering) of the products, yields signal components proportional to $\cos\phi_i$ and $\sin\phi_i$, respectively. Because these components assume values $\pm 1/\sqrt{2}$, the information is extracted from the sign of the decision variable.

Demodulation of DQPSK signals is carried out by means of differentially coherent detection. This detection scheme differs from coherent detection in that the recovered carrier (used in the latter) is replaced by the received signal being delayed by one symbol interval. Consider a received signal $s_i(t)$

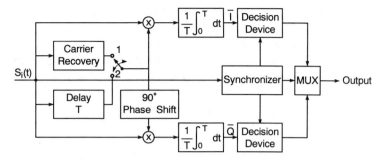

**Figure 9.7.**  QPSK demodulator (switch at position 1) and DQPSK demodulator (switch at position 2).

as given by Equation 9.6 and its delayed version $s_{i-1}(t) = A\cos(2\pi f_c + \phi_{i-1})$. By performing the multiplication $s_i(t)s_{i-1}(t)$ followed by integration (low-pass filtering), a signal component proportional to $\cos(\phi_i - \phi_{i-1})$ is obtained. Because, from Equation 9.9, $\phi_i - \phi_{i-1} = -\Delta\phi_i$ and the phase change $\Delta\phi_i$ assumes values as given by Table 9.2 (i.e., $\Delta\phi_i = \pm\pi/4, \pm 3\pi/4$), then $\cos\Delta\phi_i = \pm 1/\sqrt{2}$. Therefore, the in-phase component of the information is extracted from the sign of the decision variable. Likewise, the quadrature component can be obtained by multiplying $s_i(t)$ and the delayed signal $s_{i-1}(t)$, the latter with a phase shift of $\pi/2$ rad. The resultant signal (after the low-pass filtering) is proportional to $\sin(\phi_i - \phi_{i-1})$, assuming values of $\pm 1/\sqrt{2}$. Figure 9.7 shows a block diagram of the demodulator.

### 9.7.2  Power Spectra of QPSK and DQPSK Signals

Consider initially the QPSK modulation scheme. It is assumed that the information data at the modulator input is random, with both symbols (1 and 0) considered to be statistically independent and equally likely. During the signalling interval $0 \le t \le T$, the amplitudes of the quadrature carriers may assume the values $+g(t)$ or $-g(t)$, where $g(t)$ is the symbol shaping function. In particular, as seen before, $g(t) = A/\sqrt{2}$. Therefore, the in-phase and quadrature components can be viewed as random binary waves, each of which has a power spectral density $S_g(f)$ equal to $(A^2T/2)\mathrm{sinc}^2(Tf)$, where $T$ is the symbol duration and $\mathrm{sinc}(x) = [\sin(\pi x)]/(\pi x)$ (see Appendix 9A). The power spectral densities $S_Q(f)$ and $S_i(f)$ of the respective modulated components $\pm g(t)\cos(2\pi f_c t)$ and $\pm g(t)\sin(2\pi f_c t)$, generating $s_i(t)$, are given by (see Appendix 9A)

$$S_Q(f) = S_I(f) = \tfrac{1}{4}\big[S_g(f - f_c) + S_g(f + f_c)\big] \tag{9.11}$$

Because these components are statistically independent, the resultant power spectral density, $S_s(f)$, of the band-pass signal $s_i(t)$ is the sum of their individual power spectral densities, i.e., $S_s(f) = 2S_Q(f) = 2S_I(f)$. However, for convenience, we shall use the one-sided (positive-frequency) representation, $S(f)$, of the spectrum, so that $S(f) = 2S_s(f)$. Therefore,

$$S(f) = (A^2/2)T \, \text{sinc}^2[T(f - f_c)]$$

$$= 2(A^2/2)T_b \, \text{sinc}^2[2T_b(f - f_c)] \tag{9.12}$$

where $T_b$ is the bit duration ($T = 2T_b$). The power spectral density of the QPSK signal is plotted in Figure 9.11 (see Section 9.8.2), normalized with respect to $T$ and the carrier power $A^2/2$.

The differential encoding does not modify the power spectral density, so that Equation 9.12 also applies to DQPSK signals.[1]


### 9.7.3   BER Performance Over a Gaussian Channel

#### 9.7.3.1 *QPSK Signal*

The QPSK signal, $s_i(t)$ arrives at the receiver as $s_i(t)$ + noise. The demodulator estimates the mean value of each bit composing the symbol and decides for level 1 if the detected bit is positive and for level 0 otherwise. An error occurs if at least one bit is erroneously detected.

*Probability of Erroneous Decision.* Because the demodulator presents two symmetrical detection branches, we shall concentrate our analysis on only one of them. Consider, for instance, the in-phase detection branch and a locally generated carrier of $\sqrt{2} \, \cos(2\pi f_c t)$.* The estimated mean value of the received signal is (see Figure 9.7).

$$\bar{I} = \frac{1}{T} \int_0^T [s_i(t) + \text{noise}] \sqrt{2} \, \cos(2\pi f_c t) \, dt$$

$$= \frac{A}{\sqrt{2}} \cos\left[(2i - 1)\frac{\pi}{4}\right] + n(t) \tag{9.13}$$

where $n(t)$ is a Gaussian noise with a mean value equal to zero and a variance equal to $N_0/2$. Without loss of generality we may assume $i = 1$, corresponding to bit 1 being transmitted. Accordingly, the estimated value $\bar{I}$ is $A/2 + n(t)$. If the noise exceeds $A/2$, an erroneous decision is made.

---

*The signal $\sqrt{2} \, \cos(2\pi f_c t)$ is chosen for the optimum detection criterion.

Define a decision variable $v(t)$ such that

$$v(t) = n(t) - A/2 \tag{9.14}$$

Because $n(t)$ is a Gaussian variable with distribution $p(n)$, $v(t)$ is also a Gaussian variable with a distribution $p(v)$, such that $p(n)|dn| = p(v)|dv|$. Therefore,

$$p(v) = \frac{1}{\sqrt{\pi N_0}} \exp\left[ -\left( \frac{v + A/2}{\sqrt{N_0}} \right)^2 \right] \tag{9.15}$$

An erroneous decision is taken if $v(t) = n(t) - A/2 > 0$, occurring with a probability $P_W$, such that

$$P_W = \text{prob}(v > 0) = \int_0^\infty p(v)\, dv \tag{9.16}$$

Using the transformation of variables $z(t) = (v(t) + A/2)/\sqrt{N_0}$ and the relation $p(z)|dz| = p(v)|dv|$, we obtain

$$P_W = \frac{1}{\sqrt{\pi}} \int_{A/2\sqrt{N_0}}^\infty \exp(-z^2)\, dz = \frac{1}{2} \text{erfc}\left( \frac{A}{2\sqrt{N_0}} \right) \tag{9.17}$$

where $\text{erfc}(\cdot)$ is the complementary error function.

We may express Equation 9.17 in terms of the carrier-to-noise ratio $\gamma_c$ and the signal-to-noise ratio per bit $\gamma_b$, using the relations

$$\gamma_c = A^2/2N_0 \quad \text{and} \quad \gamma_c = 2\gamma_b \tag{9.18}$$

Therefore

$$P_W = \tfrac{1}{2} \text{erfc}\left( \sqrt{\gamma_c/2} \right) = \tfrac{1}{2} \text{erfc}\left( \sqrt{\gamma_b} \right) \tag{9.19}$$

***Probability of Symbol Error.*** A symbol is correctly detected if both bits composing that symbol are correctly detected. The probability of this event is $(1 - P_W)^2$. Therefore, the probability of symbol error $P_e$ is

$$P_e = 1 - (1 - P_W)^2 \tag{9.20}$$

***Probability of Bit Error.*** The derivation of the bit error probability $P_b$ as a function of the symbol error probability is rather tedious because of its dependence on the particular mapping of symbols onto signal phases. When Gray code is used and the symbol error is acceptable small, the following approximation works reasonably well.

The most probable errors correspond to an erroneous selection of an adjacent phase to the wanted phase. Because adjacent phases are encoded with the minimum distance criterion (1 bit difference between adjacent code words), most 2-bit symbol errors contains only a single-bit error. Therefore, we may use the approximation

$$P_b \simeq \tfrac{1}{2} P_e \tag{9.21}$$

Hence from Equations 9.21, 9.20, and 9.19 we obtain

$$P_b = \tfrac{1}{2} P_e = \tfrac{1}{2} \left[ \mathrm{erfc}\!\left(\sqrt{\gamma_b}\right) - \tfrac{1}{4} \mathrm{erfc}^2\!\left(\sqrt{\gamma_b}\right) \right] \tag{9.22}$$

Equation 9.22 is plotted in Figure 9.8.

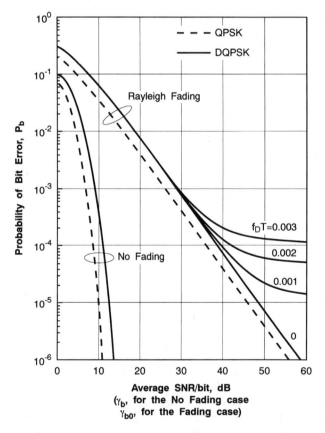

**Figure 9.8.**  Bit error rate performance of QPSK and DQPSK. (*Source*: After H. Suzuki, Canonical receiver analysis for *M*-ary angle modulations in Rayleigh fading environment, *IEEE Trans. Vehicular Tech.*, VT-31(1), 7–14, February 1982.)

### 9.7.3.2 *DQPSK Signal*

The derivation of the probability of a binary digit error for four-phase DPSK with Gray coding is rather cumbersome. We shall only state the final result and refer the interested readers to Proakis.[4]

$$
P_b = \int_b^\infty x \exp\left(-\frac{a^2 + x^2}{2}\right) I_0(ab)\, dx - \frac{1}{2} \exp\left[-\frac{1}{2}(a^2 + b^2)\right] I_0(ab)
$$

$$(9.23)$$

where $I_0(\cdot)$ is the modified Bessel function of order zero and

$$
a = \sqrt{\frac{\gamma_b}{2}} \left(\sqrt{2 + \sqrt{2}} - \sqrt{2 - \sqrt{2}}\right)
$$

$$(9.24)$$

$$
b = \sqrt{\frac{\gamma_b}{2}} \left(\sqrt{2 + \sqrt{2}} + \sqrt{2 - \sqrt{2}}\right)
$$

Equation 9.23 is plotted in Figure 9.8.

### 9.7.4 BER Performance Over a Fading Channel

In studying the BER performance over a fading channel we may distinguish between two situations, namely, slow and fast Rayleigh fading environments. In the former case we use the quasistatic approximation in which the nonselective fading phenomenon occurs (narrowband transmission). In the latter case, due to wideband transmission (high bit rate), the Doppler shift cannot be neglected, and the signal is affected by selective fading.

*Probability of Bit Error.* For a small number of diversity branches,* errors are usually assumed to occur in bursts in a fading environment. Therefore, we may consider that, for any symbol, all of the three possible symbol errors are equally likely and occur with probability $P_e/3$, where $P_e$ is the average symbol error probability. Moreover, there are $\binom{k}{2}$ ways in which $k$ bits out of 2 may be in error. Consequently, the average number of bit error per 2-bit symbol is

$$
\frac{P_e}{3} \sum_{k=1}^{2} k \binom{k}{2} = \frac{4}{3} P_e
$$

---

* In our analysis we shall consider only one diversity branch.

The average bit error probability is obtained by dividing the above result by 2, the number of bits per symbol. Hence,

$$P_b = \tfrac{2}{3} P_e = \tfrac{2}{3} \left[ 1 - (1 - P_W)^2 \right] \tag{9.25}$$

### 9.7.4.1 *QPSK Signal (Nonselective Fading)*

The probability of an erroneous decision is obtained by averaging Equation 9.19 over the probability density function of $\gamma_b$ (or $\gamma_c$). We shall explore the SNR per bit ($\gamma_b$) case, knowing that the calculations for $\gamma_c$ follow exactly the same procedure. In a Rayleigh environment the distribution of $\gamma_b$ is

$$p(\gamma_b) = \frac{1}{\gamma_{b0}} \exp\left( -\frac{\gamma_b}{\gamma_{b0}} \right) \tag{9.26}$$

where $\gamma_{b0}$ is the average SNR per bit. Accordingly

$$P_W = \int_0^\infty \tfrac{1}{2} \operatorname{erfc}(\gamma_b) p(\gamma_b) \, d\gamma_b \tag{9.27}$$

With Equation 9.26 in Equation 9.27 we obtain

$$P_W = \tfrac{1}{2}(1 - F) \tag{9.28}$$

where

$$F = \sqrt{\frac{\gamma_{b0}}{1 + \gamma_{b0}}} \tag{9.29}$$

Therefore, with Equations 9.29, 9.28, and 9.25, the bit error probability of a QPSK fading signal can be calculated. The corresponding curve is plotted in Figure 9.8. It is straightforward to show that, for large $\gamma_{b0}$ ($\gamma_{b0} \gg 1$), the BER can be well approximated by

$$P_b \simeq \frac{1}{3\gamma_{b0}} = \frac{2}{3\gamma_{c0}} \tag{9.30}$$

where $\gamma_{c0}$ is the average carrier-to-noise ratio.

### 9.7.4.2 *DQPSK Signal (Nonselective Fading)*

The bit error probability for a DQPSK fading signal can be estimated by averaging Equation 9.23 over the distribution of $\gamma_b$, given by Equation 9.26. Equivalently, the corresponding probability of an erroneous decision can be averaged over the distribution of $\gamma_b$ and used in the appropriate equations.

Again in this case we shall simply state the result.[5] It has been found that

$$F = \frac{\gamma_{b0}}{\sqrt{(\gamma_{b0} + 1)^2 - 1/2}} \qquad (9.31)$$

With Equations 9.31, 9.28, and 9.25 the bit error probability of DQPSK fading can be calculated. This is plotted in Figure 9.8. For $\gamma_{b0} \gg 1$ a good approximation for BER is easily found to be

$$P_b \simeq \frac{2}{3\gamma_{b0}} = \frac{4}{3\gamma_{c0}} \qquad (9.32)$$

### 9.7.4.3  *DQPSK Signal (Selective Fading)*

In a fast Rayleigh fading environment the envelope correlation of the signal is different from unity. It is shown that,[5] in this case,

$$F = \frac{J_0(2\pi f_D T)}{\sqrt{2(1 + 1/\gamma_{b0})^2 - J_0^2(2\pi f_D T)}} \qquad (9.33)$$

where $J_0(2\pi f_D T)$, the normalized envelope correlation, is the zeroth-order Bessel function of the first kind, and $f_D$ is the maximum Doppler frequency.

With Equations 9.33, 9.28, and 9.25, the BER of the DQPSK signal in a fast-fading environment can be estimated. This is plotted in Figure 9.8 for several values of the parameter $f_D T$. Note that when $f_D T = 0$ we have $J_0(0) = 1$ and Equation 9.33 equals Equation 9.29.

## 9.8  MSK AND GMSK MODULATION SCHEMES

Minimum shift keying (MSK) and Gaussian minimum shift keying (GMSK) are special cases of binary frequency shift keying (FSK) modulation in which the phase information of the received signal is fully explored so that noise performance can be significantly improved. In particular, GMSK is an improved version of MSK, as we shall see in this section. In all of the cases each binary symbol is identified by one carrier frequency. Moreover, the changes of frequencies of the modulated carrier, keyed by the binary input, do not affect the carrier phase. This type of FSK modulation is known as continuous-phase frequency-shift keying (CPFSK).

### *MSK Signal*

Let $s(t)$ be a CPFSK signal defined within the time interval $0 \le t \le T_b$ so that

$$s(t) = A \cos[2\pi f_c t + \theta(t)] \qquad (9.34)$$

TABLE 9.3.   Generated Signals in the MSK Modulation

| Previous Symbols | Present Symbol | $s(t)$ |
|:---:|:---:|:---:|
| 0 | 0 | $A \cos(2\pi f_0 t)$ |
| 0 | 1 | $A \cos(2\pi f_1 t)$ |
| 1 | 0 | $-A \cos(2\pi f_0 t)$ |
| 1 | 1 | $-A \cos(2\pi f_1 t)$ |

where $T_b$ is the bit interval, $\theta(t)$ is the phase of $s(t)$, $f_c$ is the nominal carrier frequency, and $A$ is the carrier amplitude.

The carrier frequency is chosen as

$$f_c = \tfrac{1}{2}(f_0 + f_1) \tag{9.35}$$

where $f_0$ and $f_1$ are frequencies used to transmit symbols 0 and 1, respectively. The phase $\theta(t)$ is a linear function of time as given by

$$\theta(t) = \theta(0) \pm \pi \Delta f t, \qquad 0 \leq t \leq T_b \tag{9.36}$$

where $\Delta f = f_0 - f_1$ is the frequency deviation and $\theta(0)$ is the carrier initial phase. The initial phase $\theta(0)$ assumes the value 0 or $\pi$ depending on the past history of the modulation process. Note from Equation 9.36 that the plus and minus signs correspond to the transmission of symbols 0 and 1, respectively.

When the frequency deviation $\Delta f$ is chosen to be half of the input bit rate, i.e., $\Delta f = \tfrac{1}{2}T_b$, the phase of the signal $s(t)$ can take on only* the values $+\pi/2$ or $-\pi/2$ at odd multiples of $T_b$, and the values 0 or $\pi$ at even multiples of $T_b$. Because $\Delta f = \tfrac{1}{2}T_b$ is the minimum frequency spacing that enables the symbols 0 and 1 to be detected without mutual interference, this special case of a CPFSK modulation is referred to as minimum shift keying (MSK). Note that MSK is a binary digital FM with a modulation index of 0.5.

Considering $\Delta f = \tfrac{1}{2}T_b$ and using Equations 9.36 and 9.35 in Equation 9.34, we conclude that

$$s(t) = \begin{cases} \pm A \cos(2\pi f_0 t) & \text{for symbol 0} \\ \pm A \cos(2\pi f_1 t) & \text{for symbol 1} \end{cases} \tag{9.37}$$

where the plus sign corresponds to $\theta(0) = 0$ and the minus sign corresponds to $\theta(0) = \pi$. We may condition $\theta(0)$ to be 0 if the previous symbol was 0 and to be $\pi$ if the previous symbol was 1. Therefore, with this assumption, the generated signals for each combination of symbols are as given by Table 9.3.

---

*Note that the phase shifts are modulo-$2\pi$.

With these considerations and using standard trigonometric identities, Equation 9.34 may be rewritten as

$$s(t) = g(t)\cos(2\pi f_c t) + g(t - T_b)\sin(2\pi f_c t) \qquad (9.38)$$

where

$$g(t) = A\cos[\theta(0)]\cos\left(\frac{\pi}{2T_b}t\right), \qquad -T_b \le t \le T_b \qquad (9.39)$$

is the symbol shaping function.

### GMSK Signal

From the preceding discussion we understand that MSK modulation yields a constant envelope and has coherent detection capability. Moreover, as we shall see, the bandwidth of the modulated wave is relatively narrow. On the other hand, its out-of-band radiation cannot be neglected, constituting a serious constraint for mobile radio applications.

The output power spectrum of MSK can be controlled by low-pass filtering the binary input data prior to modulation. This guarantees the constant envelope property, an essential feature that renders the mobile signal robust against fading. However, in order to accomplish this, the low-pass filter (LPF) must present the following properties[6]:

- Narrow bandwidth and sharp cutoff, to suppress high-frequency components
- Low overshoot impulse response, to protect against excessive instantaneous frequency deviation
- Preservation of the filter output pulse area, corresponding to a phase shift of $\pi/2$, for simple coherent detection

These conditions are satisfied by a Gaussian LPF, and the modified MSK modulation is referred to as Gaussian MSK or GMSK.

### 9.8.1 Generation and Detection of MSK and GMSK Signals

#### 9.8.1.1 Modulation

An orthogonal modulator, as described by Equation 9.38, can be designed to implement the MSK function. Because the spectral density of an orthogonal FSK signal depends on its symbol shaping functions, the latter can be manipulated to yield the desired output power spectrum. Consequently, several related modulation techniques can be derived by simply using an appropriate symbol shaping function. Tamed frequency modulation (TFM)

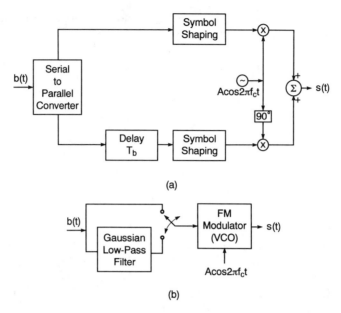

(a)

(b)

**Figure 9.9.** MSK and GMSK modulators: (a) orthogonal modulator; (b) MSK modulator (switch at position 1); GMSK modulator (switch at position 2).

and minimum shift keying are some of the many existing techniques that use an orthogonal modulator, as shown in Figure 9.9a.

However, the simplest way of generating an MSK signal or a GMSK signal is by using an FM modulator. In the case of MSK, the information bit stream is used to modulate the frequency of a VCO. As for GMSK, the information bit stream is first low-pass filtered through a Gaussian filter and then used to modulate the frequency of a VCO. This scheme is shown in Figure 9.9b. The weak point of this solution is the difficulty of keeping the center frequency within a range where the linearity and sensitivity of the FM modulator are still kept. More elaborate solutions, such as the orthogonal modulator or the use of PLL circuitry, can be used to minimize this problem.

### 9.8.1.2 Demodulation

Like the QPSK systems, MSK and also GMSK can be coherently demodulated in quadrature channels, as shown in Figure 9.10a. Note that the integration interval is $2T_b$ and that the quadrature channel is delayed by one bit interval $(T_b)$ with respect to the in-phase channel. After the threshold detection (decision device) the bits are combined to restore the original information.

Noncoherent detection can be accomplished by the use of a limiter and frequency discriminator as shown in Figure 9.10b. This is an easy and economical way of performing demodulation at the expense of a system

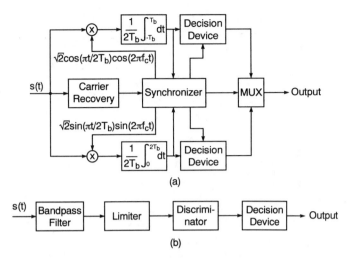

**Figure 9.10.**  MSK and GMSK signal detector; (a) orthogonal coherent detector; (b) noncoherent detector.

performance degradation. The limiter supplies the discriminator with a constant-envelope signal. The discriminator has an output proportional to the derivative of the instantaneous phase of the input signal, corresponding to the signal frequency. The decision device samples the detected frequency and compares it with a set of thresholds to decide the transmitted frequency.

### 9.8.2  Power Spectra of MSK and GMSK Signals

The power spectral density is determined on the assumption that the input binary wave is random, with symbols 1 and 0 equally likely and transmitted during statistically independent time slots. The power spectral density of the band-pass signal is basically dependent on its symbol shaping function $g(t)$, given by Equation 9.39, and $g(t - T_b)$. These functions have the same energy spectral densities $\psi_g(f)$ given by $|G(f)|^2$, where $G(f)$ is the Fourier transform of $g(t)$. Therefore

$$\psi_g(f) = 16\frac{AT_b^2}{\pi^2}\left[\frac{\cos 2\pi T_b f}{1 - (4T_b f)^2}\right]^2 \tag{9.40}$$

The power spectral densities of the in-phase and quadrature components are given by $\psi_g(f)$ averaged over the symbol shaping function duration $2T_b$, i.e., $S_Q(f) = S_I(f) = \psi_g(f)/2T_b$ (see Appendix 9A). Because these components are statistically independent, by following the same reasoning as in the QPSK case (Section 9.7.2) we find that the one-sided power spectral density

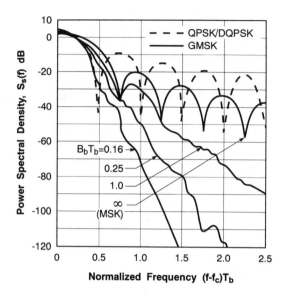

**Figure 9.11.** Power spectra of MSK, GMSK, QPSK, and DQPSK. (*Source*: After K. Murota and K. Hirade, GMSK modulation for digital mobile radio telephony, *IEEE Trans. Commun.*, COM-29(7), 1044–1050, July 1981.)

$S(f)$ of the MSK signal $s(t)$ is

$$S(f) = \frac{16(A^2/2)}{\pi^2} T_b \left\{ \frac{\cos[2\pi T_b(f - f_c)]}{1 - [4T_b(f - f_c)]^2} \right\}^2 \tag{9.41}$$

The spectral density of an MSK signal given by Equation 9.41 can be seen in Figure 9.11, normalized with respect to $T_b$ and the carrier power $A^2/2$. Note that the MSK spectrum falls off at a higher rate than that of QPSK. On the other hand, the main lobe of the MSK spectrum is wider than that of QPSK, the respective first nulls occurring for $T_b(f - f_c) = 0.75$ and 0.5.

As for the GMSK signal, its power spectrum corresponds to that of the MSK signal but modified by the effects of the premodulation Gaussian low-pass filtering. This is shown in Figure 9.11 (from Reference 6) where the normalized 3-dB down bandwidth, $B_b T_b$, of the Gaussian LPF is a parameter. Note that, when $B_b T_b$ tends to infinity, the power spectrum of GMSK coincides with that of MSK. As $B_b T_b$ increases, so does the rate at which the power spectrum falls. The parameter $B_b T_b$ has a great influence on the spectrum efficiency and must be carefully selected by the system designer. Murota[6] has shown that an optimum condition for maximizing the spectral efficiency of a GMSK cellular radio system is obtained when $B_b T_b = 0.25$. For this situation, (1) with a channel spacing of 25 kHz, (2) with a bit rate of $1/T_b = 16$ kbit/s, and (3) with a predetection rectangular band-pass $B_i T_b =$

1, the ratio of out-of-band radiation power to total power is less than $-60$ dB, as specified by CCIR.[7]

### 9.8.3 BER Performance Over a Gaussian Channel

In this section we shall consider the bit error rate performance of MSK and GMSK moduations in the presence of additive white Gaussian noise using coherent detection.

#### 9.8.3.1 *MSK Signal*

The detection of an MSK signal $s(t)$ is carried out in a way similar to that of QPSK. The mean value of the received signal is evaluated at each branch of the demodulation as indicated in Figure 9.10a. These estimated values are

$$\bar{I} = \frac{1}{2T_b} \int_{-T_b}^{T_b} [s(t) + \text{noise}] \sqrt{2} \cos\left(\frac{\pi}{2}T_b\right)\cos(2\pi f_c t)\, dt$$

$$= \bar{Q} = \frac{1}{2T_b} \int_{0}^{2T_b} [s(t) + \text{noise}] \sqrt{2} \sin\left(\frac{\pi}{2}T_b\right)\sin(2\pi f_c t)\, dt$$

$$= \frac{A}{2} + n(t) \tag{9.42}$$

Following the same procedure as for the QPSK, we find that both techniques present the same bit error probability $P_b$,

$$P_b = \frac{1}{2}\left[\text{erfc}\left(\sqrt{\gamma_b}\right) - \frac{1}{4}\text{erfc}^2\left(\sqrt{\gamma_b}\right)\right] \tag{9.43}$$

Equation 9.43 is plotted in Figure 9.12.

#### 9.8.3.2 *GMSK Signal*

The GMSK modulation may be viewed as a modified version of MSK, both MSK and GMSK constituting binary digital modulations. As such, the BER performance, for high signal-to-noise ratio $\gamma_b$, may be approximated by

$$P_b \simeq \frac{1}{2}\text{erfc}\left(\sqrt{\alpha\gamma_b}\right) \tag{9.44}$$

where the parameter $\alpha$ depends on the premodulation Gaussian low-pass filter. In other words, $\alpha$ is a function of the parameter $B_bT_b$. For $B_bT_b = \infty$, $\alpha$ is found to be equal to 1, corresponding to the case of the MSK modulation. For the optimum condition $B_bT_b = 0.25$, the parameter $\alpha$ is found to be

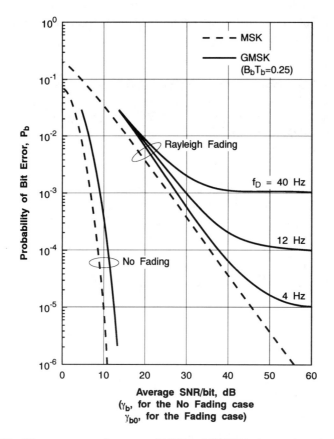

**Figure 9.12.**    Bit error rate performance of MSK and GMSK (coherent detection). (*Source*: After K. Murota and K. Hirade, GMSK modulation for digital mobile radio telephony, *IEEE Trans. Commun.*, COM-29(7), 1044–1050, July 1981.)

approximately equal to 0.68.[6] The probability of the bit error for this condition is plotted in Figure 9.12.

### 9.8.4    BER Performance Over a Fading Channel

#### 9.8.4.1    *MSK Signal (Nonselective Fading)*

As we remarked before, the BER performance of the MSK modulation equals that of QPSK. Therefore, its bit error probability can be estimated by using Equation 9.29 in Equation 9.28 and the resultant equation in Equation 9.25. The corresponding curve is reproduced in Figure 9.12.

A thorough study of the performance of MSK system in fast Rayleigh fading environment (selective fading) with differential and discriminator detection can be found in the publications by Hirade and co-workers[8,9] and in the book edited by Feher.[1]

### 9.8.4.2 *GMSK Signal (Nonselective Fading)*

An approximate BER performance can be estimated by averaging Equation 9.43 over the distribution of $\gamma_b$ given by Equation 9.26. The resultant bit error probability, neglecting the quadratic term $\text{erfc}^2(\cdot)$ is

$$P_b \simeq \int_0^\infty \frac{1}{2} \, \text{erfc}\left(\sqrt{\alpha\gamma_b}\right) p(\gamma_b) \, d\gamma_b$$

$$\simeq \frac{1}{2}\left(1 - \sqrt{\frac{\alpha\gamma_{b0}}{1 + 2\gamma_{b0}}}\right) \tag{9.45}$$

where the parameter $\alpha$ depends on the premodulation Gaussian LPF bandwidth. This BER is plotted in Figure 9.12 for $\alpha = 0.68$ ($B_b T_b = 0.25$).

### 9.8.4.3 *GMSK Signal (Selective Fading)*

The BER performance of the GMSK modulator in a fast Rayleigh fading environment has been evaluated experimentally by means of a Rayleigh fading simulator.[6] The corresponding curves for various values of the maximum Doppler frequency $f_D$ is shown in Figure 9.12.[6]

## 9.9 COMBINED TECHNIQUES

In the introduction to this chapter we remarked that the main processes composing a digital communication system are A/D conversion and modulation. In fact, this is a simplified model of a digital system, adopted as such for convenience. A more complete model would include a channel coding block in between A/D conversion and modulation. The channel coding is responsible for introducing redundancy into the speech signal after the A/D conversion process. It may sound curious (or rather senseless) to reintroduce redundancy into a signal from which most of it had been extracted in a previous step. The interesting point is that the extracted redundancy has an unpredictable characteristic, whereas the reintroduced redundancy is totally controllable. The signal without the "unpredictable" redundancy is very susceptible to a disastrous (abrupt) degradation with increasing noise in the system. With the inclusion of the controllable redundancy, the signal may suffer a graceful degradation, instead.

The use of controllable redundancy (error-correcting codes) improves the error rate performance of the system at the expense of extra bits included in the data stream. Accordingly, the modulator must operate at higher bit rate, requiring extra channel bandwidth. Higher-order modulation schemes can be used to save bandwidth, but larger signal power is required to maintain the same error probability. Some alternative solutions have recently been proposed and they are briefly described next.

## 9.9.1 TCM—Combined Modulation and Coding

In trellis coded modulation (TCM), modulation and coding are treated not as individual entities, but rather as a unique process. In TCM a higher-order modulation is combined with convolutional code to improve the error rate performance while keeping the bandwidth unaltered. At the receiver end, instead of performing demodulation and then decoding, both processes are carried out in a single step. The performance of the system is dictated by the free Euclidean distance between the transmitted signal sequences, instead of the free Hamming distance of the convolutional code. Therefore, the choice of code and signal constellation must be performed jointly. The detection process should involve soft decision rather than hard decision, improving the performance.

This combined scheme was first explored by Ungerboeck,[22] using a modulator constellation twice as large as that necessary for uncoded transmission. Although the bandwidth was kept the same, a power gain was accomplished. The asymptotic coding gain of Ungerboeck codes is given by

$$G = 10 \log \left( \frac{d_{\text{free}}}{d_{\text{ref}}} \right)^2 \tag{9.46}$$

where $d_{\text{free}}$ is the free Euclidean distance of the code, and $d_{\text{ref}}$ is the minimum Euclidean distance of an uncoded modulation scheme operating with the same energy per bit.

### 9.9.1.1 *Ungerboeck Codes*

Trellis coded modulation presents two basic characteristics:

1. The signal constellation contains more signal points than what is required for the modulation scheme of interest with the same data rate. For instance, if initially an $N$-ary modulation is required, in TCM an $M$-ary scheme, with $M > N$, must be used.
2. Convolutional coding is used.

The additional points in the signal constellation allow for redundancy and can be used for detecting and correcting errors. Convolutional coding introduces dependence between successive signal points, when only certain patterns are permitted. TCM principles will be illustrated in this section using a two-dimensional signal constellation. In particular, we use $M$-ary PSK schemes where the constellation presents a circular grid.

Consider, initially, a quadrature modulation with $b$ bits per symbol, corresponding to a constellation of $2^b$ signal points. In TCM we choose a parameter $a$, with $a \geq 1$, such that a modulation scheme of a constellation with $2^{b+a}$ signal points is used. This constellation is then partitioned into $s$ subsets, each of which contains $(2^{b+a})/s$ equally spaced signal points. Assume that $k$ out of $b$ bits per symbol enter a rate-$k/n$ binary convolutional encoder such that $2^n = s$. The resulting $n$ coded bits per symbol selects one out of the $s$ subsets. The remaining uncoded $(b - k)$ bits select a particular point of that subset, with $2^{b-k} = (2^{b+a})/s = 2^{b+a-n}$. The class of trellis codes implementing this is known as *Ungerboeck codes*.

### 9.9.1.2 A TCM for 8-PSK

The set of an octaphase signalling waveform (8-PSK) is given by

$$s_i(t) = A \cos\left(2\pi f_c t + (i - 1)\frac{\pi}{4} + \lambda\right) \tag{9.47}$$

where $i = 1, 2, \ldots, 8$, $0 \leq t \leq T$, $T$ is the symbol duration, $A$ is the carrier amplitude, $f_c$ is the carrier frequency, and $\lambda$ is the initial phase. Following the same steps as in Section 9.7 we rewrite Equation 9.47 as follows:

$$s_i(t) = I_i A \cos(2\pi f_c t) - Q_i A \sin(2\pi f_c t) \tag{9.48}$$

where

$$I_i = \cos \phi_i, \quad Q_i = \sin \phi_i \quad \text{and} \quad \phi_i \triangleq (i - 1)\frac{\pi}{4} + \lambda \tag{9.49}$$

The signal constellation is composed of eight points given by $(I_i, Q_i)$, $i = 1, 2, \ldots, 8$, equally spaced on a circular grid of radius 1.

Suppose that a transmission of 2 bits per symbol ($b = 2$) is required, for which a QPSK signalling is of interest (constellation with $2^2$ signal points). If, however, in TCM an 8-PSK scheme is used, a constellation twice as large is achieved ($2^{b+a} = 8$, for which $a = 1$). Moreover, assume that a rate-1/2 ($k = 1$, $n = 2$) convolutional encoder is used. Therefore, we find that the signal constellation is partitioned into $2^n = s = 4$ subsets, each of which contains $2^{b-k} = 2$ equally spaced signal points. In other words, each subset contains two antipodal signal points (signals with a phase difference of 180°).

Let us estimate the asymptotic coding gain of this configuration. From Equation 9.46 we must determine $d_{\text{ref}}$ and $d_{\text{free}}$. The parameter $d_{\text{ref}}$ is the minimum Euclidean distance of an uncoded QPSK and is given by the minimum distance between any pair of signal points $(I_i, Q_i)$ described by Equation 9.8. Therefore, from Table 9.2 (or, equivalently, from Equation 9.8), $d_{\text{ref}} = \sqrt{2}$. The parameters $d_{\text{free}}$—the free Euclidean distance—can be no larger than the Euclidean distance between the antipodal signal points of each subset. This is given by the distance between any pair of points $(I_j, Q_j)$ and $(I_{j+4}, Q_{j+4})$, $j = 0, 1, 2, 3$, where $(I_i, Q_i)$ are described by Equation 9.49. Accordingly, $d_{free} = 2$. Finally, from Equation 9.46, $G_a = 10 \log_{10} 2 = 3$ dB.

### 9.9.2   Unequal Error Protection

Redundancy can be more efficiently utilized if the bits with more-relevant information of the speech signal are better protected. This technique is known as unequal error protection.

## 9.10   SUMMARY AND CONCLUSIONS

The main appeal of digital communication is that it renders the system more flexible to accommodate technological evolution. Following the world-wide trend of network digitization, the second generation of mobile radio has adopted the fully digital approach. Digital technology is rather attractive but requires special attention when radio transmission is involved. The problems become more challenging in the case of mobile radio.

One of the greatest challenges of using digital radio communication is the reduction of the required bandwidth. As we know, digital transmission requires a far greater bandwidth than analog transmission. As far as digital transmission is concerned, bandwidth reduction is directly related to bit rate reduction.

The use of appropriate speech-coding algorithm can drastically reduce the number of bits required to encode the speech. These algorithms are based on the fact that human speech contains a great deal of redundancy. If the redundancies are appropriately extracted, the resultant speech will still present acceptable quality and can be represented with fewer bits. From the conventional 64-kbit/s PCM, the speech-coding algorithms have evolved to a bit rate of less than 8 kbit/s, still keeping toll quality.

Higher-order modulation schemes can also be used to reduce bandwidth. Given that a higher-order modulation is used, the other requirement is that it should present an almost constant envelope for robustness against fading. Moreover, "insensitivity" to nonlinear amplification is also a requisite. The great challenge, however, is to reconcile power efficiency and spectral efficiency. Power efficient techniques present low bit error rate, obtained at the expense of a wider spectrum. Spectrally efficient modulation techniques, on the other hand, offer a poor BER performance.

A promising approach is to combine modulation and coding to be treated as a unique entity. In the combined approach it is possible to introduce redundancy (error-correcting code) to render the transmission robust while keeping the same bandwidth. The usual approach is known as trellis coded modulation (TCM). It is also possible to extract the redundancy of speech by means of a trellis coded quantization (TCQ)[23] and then combine TCQ and TCM to make large mean-square error in the source coding very unlikely.[24]

## APPENDIX 9A   POWER SPECTRAL DENSITY

### 9A.1   Power Spectral Density

Let $g(t)$ be a power signal* and let $g_T(t)$ be its truncated version, so that $g_T(t) = g(t)$ for $-T \leq t \leq T$ and $g_T(t) = 0$, otherwise. Therefore, the average signal power is

$$P = \lim_{T \to \infty} \frac{1}{2T} \int_{-T}^{T} g^2(t)\, dt = \lim_{T \to \infty} \frac{1}{2T} \int_{-\infty}^{\infty} g_T^2(t)\, dt \qquad (9A.1)$$

Let $G_T(f)$ be the Fourier transform of $g_T(t)$, i.e., $G_T(f) = \mathfrak{F}[g_T(t)]$. Using the convolution property, we note that

$$\mathfrak{F}[g_T^2(t)] = G_T(f) * G_T(f) = \int_{-\infty}^{\infty} G_T(\lambda) G_T(f - \lambda)\, d\lambda \qquad (9A.2)$$

Using the definition of the Fourier transform we have

$$\mathfrak{F}[g_T^2(t)] = \int_{-\infty}^{\infty} g_T^2(t) \exp(-2\pi ft)\, dt \qquad (9A.3)$$

Equating Equations 9A.2 and 9A.3 for $f = 0$ we obtain

$$\int_{-\infty}^{\infty} g_T^2(t)\, dt = \int_{-\infty}^{\infty} G_T(\lambda) G_T(-\lambda)\, d\lambda = \int_{-\infty}^{\infty} |G_T(\lambda)|^2\, d\lambda \qquad (9A.4)$$

Replacing Equation 9A.4 into Equation 9A.1 with a convenient change of variables we obtain

$$P = \int_{-\infty}^{\infty} S_g(f)\, df$$

where

$$S_g(f) = \lim_{T \to \infty} \frac{1}{2T} |G_T(f)|^2 \qquad (9A.5)$$

is called the power spectral density or power spectrum of $g(t)$.

---

*Power signal has infinite energy.

## 9A.2   Correlation Function

We can rewrite Equation 9A.5 in terms of $\overline{G}_T(f)$, the complex conjugate of $G_T(f)$. Hence

$$S_g(f) = \lim_{T \to \infty} \frac{1}{2T} G_T(f) \overline{G}_T(f) \tag{9A.6}$$

Using the convolution property we find that

$$\mathfrak{F}\left[ g_T^2(t) \right] = \lim_{T \to \infty} \frac{1}{2T} g_T(\tau) * g_T(-\tau)$$

$$= \lim_{T \to \infty} \frac{1}{2T} \int_{-\infty}^{\infty} g_T(t) g_T(t - \tau)\, dt \triangleq R_g(\tau) \tag{9A.7}$$

The function on the right-hand side, $R_g(\tau)$, is called autocorrelation function of the power signal. This function can be redefined in terms of the power signal as

$$R_g(\tau) = \lim_{T \to \infty} \frac{1}{2T} \int_{-\infty}^{\infty} g_T(t) g_T(t - \tau)\, dt \tag{9A.8}$$

From Equation 9A.7 we conclude that the power spectral density is the Fourier transform of the autocorrelation function, i.e.,

$$S_g(f) = \int_{-\infty}^{\infty} R_g(\tau) \exp(-2\pi \tau f)\, d\tau \tag{9A.9}$$

Now consider a stationary process $X(t)$. Its autocorrelation function is defined as

$$R_x(\tau) = E[X(t)X(t - \tau)] \tag{9A.10}$$

## 9A.3   Modulated Wave

Let $y(t)$ be a modulated wave

$$y(t) = g(t)\cos(2\pi f_c t) \tag{9A.11}$$

Using Equation 9A.5,

$$S_y(f) = \lim_{T \to \infty} \frac{1}{2T} |Y_T(f)|^2 \tag{9A.12}$$

where $Y_T(f) = \Im[y_T(t)]$ and

$$y_T(t) = g_T(t)\cos(2\pi f_c t) \tag{9A.13}$$

Hence,

$$Y_T(f) = \tfrac{1}{2}[G_T(f - f_c) + G_T(f + f_c)] \tag{9A.14}$$

With Equation 9A.14 in Equation 9A.12 we find

$$S_y(f) = \lim_{T \to \infty} \tfrac{1}{4}\Big[|G_T(f - f_c)|^2 + |G_T(f + f_c)|^2\Big]$$

$$= \tfrac{1}{4}[S_g(f - f_c) + S_g(f + f_c)] \tag{9A.15}$$

because $G_T(f - f_c)$ and $G_T(f + f_c)$ are nonoverlapping spectra (their cross product is zero).

The same result is obtained for a modulated stationary process $Y(t)$ such that

$$Y(t) = X(t)\cos(2\pi f_c t + \theta) \tag{9A.16}$$

where $X(t)$ is a stationary process and $\theta$ is a random variable uniformly distributed over 0 and $2\pi$. In this case

$$R_y(\tau) = E[Y(t + \tau)Y(t)]$$

$$= \tfrac{1}{2}E[X(t + \tau)X(t)]E[\cos(2\pi f_c \tau) + \cos(4\pi f_c t + 2\pi f_c \tau + 2\theta)]$$

$$= \tfrac{1}{2}R_x(\tau)\cos(2\pi f_c \tau) \tag{9A.17}$$

Taking the Fourier transform of Equation 9A.17, we find

$$S_y(f) = \tfrac{1}{4}[S_x(f - f_c) + S_x(f + f_c)] \tag{9A.18}$$

## 9A.4    Random Binary Wave[11]

Let $x(t)$ be a sample function of a random binary process $X(t)$ assuming amplitude levels of $-A$ and $+A$ with equal probability. Let $T$ be the duration of each pulse. Consider $t_d$, the sample value of a uniformly distributed variable $T_d$, as the starting time of the first pulse. Hence the distribution of $T_d$ is

$$P_{T_d}(t_d) = \frac{1}{T}, \qquad 0 \le t_d \le T \tag{9A.19}$$

We want to find the autocorrelation function $R_x(t_i, t_j)$, where $t_i$ and $t_j$ are the times of observations of $X(t)$.

If $|t_i - t_j| > T$, then $X(t_i)$ and $X(t_j)$ occur in different pulse intervals. Therefore, they are independent and

$$E[X(t_i)X(t_j)] = E[X(t_i)]E[X(t_j)] = 0$$

because

$$E[X(t)] = 0 \quad \forall \, t$$

Let $|t_i - t_j| < T$, with $t_i = 0$ and $t_j < t_i$. In this case $X(t_i)$ and $X(t_j)$ occur in the same pulse interval if $t_d < T - |t_i - t_j|$. Therefore,

$$E[X(t_i)X(t_j)|t_d] = A^2, \qquad t_d < T - |t_i - t_j|$$

Then

$$R_x(\tau) = E[X(t_i)X(t_j)] = \int_0^{T-|\tau|} A^2 P_{T_d}(t_d)\, dt_d$$

$$= A^2\left(1 - \frac{|\tau|}{T}\right), \qquad |\tau| < T \tag{9A.20}$$

where $\tau = |t_i - t_j|$.

Taking the Fourier transform of $R_x(\tau)$ we obtain

$$S_x(f) = A^2 T \, \text{sinc}^2(fT) \tag{9A.21}$$

where $\text{sinc}(x) = \sin(\pi x)/(\pi x)$.

If we estimate the energy spectral density $\psi_g(t)$ of a rectangular pulse $g(t)$ of amplitude $A$ and duration $T$ we obtain

$$\psi_g(f) = |G(f)|^2 = A^2 T^2 \, \text{sinc}^2(fT) \tag{9A.22}$$

Therefore we may state that the power spectral density $S_x(f)$ of a random binary wave assuming values $+g(t)$ and $-g(t)$ is given by

$$S_x(f) = \frac{\psi_g(f)}{T} \tag{9A.23}$$

where $T$ is symbol duration and $g(t)$ is the symbol shaping function.


## REFERENCES

1. Feher, K., Ed., *Advanced Digital Communications—Systems and Signal Processing Techniques*, Prentice-Hall, Englewood Cliffs, NJ, 1987.
2. Haykin, S., *Digital Communications*, John Wiley & Sons, New York, 1988.
3. Feher, K., *Digital Communications, Satellite/Earth Station Engineering*, Prentice-Hall, Englewood Cliffs, NJ, 1981.

4. Proakis, J. G., *Digital Communications*, McGraw-Hill, New York, 1989.

5. Suzuki, H., Canonic receiver analysis for *M*-ary angle modulations in Rayleigh fading environment, *IEEE Trans. Vehicular Tech.*, VT-31(1), 7–14, February 1982.

6. Murota, K. and Hirade, K., CMSK modulation for digital mobile radio telephony, *IEEE Trans. Commun.*, COM-29(7), 1044–1050, July 1981.

7. CCIR Rec. 478-3, Technical Characteristics of Equipment and Principles Governing the Allocation of Frequency Channels Between 25 and 1000 MHz for the Land Mobile Service, 13, 1982.

8. Hirade, K., Ishizuka, M., and Adachi, F., Error-rate performance of digital FM with discriminator-detection in the presence of cochannel interference under fast-Rayleigh environment, *Trans. IECE Japan*, 61-E, 704, September 1978.

9. Hirade, K., Ishizuka, M., Adachi, F., and Ohtani, K., Error-rate performance of digital FM with differential detection in land mobile radio channel, *IEEE Trans. Vehicular Tech.*, VT-28, 204, August 1979.

10. Adachi, F. and Ohno, K., Performance analysis of GMSK frequency detection with decision feedback equalization in digital land mobile radio, *IEE Proc.*, 135(3), Part F, 199–207, June 1988.

11. Haykin, S., *Digital Communications*, John Wiley & Sons, New York, 1988.

12. O'Shaughnessy, D., *Speech Communication—Human and Machine*, Addison-Wesley, Reading, MA, 1987.

13. Vary, P., Hellwig, K., Hofmann, R., Sluyter, R. J., Galand, C., and Rosso, M., Speech Codec for the European mobile radio system, in *Proc. Int. Conf. Acoustics Speech and Signal Processing*, ICASSP-1988, 227–230, 1988.

14. Vary, P., GSM speech Codec, in *Proc. Digital Cellular Radio Conference*, Hagen, FRG, October 1988.

15. Atal, B. S. and Schroeder, M. R., Stochastic coding of speech signals at very low bit rates, *Proc. IEEE Int. Conf. Communications*, 48.1, May 1984.

16. Schroeder, M. R. and Atal, B. S., Code-excited linear prediction (CELP): high quality speech at very low bit rates, in *Proc. IEEE Int. Conf. Acoustics, Speech, Signal Processing*, 937–940, March 1985.

17. EIA/TIA, Project Number 2215, Cellular System–Dual-Mode Mobile Station–Base Station Compatibility Standard, IS-54, December 1989.

18. Adoul, J. P., Mabilleau, P., Delprat, M., and Morissette, S., Fast CELP coding based on algebraic codes, in *Proc. IEEE Int. Conf. Acoustics, Speech, Signal Processing*, pp. 1957–1960, 1987.

19. Adoul, J. P. and Lamblin, C., A comparison of some algebraic structures for CELP coding of speech, in *Proc. IEEE Int. Conf. Acoustics, Signal Processing*, pp. 1953–1956, 1987.

20. Bellamy, J. C., *Digital Telephony*, John Wiley & Sons, New York, 1982.

21. Calhoun, G., *Digital Cellular Radio*, Artech House, Norwood, MA, 1988.

22. Ungerboeck, G., Channel coding with multilevel/phase signals, *IEEE Trans. Inform. Theory*, IT-28, 55–67, January 1982.

23. Marcellin, M. W. and Fischer, T. R., Trellis coded quantization of memoryless and Gauss–Markov sources, *IEEE Trans. Commun.*, COM-38, 82–93, January 1990.

24. Fischer, T. R. and Marcellin, M. W., Joint trellis coded quantization/modulation, *IEEE Trans. Commun.*, COM-39(2), 172–176, February 1991.

# PART V

# Multiple Access

CHAPTER **10**

# Multiple-Access Architecture

This chapter examines the various transmission technologies that can be used in a mobile radio system. The different approaches can be broadly grouped into two categories: narrowband and wideband transmission. Narrowband and wideband systems are then examined and their features and limitations are discussed. The bulk of the chapter is dedicated to the three multiple-access schemes, namely, FDMA, TDMA, and CDMA. These methods are analyzed independently, according to their own characteristics. The basics of the three access schemes and their advantages and disadvantages are also considered. In particular, for the FDMA case we discuss the effects of nonlinearities of the radio equipment circuitry. In TDMA systems the main issues are timing hierarchy, synchronization, and efficiency measures. CDMA systems are then analyzed in the light of information of the spread-spectrum technology. Accordingly, a great deal of the corresponding section is steered toward providing the background for spread-spectrum fundamentals and techniques. Finally, we discuss the modes of providing two-way communications in a fully trunked system. These techniques are named frequency-division duplex and time-division duplex and their main characteristics are briefly analyzed.

## 10.1  INTRODUCTION

Resource-sharing can be a very efficient way of achieving high capacity in any communication network. As far as mobile radio systems are concerned the resources are the channels or, more generically, the bandwidth. To be more efficient, the access mode must allow any terminal (the mobiles) to be able to use the resources in a fully trunked system. If the channels are assigned on demand, the procedure is called demand-assigned multiple access (DAMA), or simply multiple access.

Depending on how the available spectrum is utilized, the system can be broadly classified into narrowband or wideband. In the narrowband architecture the total frequency band is split into several narrowband channels, whereas in the wideband architecture the whole (or a significant amount of) the spectrum is available to all the users.

There are basically three access methods, classified according to the means (frequency, time, code) used to implement them:

- Frequency-division multiple access (FDMA)
- Time-division multiple access (TDMA)
- Code-division multiple access (CDMA)

FDMA is intrinsically a narrowband architecture whereas CDMA is wideband. TDMA, on the other hand, can be implemented either as a narrowband or as a wideband system.

When two-way communications are involved, the full-duplex connection can be provided by means of frequency or time division. In the first case we have frequency-division duplex (FDD) and in the second case we have time-division duplex (TDD).

All these issues will be examined in this chapter.

## 10.2  NARROWBAND AND WIDEBAND ARCHITECTURES[3]

### 10.2.1  Narrowband Architecture

Narrowband systems divide the total frequency spectrum into as many channels as the technology allows. Actually, each channel comprises a set of two carrier frequencies used for two-way communication. Frequencies used for the uplink (mobile to base station) are named *reverse channels*, whereas frequencies of the downlink (base station to mobile) are called *forward channels*. Forward and reverse channels are positioned in the spectrum so as to have maximum frequency separation between them, to avoid interference. As an example, assuming that the available bandwidth is 40 MHz and that each one-way channel can be made as narrow as 25 kHz, then the total number of channels for the two-way communication is $40 \times 10^6 / 50 \times 10^3 = 800$. If the available 40 MHz is contiguous, then the maximum possible separation between a forward channel and the corresponding reverse channel is 20 MHz. Usually this band can be split into two 20-MHz contiguous bands separated from each other by a certain frequency band within which other services are allocated. This obviously increases the frequency separation between forward and reverse channels.

Narrowband channels are usually associated with high-capacity* systems but sometimes with low-quality transmission. Accordingly, a lot of effort has been invested in the modulation technology so as to reduce the channel band and still maintain acceptable voice quality. Moreover, narrow and sharp filters are required in the radio equipment in order to minimize adjacent-

---

*The narrower the channel, the more channels are accommodated within the same bandwidth.

channel interference. This contributes to substantially increasing the costs of the apparatus.

Another aspect associated with narrowband systems is that the transmission experiences nonselective fading. This means that when fades occur, all of the information (the whole channel) is affected. As explained in Chapter 5, selective fading is likely to occur in a channel band exceeding the coherence bandwidth, given by $1/2\pi\overline{T}$, where $\overline{T}$ is the delay spread. With $\overline{T} = 0.5$ $\mu$s the corresponding coherence bandwidth is approximately 320 kHz, which is much larger than the channel band used in narrowband mobile systems.

Occurrence of call blocking is another characteristic of the narrowband system. The establishment of a call is based on the availability of a channel. Once the channels are all being used, new calls will be blocked. Methods of improving the traffic performance of the system are described in Chapter 12.

FDMA and narrowband TDMA are considered as narrowband architectures.

### 10.2.2 Wideband Architecture

The main feature of wideband systems is that either all the spectrum available or a considerable portion of it is used by each carrier. This is exactly the case for wideband TDMA systems, whereas CDMA systems frequently use the whole frequency spectrum.

The great advantage of wideband systems is that the transmission bandwidth always exceeds the coherence bandwidth for which the signal experiences only selective fading. That is, only a small fraction of the frequencies composing the signal is affected by fading. Although fades are quite frequent and deep (20–40 dB) in narrowband systems, the spreading of the signal throughout the whole bandwidth in wideband systems works toward minimizing their effects, so that these same fades would provoke a much milder effect (2–3 dB).[14] Likewise, both deliberate and unintentional interferences are reduced with the use of wideband techniques.

In particular, CDMA systems do not experience blocking probability of any kind. Because all the subscribers can use all the spectrum at the same time (see Section 10.5), in theory, new calls are always possible. The corresponding signals will be spread throughout the spectrum in the same way as the signals of the calls already established. The only consequence is the degradation of the signal-to-interference ratio, because any information, other than that of the wanted signal, works as an interfering (unwanted) signal. Accordingly, wideband systems are claimed to have a graceful performance degradation with the increase of the number of users.

### 10.3 FREQUENCY-DIVISION MULTIPLE ACCESS

The most conventional method of multiple access is frequency-division multiple access (FDMA). In FDMA the signals (from the mobiles or base

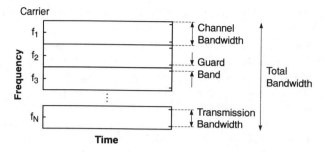

**Figure 10.1.** SCPC/FDMA architecture.

stations) are transmitted on carriers using different RF center frequencies. The simplest arrangement within the FDMA architecture is the one where a separate carrier for each channel is provided. This scheme, known as single-channel per carrier (SCPC), is very efficient in that the channels can be used in a demand-assigned mode.

Within a cell all the channels are available to all the mobiles and the channel assignment is carried out on a first-come–first-served basis. The number of channels, given a frequency spectrum, depends not only on the modulation technique but also on the guard bands between the channels. These guard bands allow for imperfect filters and oscillators and can be used to minimize adjacent-channel interference.

Among all the available channels some are dedicated to control purposes. The number of control channels varies with the size of the system, but this usually constitutes a small proportion of the total number. Note that analog as well as digital systems can use SCPC/FDMA, the only requirement being the use of only one channel per carrier. Figure 10.1 depicts the SCPC/FDMA format.

## 10.3.1  Main Features of the FDMA Architecture

In this section we consider some of the main features of the FDMA architecture, including digital FDMA.[3]

1. *SCPC/FDMA*—In a single-channel-per-carrier (SCPC) FDMA, a separate frequency is alloted to each telephone circuit.
2. *Continuous transmission*—The channels, once assigned, are used on a non–time-sharing basis. This means that both subscriber and base station can use their corresponding allotted channels continuously and simultaneously.

3. *Narrow bandwidth*—Because each allotted channel is used by only one subscriber, the required bandwidth is reasonably narrow. The existing analog cellular systems use 25–30-kHz channels. Digital FDMA systems can make use of low bit rate speech-coding techniques to reduce the channel band even more.

4. *Low ISI*—Intersymbol interference (ISI) is an intrinsic problem of digital systems and is particularly critical in a mobile radio environment. If the delay spread is not negligible compared to the symbol duration, then a great deal of digital processing (adaptive equalization) is required in order to minimize the ISI effects (see Section 6.11). Digital FDMA systems, using constant-envelope digital modulation, may require bit rates around 1 bit/Hz.[5] Therefore, a 25-kHz channel can support a transmission rate of 25 kbit/s, corresponding to a symbol duration of 40 $\mu$s if one bit per symbol is used. Because delay spreads vary from tenths of microseconds to a few microseconds (delay spreads of 5 $\mu$s are quite unusual), ISI is likely to be very low.

5. *Low overhead*—Speech channels carry overhead messages for control purposes. In analog FDMA these messages comprise, for instance, hand-off signalling. In digital FDMA they may also include bits for synchronization, the framing, etc. Because the allotted channels can be used continuously, fewer bits need be dedicated for control compared to TDMA channels. Therefore, more bits can be used to improve the transmission quality (error-correcting codes, etc.).

6. *Simple hardware*—Given that (1) very little or even no digital processing to combat ISI may be required and (2) ease of framing and synchronization can be achieved, besides other reasons, less-complex hardware can be used in both mobile unit and base station.

7. *Use of Duplexer*—Because (1) the system operates on a full-duplex basis, (2) only one antenna is used for transmission and reception, and (3) both transmitter and receiver are on continuously, the use of a duplexer (filters between transmitter and receiver) is mandatory in order to avoid interference. The cost of the duplexer represents as much as 10% of the total cost of the mobile unit equipment.[3]

8. *High base-station cost*—The SCPC/FDMA architecture requires the use of one transmitter, one receiver, two codecs, and two modems per channel. If more than one channel, say *n*, could share the same carrier as in TDMA, the number of such equipment at the base station would be divided by *n*. If on the one hand the use of more than one channel per carrier increases the complexity and therefore the costs of the equipment, on the other hand this renders the system more capable of accommodating more subscribers, making the equipment per subscriber less costly.

9. *Perceptible hand-off*—The continuous transmission on FDMA channels makes the transition from one channel to another, when a hand-off is to take place, a rather perceptible phenomenon. In fact, the interruption of transmission can be long enough to provoke a "click" or several clicks in

case many hand-offs are necessary (e.g., in small-cell systems, etc.). In TDMA systems the idle time slots can be used for this, so that the interruption of transmission can be well worked out to reduce these clicks.

## 10.3.2 Effects of Nonlinearities on FDMA Systems

The $N$ channels available in the cell can share the same antenna at the base station in two possible ways:

1. Each channel has its own power amplifier and the $N$ amplifiers have their outputs connected to a power combiner, which feeds a common antenna.
2. All the $N$ channels share a common power amplifier, with its output connected to the antenna.

Both the power combiner and the power amplifier present nonlinear properties, usually in the form of a saturation at high power levels. There are many effects due to these nonlinearities:

1. Spectral spreading
2. Modulation transfer
3. Signal suppression
4. Intermodulation

We shall describe each of these effects.

### 10.3.2.1 *Spectral Spreading*

The desired signal exceeds its own bandwidth, causing adjacent-channel interference.

### 10.3.2.2 *Modulation Transfer*

Amplifying devices may produce AM/PM (phase modulation) conversion, where a change in the envelope of a multicarrier input causes a change in the output phase of each signal component. Let $A(t)$ be the input envelope and let $\theta(A)$ be the phase modulation induced by the envelope fluctuation. For small input drive levels the following relationship holds true[2]:

$$\theta(A) \simeq KA^2(t) \tag{10.1}$$

where $K$ is a constant.

Consider that the input $x(t)$ is a summation of $N$ sinusoids, i.e.,

$$x(t) = \sum_{i=1}^{N} A_i \cos[\omega_0 t + \phi_i] \triangleq A(t)\cos[\omega_0 t + \phi(A)] \tag{10.2}$$

where

$$A^2(t) = \left( \sum A_i \cos \phi_i \right)^2 + \left( \sum A_i \sin \phi_i \right)^2 \tag{10.3}$$

is the squared envelope, and

$$\phi(A) = \tan^{-1} \left( - \frac{\sum A_i \sin \phi_i}{\sum A_i \cos \phi_i} \right) \tag{10.4}$$

is the resultant phase.

The AM/PM output $y(t)$ is also a summation of sinusoids, each shifted by $\theta(A)$, that is

$$y(t) = \sum_{i=1}^{N} A_i \cos[\omega_0 t + \phi_i + \theta(A)] \tag{10.5}$$

For $\theta(A) \ll 1$, $\cos[\theta(A)] \simeq 1$ and $\sin[\theta(A)] \simeq \theta(A)$. Therefore

$$y(t) \simeq \sum_{i=1}^{N} A_i \cos(\omega_0 t + \phi_i) - \theta(A) \sum_{i=1}^{N} A_i \sin(\omega_0 t + \phi_i)$$

$$= A(t)\cos[\omega_0 t + \phi(A)] + d(t)$$

$$= x(t) + d(t) \tag{10.6}$$

Using Equation 10.1, the distortion term $d(t)$ can be expressed by

$$d(t) \simeq -KA^2(t)A(t)\sin[\omega_0 t + \phi(A)]$$

$$\simeq -KA^3(t)\sin[\omega_0 t + \phi(A)] \tag{10.7}$$

Compare the signals $x(t)$ and $d(t)$ (Equations 10.2 and 10.7, respectively). Note that the distortion occurs at precisely the same frequencies, has a different amplitude, and is shifted 90° in phase.

### 10.3.2.3 Signal Suppression

When the amplifier operates near its linear region, the output voltage $g(x)$ is proportional to the input voltage $x(t)$, i.e.,

$$g(x) = Gx(t) \tag{10.8}$$

where $G$ is known as small-signal voltage gain. Outside the linear region

Equation 10.8 does not hold true and

$$g(x) < Gx(t) \tag{10.9}$$

Consider an envelope sinusoid of amplitude $B$ received by the mobile, in the presence of a large Gaussian interference. Let the Gaussian term be a large number of other independent sinusoidal signals with a resultant envelope $A$. The interference $A$ has a Rayleigh density $p(A)$, such that

$$p(A) = \frac{A}{\sigma^2} \exp\left(\frac{-A^2}{2\sigma^2}\right) \tag{10.10}$$

It is shown by Spilker[2] that the ratio between the output signal-to-noise ratio $\text{SNR}_O$ and the input signal-to-noise ratio $\text{SNR}_I$ is

$$\frac{\text{SNR}_O}{\text{SNR}_I} = R \tag{10.11}$$

where

$$R = \frac{\left[\int_0^\infty Ag(A)p(A)\, dA\right]^2}{\left[\int_0^\infty g^2(A)p(A)\, dA\right]\left[\int_0^\infty A^2 p(A)\, dA\right]} \tag{10.12}$$

and $g(A)$ is the envelope output.

Using the Schwarz inequality (see Appendix 5B), we conclude that $R$, the signal-suppression ratio, is less than or equal to 1 ($R \leq 1$). Accordingly, there can be no signal enhancement, no matter what the nonlinearity $g(A)$ should be.

### 10.3.2.4 *Intermodulation*

Intermodulation (IM) is the presence of unwanted signal-dependent spectral components. In this section we shall illustrate the fundamentals of IM effects and calculate the magnitude of the intermodulation produces (IP) through an example.

Let $g(x)$ be the voltage transfer function of a nonlinear device, such that

$$g(x) = k_1 x + k_2 x^2 + k_3 x^3 \tag{10.13}$$

where $x$ is the input voltage and $k_i$ are constants. Consider that the input $x$ is composed of three narrowband band-pass signals,

$$x = A \cos a + B \cos b + C \cos c \tag{10.14}$$

with $a = (\omega_c - \Delta\omega)t$, $b = \omega_c t$, $c = (\omega_c + \Delta\omega)t$, where $\omega_c$ is the carrier frequency and $\Delta\omega$ is the separation between adjacent carrier frequencies.

Under the conditions that (1) $\omega_c \gg \Delta\omega$ and (2) a zonal filter is placed after the nonlinear device to remove out-of-band harmonics, with Equation 10.14 in Equation 10.13 we obtain*

$$g(x) = \underbrace{C_1 \cos a + C_2 \cos b + C_3 \cos c}_{\text{wanted signal}}$$

$$+ \underbrace{C_4 \cos(2b - a) + C_5(2b - c) + C_6 \cos(a - b + c)}_{\text{intermodulation products}} \quad (10.15)$$

The values of the constants in Equation 10.15 are

$C_1 = k_1 A + \frac{3}{4} k_3 A(A^2 + 2B^2 + 2C^2)$, linear at $a = \omega_c - \Delta\omega$

$C_2 = k_1 B + \frac{3}{4} k_3 B(2A^2 + B^2 + 2C^2)$, linear at $b = \omega_c$

$C_3 = k_1 C + \frac{3}{4} k_3 C(2A^2 + 2B^2 + C^2)$, linear at $c = \omega_c + \Delta\omega$

$C_4 = \frac{3}{4} k_3 AB^2$, intermodulation at $2b - a = \omega_c + \Delta\omega$

$C_5 = \frac{3}{4} k_3 B^2 C$, intermodulation at $2b - c = \omega_c - \Delta\omega$

$C_6 = \frac{3}{2} k_3 ABC$, intermodulation at $a - b + c = \omega_c$

It is possible to avoid completely third-order or third- and fifth-order IM products by widening the bandwidth and assigning the channels appropriately. Consider that $N$ sinusoids occupy a bandwidth of $NB$, where $B$ is assigned to each signal channel. The total required bandwidth in order to avoid IM is $MB$, where $MB \geq NB$. The relation between $N$ and $M$ is shown in Figure 10.2.[2] As an example, if $N = 4$, then $M = 7$ or $M = 20$ to avoid third-order or third- and fifth-order IM products, respectively.

The appropriate assignment of the $N$ channels out of the $M$ required channels was obtained by Spilker[2] and is shown in Table 10.1.

The upper part of the table shows the frequency plan with IM products spreading limited to $3B$, and the lower part shows this without IM product spreading.

---

*In this case the out-of-band terms are of type $dc$, $2a$, $2b$, $2c$, $a + b$, $a - b$, $a + c$, $a - c$, $b + c$, $b - c$, $2a - b$, $2a - c$, $2c - a$, $2c - b$, $a + b - c$, and $-a + b + c$, exceeding the range $\omega_c - \Delta\omega < \omega < \omega_c + \Delta\omega$.

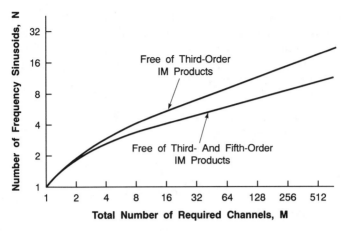

**Figure 10.2.** Required number of channels versus number of carriers to avoid IM products. (*Source*: W. C. Y. Lee, *Mobile Communications Design Fundamentals*, Howard W. Sams, Indianapolis, IN, 1986. Reprinted with permission of Prentice-Hall Computer Publishing.)

**TABLE 10.1.   Frequency Plans to Avoid Third-Order IM Products**

| IM Product Spreading | Number of Carriers $N$ | Number of Required Channels $M$ | Frequencies $F_i$ |
|---|---|---|---|
| Yes | 3 | 4 | 1, 2, 4 |
| (limited) | 4 | 7 | 1, 2, 5, 7 |
| to $3B$ | 5 | 12 | 1, 2, 5, 10, 12 |
| | 6 | 18 | 1, 2, 5, 11, 13, 18 |
| | 7 | 26 | 1, 2, 5, 11, 19, 24, 26 |
| | 8 | 35 | 1, 2, 5, 10, 16, 23, 33, 35 |
| | 9 | 46 | 1, 2, 5, 14, 25, 31, 39, 41, 46 |
| | 10 | 62 | 1, 2, 8, 12, 27, 46, 48, 57, 60, 62 |
| No | 3 | 7 | 1, 3, 7 |
| | 4 | 15 | 1, 3, 7, 15 |

*Source*: W. C. Y. Lee, *Mobile Communications Design Fundamentals*, Howard W. Sams, Indianapolis, IN, 1986. Reprinted with permission of Prentice-Hall Computer Publishing.

## 10.4   TIME-DIVISION MULTIPLE ACCESS

Time-division multiple access (TDMA) is the primary alternative to frequency-division multiple access (FDMA). TDMA takes advantage of the sampling (Nyquist) theorem in accommodating more information within a radio frequency by increasing the bit transmission rate. Accordingly, a carrier frequency can be shared by several (say, $n$) terminals, each of which makes use of this carrier in separate nonoverlapping time slots. The number of time slots per carrier depends on many factors, such as modulation technique,

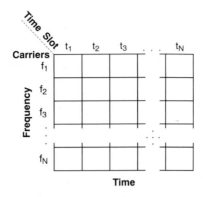

**Figure 10.3.** TDMA architecture.

available bandwidth, etc., characterizing the narrowband TDMA or the wideband TDMA. An illustration of the TDMA architecture is shown in Figure 10.3.

In digital cellular radio, the users have access to all frequencies within the cell and to all time slots within the frequencies. Note, therefore, that such an access mode is, in fact, a combination of FDMA and TDMA.

Transmission from or to a given mobile unit is carried out in a noncontinuous mode, characterizing the buffer-and-burst communication. Once the mobile has accessed a carrier, transmission and reception are performed in distinct time slots, separated from one another by some guard time slots, which can be used by other mobiles. The use of the buffer-and-burst mode implies that the channel transmission rate must be faster than the coding (or coding plus overhead) rate by a factor equal to or greater than the number of time slots per carrier.

### 10.4.1 Main Features of the TDMA Architecture

In this section we shall examine the main characteristics of the TDMA architecture.[3]

1. *Multiple channels per carrier*—The total time of a carrier is divided into time slots so that it can be shared by more than one user. In the European system (GSM) the number of voice channels per carrier is eight whereas in the American system this is three.
2. *Burst transmission*—The channels are used on a time-sharing basis. Therefore, the signal transmission occurs during specific time slots.
3. *Narrow or wide bandwidth*—The bandwidths of TDMA systems depends on various factors, including the modulation scheme. Carrier spacing varies from tens of kilohertz to hundreds of kilohertz. The European

system uses 200-kHz carrier spacing with a 271-kbit/s transmission rate. The corresponding figures for the American system are 30 kHz and 48.6 kbit/s.

4. *High ISI*—Intersymbol interference is intrinsically related to the symbol rate and not to TDMA architecture. However, TDMA systems usually take advantage of the digital transmission to increase the symbol rate, resulting in high ISI. Note that for the GSM system the symbol duration is 3.6 $\mu$s, which is already comparable to the delay spread in urban areas. For the American system this is not as critical, although both systems make use of adaptive equalization to combat ISI.

5. *High overhead*—Because of the burst transmission characteristic of TDMA systems, synchronization can be a complicated issue. Therefore, a reasonable amount of the total transmitted bits must be dedicated to synchronization purposes. This, in connection with the guard time between time slots, can represent 20–30% of the channel bits.

6. *Complex hardware*—The use of digital technology permits the inclusion of several facilities in the mobile unit, increasing its complexity. One example of this is the use of slow frequency hopping to counteract multipath fading, as proposed by the GSM system.

7. *No duplexer used*—Because transmission and reception are carried out on different time slots, the duplexer can be completely avoided. Instead, a switching circuit can be used to turn the transmitter and receiver on or off at the convenient times.

8. *Low base-station cost*—The use of multiple channels per carrier provides a proportional reduction of the equipment at the base station.

9. *Efficient hand-off*—TDMA systems can take advantage of the fact that the transmitter is switched off during idle time slots to improve the hand-off procedure.

10. *Efficient power utilization*—FDMA systems require a 3- to 6-dB power back-off[2] in order to minimize intermodulation effects. TDMA can achieve efficiencies of 90% or more compared to the 3- to 6-dB loss in power of FDMA. Time-division multiplex systems can operate the output amplifier at full saturation, resulting in a better power utilization.

11. *IM minimization*—Intermodulation products can be minimized if a convenient guard time between slots is used.

## 10.4.2  Timing Hierarchy

Timing hierarchy is an ordered set of time duration or intervals used to control and configure a TDMA system. Figure 10.4 illustrates a typical timing hierarchy where not all levels are necessarily used. The fields shown in Figure 10.4 are described next.

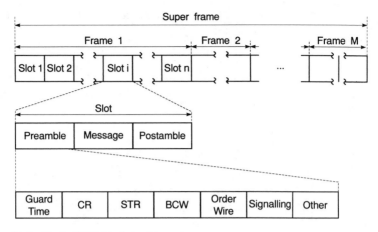

**Figure 10.4.** Typical TDMA timing hierarchy.

1. *Superframe*—number of sequential frames organized to distribute system and network control or signalling information
2. *Frame*—time interval over which the signal format is established and then repeated indefinitely
3. *(Time) slot*—a subdivision of a frame
4. *Burst time*—an integer number of time slots
5. *Burst*—an accessing signal that uses a certain number of slots in the frame
6. *Guard time*—a time interval between bursts to assure that no overlap occurs
7. *Preamble*—initial portion of a burst, consisting of the following elements
   a. CR—carrier recovery, used in coherent demodulating systems
   b. STR—symbol-timing recovery, also known as clock or bit timing recovery (BTR)
   c. UW—unique word or burst code word (BCW) for burst synchronization
   d. HKS—housekeeping symbols such as signalling bits, order wire for voice or data, command and control signalling and error-monitoring symbols
8. *Message*—portion of a burst containing the desired data
9. *Postamble*—end portion of a burst, used for initialization of the next burst

### 10.4.3  Measures of Efficiency

There are several measures of efficiency that can be used in a TDMA system. We shall consider three types.

*Frame efficiency* ($\eta_F$) is the ratio of the portion of the frame available for messages to the total frame length:

$$\eta_F = 1 - \frac{\left(S + \sum_{i=1}^{n}(G_i + P_i + Q_i)\right)T_s}{F} \qquad (10.16)$$

where  $F$ = frame length (in microseconds)
  $S$ = number of symbols in synchronization bursts
  $G_i$ = ratio of the guard time and symbol duration of access $i$
  $P_i$ = number of symbols in preamble of access $i$
  $Q_i$ = number of symbols in postamble of access $i$
  $T_s$ = symbol duration (in microseconds)
  $n$ = number of accesses.

*Burst efficiency* ($\eta_B$) is the ratio of useful message information to the total number of bits transmitted

$$\eta_B = \frac{rM}{P + M + Q} \qquad (10.17)$$

where  $r$ = coding rate of the codec;
  $M$ = message symbols per burst;
  $P$ = number of symbols in preamble;
  $Q$ = number of symbols in postamble.

*System efficiency* ($\eta_S$) is the ratio of the useful capacity (paying traffic) to the available capacity. For a Gaussian channel,

$$\eta_S = B = \log_2\left(1 + \frac{E_s T_s}{N_0 B}\right) \qquad (10.18)$$

where  $B$ = channel bandwidth;
  $E_S$ = energy per symbol;
  $T_S$ = symbol length;
  $N_0$ = spectral density of white Gaussian noise.

## 10.4.4  Time Alignment and Synchronization

The TDMA architecture requires that the mobiles using a common carrier must have their transmitted signals reach the base station receiver at exactly the right time so that signal overlapping does not occur. If the base station provides a reference signal, those mobiles nearer the base station will respond earlier than those located further away. The nonoverlap control can be done by dimensioning the guard time appropriately. For instance, if a mobile is 35 km away from the base station, the time taken for a radio signal to travel the 70 km to and back from the mobile is $70 \times 10^3/3 \times 10^8 = 0.23$ ms. Hence 0.23 ms is the guard period to be provided on each time slot.

A more clever solution is the use of the following time alignment process, which has been adopted by both the GSM and American systems: The mobile is informed how much it must advance or retard the transmit burst in order to be correctly synchronized at the base station. By using this procedure in the preceding example, the GSM system[5] claims to reduce the guard time to 30 $\mu$s.

The mobile station derives timing from a common source, which tracks the base station symbol rate as perceived at the mobile receiver. The frequency tracking is maintained over all specified operating conditions. The synchronization word must have good autocorrelation properties to facilitate synchronization and training.

## 10.5 CODE-DIVISION MULTIPLE ACCESS

Code-division multiple access (CDMA) is an access method where all the users (1) are permitted to transmit simultaneously, (2) operate at the same nominal frequency, and (3) use the entire system bandwidth. One of the main feature of CDMA is that very little dynamic coordination is required, as opposed to FDMA and TDMA, where frequency and time management is critical.

Because all the users can transmit simultaneously throughout the whole system frequency spectrum, a private code must be assigned to each user so that his (or her) transmission can be identified. This privacy is achieved by the use of codes with very low cross-correlation among themselves (orthogonal codes). CDMA systems work asynchronously, in that each user can initiate the transmission at any time instant. Therefore, as far as digital communication is concerned, transition times of a user's message symbols may not coincide with the other users. This simplifies the network synchronization but can complicate the design of good codes.

Note that in CDMA systems, in theory, "there is no limit" on the number of users. What happens is a graceful degradation of performance as the number of simultaneous users increases. Note also that multipath interference is substantially reduced because the signal is spread in the whole spectrum and usually only part of it is affected. In the same way, CDMA systems offer a high resistance to deliberate jamming because transmission is carried out in an encoded way, only known by the users involved.

To accomplish CDMA, spread-spectrum (SS) techniques must be used. In this section we shall present a notion of spread spectrum and the essential characteristics of the techniques used. Although spread spectrum can be used in both analog and digital communications, we shall stress its application in the latter.

### 10.5.1    Spread-Spectrum Fundamentals

#### 10.5.1.1 *General Considerations*

Spread spectrum is defined as a communication technique in which the intended signal is spread over a bandwidth in excess of the minimum bandwidth required to transmit the signal. This is accomplished by the use of a wideband encoding signal at the transmitter, which is required to operate in synchronism with the receiver, where the encoding signal is also known. By allowing the intended signal to occupy a bandwidth far in excess of the minimum bandwidth required to transmit it, the signal will have a noiselike appearance. This works as a "disguise", rendering the spread-spectrum communication able to reject interference.

A typical spread-spectrum system is depicted in Figure 10.5. Note that the spectrum-spreading function is an additional block that is included in a conventional communication system, corresponding to a "second-level" modulation. Therefore, generating a spread-spectrum signal involves two steps: First the carrier is modulated by the baseband digital information with rate $R_b = 1/T_b$, where $T_b$ is the baseband symbol duration. The modulated signal $s_i(t)$ is used to modulate a wideband function $c_i(t)$ with rate $R_c = 1/T_c$, where $R_c$ is called *chip rate*. From Fourier analysis we know that the multiplication of two functions in the time domain corresponds to their convolution in the frequency domain. Therefore, if $s_i(t)$ is narrowband and $c_i(t)$ is wideband, the product $c_i(t)s_i(t)$ will have a spectrum that is approximately the same as that of $c_i(t)$ (wideband).

The wideband signal $s_i(t)c_i(t)$ is passed through the channel, where other wideband signals, interference $I(t)$, and additive noise $n(t)$ are superimposed on the desired signal. Assuming $M$ instantaneous users in the CDMA system, the received signal $r(t)$ is given by

$$r(t) = \sum_{j=1}^{M} c_j(t)s_j(t) + I(t) + n(t) \qquad (10.19)$$

It is assumed that the same spreading function $c_i(t)$ used at the transmitter is locally generated at the receiver. Furthermore, both transmitter and receiver are supposed to operate synchronously. The resulting "despread"

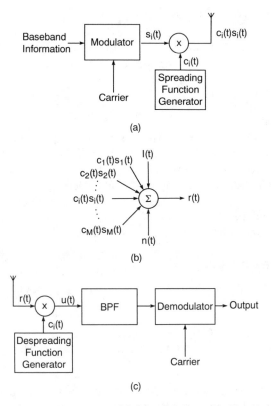

**Figure 10.5.** Spread-spectrum system model: (a) transmitter; (b) channel; (c) receiver.

signal $u(t)$ is

$$u(t) = c_i(t)\left[ \sum_{j=1}^{M} c_j(t)s_j(t) + I(t) + n(t) \right]$$

$$= c_i^2(t)s_i(t) + c_i(t)\left[ \sum_{\substack{j=1 \\ (j \neq i)}}^{M} c_j(t)s_j(t) + I(t) + n(t) \right] \quad (10.20)$$

If the set of spreading functions are chosen so that they have low cross-correlation among themselves, then only the original, modulated waveform remains after the correlation process. Other waveforms are not correlated and will be spread, appearing as noise to the modulator, that is,

$$u(t) = Ks_i(t) + \text{noise} \quad (10.21)$$

where $K \triangleq c_i^2(t) = $ constant and

$$c_i(t) \left[ \sum_{\substack{j=1 \\ (j \neq i)}}^{M} c_j(t)s_j(t) + I(t) + n(t) \right]$$

constitutes the noise component.

Note that $Ks_i(t)$ is band-pass, whereas the noise component is wideband. Thus by applying $u(t)$ to the band-pass filter (BPF) with a bandwidth just large enough to accommodate the modulated signal $Ks_i(t)$, the noise component is made band-pass, therefore with much less power. Consequently, a jammer or interference power density in the information bandwidth of the received signal will be effective only if its total power is increased by the same amount as the bandwidth expansion of the signal.

### 10.5.1.2 *Processing Gain*

Process gain $(G_p)$ is defined as the difference in decibels between output and input signal-to-noise ratios. In absolute values,

$$G_p = \frac{\text{SNR}_O}{\text{SNR}_I} \tag{10.22}$$

It represents the gain achieved by processing a spread-spectrum over an unspread signal. Clearly, this coincides with the ratio between the corresponding bandwidths, that is,

$$G_p = \frac{B_S}{B_U} \tag{10.23}$$

where $B_S$ is the bandwidth of the spread signal and $B_U$ is the bandwidth of the unspread signal.

### 10.5.1.3 *Jamming Margin*

Jamming margin $(M_j)$ gives the amount of interference above the intended signal with which the system is expected to operate. The jamming margin accounts for internal losses $L$ and is given by

$$M_j = \frac{G_p}{L \, \text{SNR}_O} \tag{10.24a}$$

or

$$M_j \, (\text{dB}) = G_p \, (\text{dB}) - [L + \text{SNR}_O] \, (\text{dB}) \tag{10.24b}$$

### 10.5.2 Spread-Spectrum Techniques

There are several spread-spectrum techniques that can be used in communication systems. Each technique is usually applied to specific fields so that very little competition between them occurs.[9] We shall give an overview of the principal techniques but the main focus will be on the two most widely used.

1. *Direct sequence or direct spread or pseudonoise systems*—The modulated carrier is further modulated by a digital code sequence whose bit rate is much higher than the information signal bandwidth.
2. *Frequency hopping or multiple frequency, code-selected, frequency-shift keying systems*—The carrier hops "randomly" from one frequency to another.
3. *Time-hopping systems*—This is exactly the pulse modulation. The code sequence keys the transmitter on and off. Because the code sequence is pseudorandom the times when the transmitter is on and off are also pseudorandom.
4. *Pulsed-FM or chirp systems*—The carrier varies (is swept) over a wide band, in some known way, during each pulse period. This technique does not necessarily employ coding.
5. *Frequency hopping and direct sequence systems*—The direct sequence modulated signal has its center frequency hopping periodically.
6. *Time and frequency hopping*—When frequency hopping alone is not enough to combat interference, then time hopping is also recommended. This combination is mainly used to minimize the near/far problem in which the stronger signals will interfere with the weaker ones. In such cases it is convenient to time all transmissions so that the wanted and unwanted signals are never transmitted at the same time.
7. *Time hopping and direct sequence*—When direct sequence alone is not enough to provide sufficiently high capacity to accommodate users in CDMA systems, time hopping is an efficient way of helping in traffic control.

Having presented the main spread-spectrum techniques, we now describe in a more-detailed way the two most widely used schemes, namely, direct sequence and frequency hopping.

#### 10.5.2.1 *Direct Sequence Systems*

Direct sequence (DS), direct spread (DS), or pseudonoise (PN) spread-spectrum (SS) uses a code sequence to modulate a carrier (or a modulated carrier). In principle, any modulation technique such as AM (pulse), FM, or PM can be used. However, the most widespread form is the binary phase-shift-keying (BPSK) modulation. A simplified model of a DS/SS system is depicted in Figure 10.6. Note that the information signal $m(t)$ and the

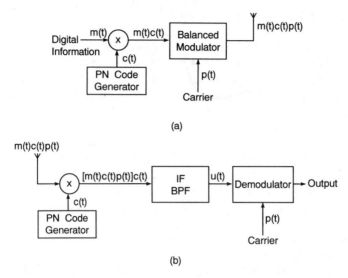

(a)

(b)

**Figure 10.6.** Simplified model of a direct sequence spread-spectrum system: (a) transmitter; (b) receiver.

spreading sequence $c(t)$ are combined before phase-modulating the carrier $p(t)$. This is convenient for digital modulation purposes.[1]

The PN code $c(t)$ is a pseudorandom binary sequence with a chip rate $R_c$ much larger than the information bit rate $R_b$. The multiplication of the PN code by the information signal results in a frequency spectrum with a bandwidth approximately equal to that of the sequence spectrum. The resultant signal is then modulated onto a carrier and transmitted. At the receiver an exact replica of the PN code is used in synchronism with the transmitter to unspread the received signal.

Note from Equation 10.20 that if the binary sequence $c_j(t)$ is composed of digits alternating between the levels $-1$ and $+1$, then $c_j^2(t) = 1 = K$. Therefore, from Equation 10.21 $u(t) = s_i(t) + \text{noise}$. Assuming a BPSK modulation in the DS/SS system of Figure 10.6, a sketch of the time-domain representation of the signals can be seen in Figure 10.7.

If the bandwidths of the spread and unspread signals are taken as being the mainlobe of their spectra, then $B_s = 2(1/T_c) = R_c$ and $B_U = 1/T_b = R_b$.* Therefore, using Equation 10.23, the processing gain of a DS/SS

---

* We are making the usual assumption that the power spectral density of a random pulse train is given by the $[(\sin x)/x]^2$ response (refer to Appendix 9A). Specifically, for a baseband signal this is $T_b[\sin(\omega T_b/2)/(\omega T_b/2)]^2$. For a modulated signal the power spectrum is $T_c\{\sin[(\omega - \omega_p)T_c/2]/[(\omega - \omega_p)T_c/2]\}^2$, where $\omega_p$ is the carrier angular frequency. Therefore, for positive frequencies $B_s = 2/T_c$ and $B_u = 1/T_b$.

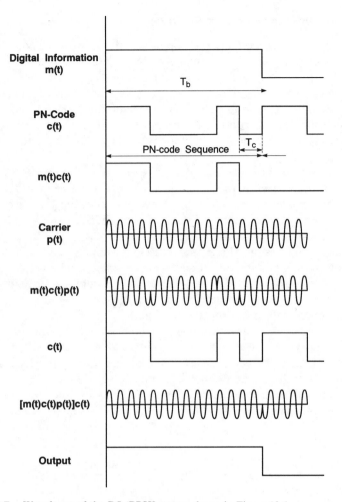

**Figure 10.7.** Waveforms of the DS/BPSK system shown in Figure 10.6.

system is

$$G_p = \frac{2T_b}{T_c} = \frac{2R_c}{R_b} \qquad (10.25)$$

Note that the larger the chip rate, the larger the processing gain.

The choice of modulation as well as code rate depend basically on (1) the available bandwidth, (2) the required process gain, and (3) the baseband information rate. Given a code rate, multilevel modulation techniques can be used to reduce the RF bandwidth. Note, however, that the processing gain is a function of this bandwidth.

### 10.5.2.2 *Frequency Hopping Systems*

In direct sequence systems the spreading of the transmission bandwidth appears in a continuous form when the PN sequence is combined with the information to modulate the carrier. The performance of spread-spectrum systems depends on the processing gain, which, in its turn, is dependent on the PN sequence chip rate. The generation of the PN code relies on the available technology to produce high chip rates. This, however, is a serious constraint, imposing practical limitations to obtaining a better processing gain.

Frequency hopping (FH), an alternative to the DS systems, is a spread-spectrum technique with the spreading of the transmission bandwidth appearing in a discrete form. This is achieved by allowing the carrier to hop from one frequency to another in a sequence dictated by the PN code. A basic FH/SS system is shown in Figure 10.8.

At a given time instant the output of FH transmitter is a single frequency. Over a period of time its output signal spectrum is ideally rectangular, with the same amount of power in every frequency, as depicted in Figure 10.9.

In practice, however, the hopping process generates spurious frequencies causing interchannel interference (cross-talk).

At the receiver the transmitted signal is mixed with a synchronous locally generated replica of the transmitter frequency sequence, offset by the intermediate frequency, $f_{IF}$. Again, any signal different from the local replica is spread by the multiplication and rejected as noise.

A usual modulation scheme for FH systems is *M*-ary frequency-shift keying and this combination is referred to as FH/MFSK. As far as demodulation is

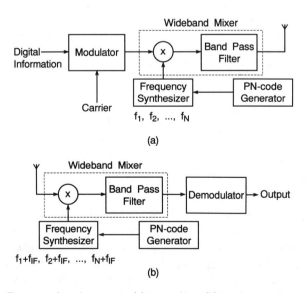

**Figure 10.8.**   Frequency hopping system: (a) transmitter; (b) receiver.

**Figure 10.9.** Frequency hopping signal spectrum.

concerned, the noncoherent approach is commonly used because it is extremely difficult to maintain phase relationships between frequency steps.

There are two basic FH systems:

- Slow frequency hopping (SFH) systems, in which several symbols of information are transmitted on each frequency hop
- Fast frequency hopping (FFH) systems, in which several hops occur during the transmission of one symbol.

In SFH systems each symbol is a chip, whereas in FFH systems the chip is characterized by a hop. An illustration of the frequency changes of SFH and FFH systems is shown in Figure 10.10.

Deliberate interference is possible if the spectral content of the transmitted signal is measured and the interfering signal is tuned to the corresponding portion of the frequency band. To combat this, the frequency hopper must hop at a sufficiently fast rate so as to skip to another frequency before the interferer can cause trouble. For mobile radio applications, this implies that the carrier frequency should hop as fast as possible because distances between wanted and interfering signals vary. Therefore, the exact time required to change to another frequency also varies.

If the frequency separation between discrete frequencies is such that it equals the message bandwidth, then from Equation 10.25

$$G_p = N \tag{10.26}$$

where $N$ is the number of available frequency choices (or channels used for hopping).

### 10.5.3 Synchronization

Synchronization is considered to be one of the most difficult parts of the CDMA architecture implementation. A rigid system timing is not required, but transmitters and receivers must work synchronously so that the despreading of the wanted information and the spreading of the unwanted signals can be accomplished.

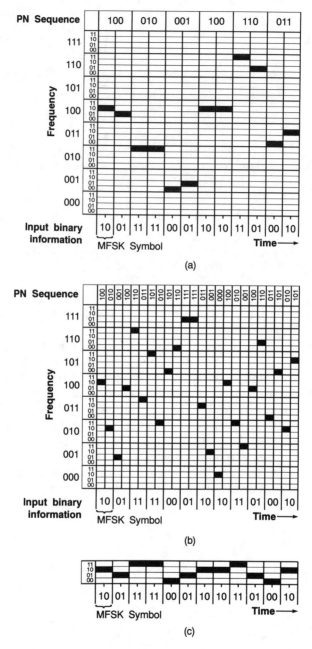

**Figure 10.10.** An illustration of frequency hopping: (a) slow frequency hopping; (b) fast frequency hopping; (c) dehopped signal.

There are basically two regions of uncertainty as far as synchronization is concerned: code phase and carrier frequency. The PN code phase uncertainty must be resolved to better than one chip duration. Ideally, once synchronization is achieved, it should continue to be so forever. There are, however, sources of errors that must be taken into account when designing a CDMA system. One such source is the Doppler effect, which affects both the carrier frequency and the code rate. Moreover, multipath propagation also imposes phase shifts in the received signal that can be really harmful at high-rate transmission. Other sources of uncertainty include the clock stability and clock rate offset.

Accordingly, the synchronization process comprises two sequential steps: initial synchronization, and tracking. The initial synchronization task is the most complicated of all. The tracking process can always be based on the knowledge of the timing gained previously. An extensive list of synchronization methods can be found in the literature[9, 10] and will not be reproduced here.

## 10.6  TWO-WAY COMMUNICATION

Two-way communication in a full-duplex fully trunked system can be implemented by means of frequency-division, time-division, and code-division methods. The latter being exactly the CDMA systems, which have been described in Section 10.5. Frequency division and time division will be characterized next.

### 10.6.1  Frequency-Division Duplex

In frequency-division duplex (FDD) communications, each direction of transmission (forward channel and reverse channel) has a separate frequency band. Consequently, simultaneous transmission in both directions is feasible. As explained in the FDMA case, a large interval between these frequency bands must be allowed so that interference can be minimized. In the same way, sharp filters with strong out-of-band rejection must be used in the radio equipments to reduce adjacent-channel interference.

### 10.6.2  Time-Division Duplex

In time-division duplex (TDD) communications, both directions of transmission use one contiguous frequency allocation, but two separate time slots. Because transmission from mobile to base and from base to mobile alternates in time, this scheme is also known as "ping-pong". As a consequence of the use of the same frequency band, the communication quality in both directions is the same. A guard time between the two time slots must be allowed in order to avoid interference. Moreover, synchronization among base stations should be provided, again for minimizing interference.

## 10.7  SUMMARY AND CONCLUSIONS

The choice of an access scheme for mobile radio systems is a rather complex task that may involve several aspects other than that of system capacity. In particular, many queries are still to be solved, remaining as matters of controversy. As far as analog systems are concerned the range of options is not so vast, so that the FDMA architecture is always used. For digital systems, however, three methods, namely, FDMA, TDMA, and CDMA, are applicable—each of which presents advantages and disadvantages. In this section we discuss some of these issues,[13] steering the analysis toward the digital approach.

*Advantages of SCPC / FDMA*

- Channel equalization is not necessary because the channels operate within the coherence bandwidth.
- Reduction of information bit rate can directly increase system capacity.
- The technology involved is well known.
- These systems are easily compatible with existing analog systems.

*Disadvantages of SCPC / FDMA*

- Growth techniques do not differ much from those used for the analog systems.
- Terminals are costly because they require narrowband filters, which are not realizable in VLSI.
- Bit transmission rate is fixed.

*Advantages of TDMA*

- Channel equalization can be used to combat fast fading.
- Bit transmission rate may be variable as a multiple or submultiple of the basic single-channel rate.
- Signal strength and bit error rates can be monitored on a frame-by-frame basis, facilitating and speeding up the hand-off process.

*Disadvantages of TDMA*

- High peak power on the uplink (mobile to base station) is required.
- A considerable amount of signal processing may be required when matched filtering and correlation detection are to be implemented.
- This increases the power consumption and imposes delay in the signal.

*Advantages of CDMA*

- Spectral density is reduced.
- Protection against jamming is provided.
- Message privacy is provided.
- Multipath effects are greatly minimized.

*Disadvantages of CDMA*

- Spread-spectrum techniques still remain to be explored for widespread commercial applications.

### 10.7.1  Final Remarks

Narrowband access schemes (FDMA and narrowband TDMA) are simple to implement but offer very little regarding protection against interference, noise, blocking, etc. Wideband systems are potentially more attractive, but require considerable advances in signal processing before they can be fully explored.

## REFERENCES

1. Bhargava, V. K., Haccoun, D., Matyas, R., and Nuspl, P., *Digital Communications by Satellite*, John Wiley & Sons, New York, 1976.
2. Spilker, J. J., Jr., *Digital Communications by Satellite*, Prentice-Hall, Englewood Cliffs, NJ, 1977.
3. Calhoun, G., *Digital Cellular Radio*, Artech House, Norwood, MA, 1988.
4. Tarallo, J. and Zysman, G. I., A digital narrowband cellular system, in *Proc. 37th IEEE Vehicular Technology Conference*, 279–280, Tampa, June 1–3, 1987.
5. Uddenfelt, J. and Persson, B., A narrowband TDMA system for a new generation cellular radio, in *Proc. 37th IEEE Vehicular Technology Conference*, 286–292, Tampa, June 1–3, 1987.
6. Bhargava, V. K., Haccoun, D., Matyas, R., and Nuspl, P., *Digital Communications by Satellite*, John Wiley & Sons, New York, 1981.
7. Balston, D. M., Pan-European cellular radio: or 1991 and all that, *Electron. Commun. Eng. J.*, 7–13, January/February 1989.
8. IS-54, EIA/TIA Project Number 2215, Cellular System: Dual-Mode Mobile Station–Base Station Compatibility Standard, December 1989.
9. Simon, M. K., Omura, J. K., Scholtz, R. A., and Levitt, B. K., *Spread Spectrum Communications*, Vol. 1, 2, 3, Rockville, MD, 1985.
10. Dixon, R. C., *Spread Spectrum System*, 2nd ed., John Wiley & Sons, New York, 1984.

11. Pickholtz, R. L., Schilling, D. L., and Milstein, L. B., Theory of spread spectrum communications—a tutorial, *IEEE Trans. Commun.*, COM-30, 855–884, May 1982.

12. Holmes, J. K., *Coherent Spread Spectrum Systems*, John Wiley & Sons, New York, 1982.

13. Cooper, G. R. and McGillem, C. D., *Modern Communications and Spread Spectrum*, McGraw-Hill, New York, 1986.

14. Parsons, J. D. and Gardiner, J. G., *Mobile Communication Systems*, Blackie & Sons, Glasgow, 1989.

15. Cooper, G. R., Nettleton, R. W., and Grybos, D. P., Cellular land-mobile radio: why spread spectrum?, *IEEE Communications Magazine*, March 1979.

16. Lee, W. C. Y., Overview of cellular CDMA, *IEEE Trans. Vehicular Tech.*, VT-40, 291–302, May 1991.

17. Gilhousen, K. S., Jacobs, I. M., Padovani, R., Viterbi, A. J., Weaver, L. A., Jr., and Wheatley, C. E., III, On capacity of a cellular CDMA system, *IEEE Trans. Vehicular Tech.*, VT-40, 303–312, May 1991.

18. Pickloltz, R. L., Milstein, L. B., and Schilling, D. L., Spread spectrum for mobile communications, *IEEE Trans. Vehicular Tech.*, VT-40, 313–322, May 1991.

19. Raith, K. and Uddenfeldt, J., Capacity of digital cellular TDMA systems, *IEEE Trans. Vehicular Tech.*, VT-40, 323–332, May 1991.

# CHAPTER 11

# Access Protocols

This chapter gives an introduction to the multiple-access protocols commonly used in multiple-access communication networks. The aim is to provide the basic concepts for the analysis of one particular protocol in a mobile radio environment. The mobile radio system constitutes a typical example of a multiple-access communication network where the control channels are shared by the mobile stations. The performance of the protocols is assessed by their throughput, and special attention is paid to the slotted ALOHA, chosen to be analyzed in a mobile radio environment. The radio channel is characterized by path loss (near/far effect) and Rayleigh fading. Moreover, the receiver is considered to work with capture effect. Contrary to what is initially expected, the combination of these phenomena greatly enhance the performance of the chosen protocol.

## 11.1 INTRODUCTION

The mobile radio system constitutes a typical example of a multiple-access communication network where the terminals, represented by the mobile stations, share the radio channels as common resources. The radio channels are divided into speech channels and control channels, the latter constituting the target of the present chapter.

Resource-sharing is intrinsically associated with system efficiency. Accordingly, it is desirable to optimize the use of the channels, keeping them busy with useful information as long as possible. However, due to the sharing, message collisions may occur in case more than one terminal attempts to transmit simultaneously. Therefore, a certain access discipline is mandatory, so that conflicts among terminals are kept to a minimum but with a maximum utilization of the channels.

## 11.2 PROTOCOL CATEGORIES

The multiple-access protocols are classified according to how much coordination is required to access the shared resources. Accordingly, there are three basic categories:

1. *Random access*—There is no coordination among the terminals. A terminal transmits a message according to its own convenience. As a consequence, message collisions are very likely to occur. The terminals are informed if they have been involved in a collision by monitoring the *acknowledgement* signal transmitted by the receiver over a separate channel.
2. *Scheduled access.* There is a total coordination among the channels. A terminal only transmits a message within a slot specially allotted for it. No collisions occur, but the channels may not be efficiently used.
3. *Hybrid access*—A certain degree of coordination is combined with the random access so that collisions are minimized.

## 11.3  PERFORMANCE EVALUATION

Many factors, such as (1) the ratio of the channel propagation delay to the transmission delay and (2) the traffic arrival process, and others, may have a great influence on the performance of the access protocol.

One important performance measure is the *throughput*, defined as the number of successfully transmitted messages per unit time. It is also important to determine the *delay* experienced by the message as a function of the throughput. However, in our analysis we shall consider only the throughput as a performance measure.

A useful definition to be used in the analysis of the protocols where collisions may occur is the *vulnerable period*. A vulnerable period $T$ of a protocol is the time interval susceptible to collisions.

If the messages are considered to have a fixed length, then the throughput is given by the number of successfully transmitted packets per packet transmission time. Accordingly, the throughput assumes values within the range 0 to 1.

Let $p$ be the probability of a successful packet transmission. Moreover, assume that the offered traffic, given in packets per packet time, is $G$. Therefore, the throughput $S$ is

$$S = Gp \qquad (11.1)$$

The packets generated in the network are constituted by new messages plus retransmitted messages. In the case where the number of packets is Poisson distributed (refer to Chapter 12), the probability that $k$ packets are generated in $T$ packet time is

$$p_k = \frac{(GT)^k}{k!} \exp(-GT) \qquad (11.2)$$

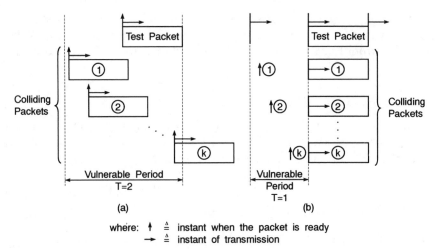

where: ⬆ ≜ instant when the packet is ready
       → ≜ instant of transmission

**Figure 11.1.** Vulnerable period of (a) ALOHA and (b) slotted ALOHA protocols.

Hereinafter we shall assume the Poisson process for the traffic generated by the terminals.

## 11.4  ACCESS PROTOCOLS

This section will focus on the various access protocols, ranging from the random protocols to the scheduled ones. Our aim is to introduce the basic concepts to be used in the analysis of a specific access protocol when this is applied to a mobile radio system.

### 11.4.1  ALOHA Protocols

The ALOHA is the best-known protocol among those belonging to the random access category. It is based on the following strategy. As soon as the message is ready for transmission, the terminal accesses the channel and sends its message. In case of collision occurrence, a random time is awaited until retransmission is attempted.

Figure 11.1a illustrates the timing of packet transmission using ALOHA. It can be seen that the vulnerable period of the packet is $T = 2$ packet times. Therefore, from Equation 11.2, the probability that an arbitrary packet is overlapped by $k$ packets is

$$p_k = \frac{(2G)^k}{k!} \exp(-2G) \tag{11.3}$$

The probability of a successful transmission in $T$ packet times is the probability that no message is generated within $T$, i.e., $p = p_0 = \exp(-GT)$. Therefore, from Equation 11.1, the throughput of the ALOHA protocol is

$$S = G \exp(-2G) \qquad (11.4)$$

Using Equation 11.4 we determine the maximum throughput of the ALOHA as being equal to $1/2e$ ($\approx 0.184$), corresponding to an offered traffic $G$ of 0.5 packet per packet time.

## 11.4.2  Slotted ALOHA

In the slotted ALOHA the time is divided into fixed-length time slots. The terminal always awaits the next slot boundary to send its packet. Consequently, the vulnerable period is $T = 1$ packet time as illustrated in Figure 11.1b. Accordingly, from Equation 11.2, the probability that a packet is overlapped by $k$ packets is

$$p_k = \frac{G^k}{k!} \exp(-G) \qquad (11.5)$$

Following the same steps as in the previous analysis, the throughput of this protocol is

$$S = G \exp(-G) \qquad (11.6)$$

Using Equation 11.6, we determine the maximum throughput of the slotted ALOHA as being equal to $1/e$ ($\approx 0.368$), corresponding to an offered load of 1 packet per packet time.

## 11.4.3  Tree Algorithm

The tree algorithm belongs to a class of protocols known as splitting algorithms. It is based on the following strategy. In the occurrence of a collision the terminals not involved in the collision go into a waiting state. Those involved are split into two groups according to a established rule (e.g., by flipping a coin). The first group is permitted to use one time slot whereas the second group can use the next time slot, in case the first group is successful. However, if another collision occurs, a further splitting is carried out until, eventually, a group with only one active terminal will be allowed to transmit successfully. The maximum throughput of this algorithm is 0.43.[7]

### 11.4.4  First-Come – First-Served Algorithm

This is another type of splitting algorithm, where packets are transmitted in a first-come–first-served (FCFS) mode. At each time slot (say time slot $k$) only the packet arriving within a specified allocation time interval (say, from $T(k)$ to $T(k) + \alpha(k)$) are entitled to be transmitted. If a collision occurs, the allocation interval is split into two equal subintervals (from $T(k)$ to $T(k) + \alpha(k)/2$ and from $T(k) + \alpha(k)/2$ to $T(k) + \alpha(k)$) and the packet arrived in the first subinterval is sent. In the case of another collision a further splitting ($\alpha(k)/4$) is required, and so on, until the transmission is successful. The throughput of this algorithm is 0.487.[8]

### 11.4.5  Carrier Sense Multiple Access Protocols

The basic characteristics of the carrier sense multiple access (CSMA) protocols is that each terminal monitors the status of the channel before transmission. If the channel is idle, the terminal transmits, otherwise the terminal awaits. There are several variations of this strategy:

1. *1-persistent CSMA*—The terminal waits until the channel is free.
2. *Nonpersistent CSMA*—The terminal waits a random time to retry the transmission.
3. *p-Persistent CSMA*—This protocol is applicable to slotted channels. In the case where the channel is idle, the packet is sent in the first possible slot with probability $p$ or in the next slot with probability $1 - p$.
4. *CSMA / CD*—In this protocol the terminal continuously monitors its own transmission for collision detection (CD) purposes. In case of collision, the transmission is aborted and a jamming signal is sent, instead. This jamming signal, sent for some units of time, is used as a "collision consensus enforcement mechanism".[1]

Although in all the CSMA protocols the channel is sensed prior to transmission, collision can still occur due to the propagation delay time. In other words a channel can be detected as idle although a remote terminal is already making use of it. Let $a$ be the propagation delay between the furthest terminals. Hence, the throughput of the nonpersistent CSMA is[9]

$$S = \frac{G \exp(-aG)}{G(1 + 2a) + \exp(-aG)} \tag{11.7}$$

### 11.4.6  Time-Division Multiple Access

In time-division multiple access (TDMA) protocol, time is divided into slots and these are clustered into frames. Each slot is dedicated to a particular user.

### 11.4.7  Frequency-Division Multiple Access

In frequency-division multiple access (FDMA) protocol the spectrum is divided into subbands and each subband constitutes a channel also dedicated to a particular user.

### 11.4.8  Code-Division Multiple Access

In code-division multiple access (CDMA) protocol, different terminals transmit with different codes. Hence, a receiver tuned to the code of one specific transmission will interpret the other transmissions as noise.

### 11.4.9  Packet-Reservation Multiple Access

Packet-reservation multiple access (PRMA)[11-13] can be viewed as a combination of TDMA and slotted ALOHA. The channel bit stream is organized in slots and frames, as in TDMA. Each slot within a frame can either be idle (available for contention) or busy (reserved). Terminals with new information to transmit contend for access to the idle slots, as in slotted ALOHA. Once a terminal is successful in the contention for a time slot in one frame, this slot is reserved for its exclusive use in subsequent frames, until it has no more packets to send.

Terminals are informed about the status (available or reserved) of each time slot by a continuous signal stream broadcast by the base station. Permission to contend for any idle time slot is granted according to a given probability, representing a design parameter in the PRMA system.

Due to its time slot reservation characteristics, the PRMA protocol is equally suitable for handling both periodic and random informations. Periodic information is represented by (1) packets belonging to a long stream of data information and (2) by the speech packets. Isolated data packets constitute the random type of information.

Although data packets can experience delays in case of system congestion, conversational speech requires prompt packet delivery. Accordingly, besides throughput, PRMA require other performance parameters, represented by the *number of simultaneous conversations* that can take place for a given *packet dropping probability* ($P_{drop}$). The $P_{drop}$ parameter is related to the speech-quality objectives. It depends on the maximum time the speech packet can be held before being discarded by its terminal.

Experiments[12] show that for a channel rate of 720 kbit/s, a source rate of 32 kbit/s, a frame duration of 16 ms, overhead of 64 bits per frame, and a maximum delay of 32 ms, the throughput is estimated as being 0.48. Moreover, for a $P_{drop} \leq 0.01$ the number of simultaneous conversation is 37. Note that under these conditions, the number of channels per carrier is $720/32 =$

22.5. Therefore, the number of simultaneous conversation per channel is $37/22.5 \approx 1.7$.

## 11.5   SOME COMMENTS ON THE PROTOCOLS

The pure ALOHA and the slotted ALOHA protocols are both susceptible to unstability. With the increase of the traffic load, the probability of collision increases substantially and eventually no packet is successfully transmitted. There are many ways of stabilizing these protocols by controlling the retrial process.

The splitting algorithms (e.g., tree and FCFS) have the advantage of better stability but are more complex to be implemented. The theoretical analysis of the FCFS protocol is particularly intricate and its performance is usually obtained by means of simulation.

The CSMA protocols are not free from collision even though the channels are sensed before transmission is allowed. This is due to the propagation delay time, which has a substantial influence on protocol performance.

The scheduled (also known as reservation-based) protocols (e.g., TDMA, FDMA, CDMA) suffer from the common disadvantage of not efficiently using the available resource. If a terminal has nothing to send, its slot is left free. This is particularly critical when messages are to be sent in bursts.

The performance of some protocols, as far as throughput is concerned, is shown in Figure 11.2. Note that the CSMA is greatly dependent on the delay parameter $a$. The CSMA/CD is a more elaborate protocol, giving better results than the others.

The PRMA protocol is still under investigation with the aim being to use it in third-generation mobile radio systems. "The vision of the third generation

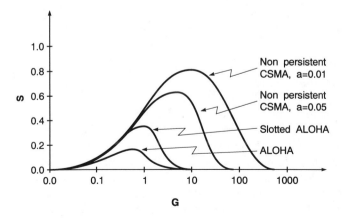

**Figure 11.2.**   Throughput of various access protocols.

is the merger of cellular and cordless technologies into a means of generalized wireless access to advanced information services".[12] By serving densely populated areas this system must handle the most diverse types of information such as voice, computer data, facsimile, and network-control messages.

We now turn our attention to the analysis of the slotted ALOHA in a mobile radio environment. Due to its simplicity and good performance, this protocol has been chosen to be used by the pan-European mobile communication system.[10] It is employed by mobile stations to provide initial access to the base station.

## 11.6  SLOTTED ALOHA IN A MOBILE RADIO ENVIRONMENT

The three main propagation characteristics in a mobile radio environment are (1) shadowing, (2) multipath propagation, and (3) the near/far phenomenon. Shadowing occurs due to obstruction. The envelope of the signal experiencing this phenomenon has lognormal distribution. Multipath propagation occurs due to the various signal scatterers randomly distributed around the mobile. The envelope of the resultant signal has a Rayleigh distribution. The near/far phenomenon is related to the different power levels experienced at the base station by the signals transmitted from mobile stations positioned at different locations in the cell. If $d$ is the distance between mobile and base station, the received mean signal power is given as a function of $d^{-\alpha}$, where $\alpha$ is a parameter in the range $2 \leq \alpha \leq 4$. The combination of these effects is usually seen as a threat to the performance of any transmission system. Accordingly, many equalization techniques are developed so as to counteract the effect of these phenomena.

However, we will show that, contrary to what is usually expected, the performance of the slotted ALOHA is greatly enhanced when employed in this "unkind" environment. We shall analyze the slotted ALOHA protocol in the presence of Rayleigh fading combined with the near/far phenomenon. The effects of shadowing are considered at the end of the chapter. The analysis that follows is based on work by Arnbak and Blitterswijk.[3]

### 11.6.1  Capture Effect

In the studies of standard ALOHA networks it is assumed that any collision provokes the destruction of all of the packets involved, leading to the retransmission of these packets.

On the other hand, it is plausible to assume that when two or more mobiles are competing for a time slot, the receiver at the base station will capture on the strongest signal. Let $W$ be the power of the wanted signal and let $I_k = \sum_{i=1}^{k} I_j$ be the joint power of the other $k$ interfering signals. Capture will occur when $W$ exceeds $I_k$ by a given threshold $R_0$. Therefore, the wanted packet is considered destroyed in a collision if and only if

$$W/I_k < R_0 \tag{11.8}$$

This implies that in our model the probability of collision given by Equation 11.5 is conditional on the fact that Inequality 11.8 holds. Hence

$$\text{prob}(k \text{ collisions given } W/I_k < R_0) = p_k = (G^k/k!)\exp(-G) \tag{11.9a}$$

Let $r_k$ be the probability that $W/I_k < R_0$, i.e.,

$$r_k = \text{prob}(W/I_k < R_0) \tag{11.9b}$$

Then the unconditional probability of collisions $q$ is

$$q = \sum_{k=1}^{\infty} p_k r_k \tag{11.10}$$

where $p_k$ and $r_k$ are given by Equations 11.9a and 11.9b, respectively. Capture will occur with probability $p$, where

$$p = 1 - q \tag{11.11}$$

Therefore, using Equation 11.10 in Equation 11.11 and the substituting Equation 11.11 into Equation 11.1, the channel throughput is obtained as

$$S = Gp = G\left[1 - \sum_{k=1}^{\infty} p_k r_k\right] \tag{11.12}$$

The probability distribution of the signal-to-interference ratio $W/I_k$ can be written as a function of their respective distributions as follows. Define the random variables $R$ and $I$ such that

$$W = RI, \qquad 0 \le R < \infty \tag{11.13a}$$

$$I = I_k, \qquad 0 \le I < \infty \tag{11.13b}$$

The joint distribution of $R$ and $I$ is

$$p(R, I) = |J| p(W, I_k)$$

where

$$|J| = \begin{vmatrix} \dfrac{\partial W}{\partial R} & \dfrac{\partial W}{\partial I} \\ \dfrac{\partial I_k}{\partial R} & \dfrac{\partial I_k}{\partial I} \end{vmatrix} = I$$

is the Jacobian of the transformation, defined by Equation 11.13. Because $W$ and $I_k$ are independent random variables

$$p(R, I) = Ip(W)p(I_k) \tag{11.14}$$

Hence, the density of $R$ is

$$p(R) = \int_0^\infty p(R, I) \, dI$$

$$= \int_0^\infty Ip(W)p(I_k) \, dI$$

$$= \int_0^\infty Ip(RI)p(I) \, dI \tag{11.15}$$

Therefore, the corresponding distribution of $R$ is $r_k$, where

$$r_k = \text{prob}(R \le R_0) = \int_0^{R_0} p(R) \, dR \tag{11.16}$$

## 11.6.2  Probability Density of the Mean Power

Let $p(W|\overline{W})$ be the distribution of the signal $W$ conditional on the mean power $\overline{W}$. Similarly, $p(I_k|\overline{I}_k)$ is the distribution of the interfering signal conditional on the mean interference $\overline{I}_k$. If we consider only multipath propagation, both $W$ and $I_k$ have a Rayleigh distribution, such that (refer to Chapter 5, Equation 5.17)

$$p(X|\overline{X}) = \frac{1}{\overline{X}} \exp\left(-\frac{X}{\overline{X}}\right) \tag{11.17}$$

where $X = W$ and $\overline{X} = \overline{W}$ or $X = I_k$ and $\overline{X} = \overline{I}_k$, as required.

The unconditional probabilities $p(W)$ and $p(I_k)$ are obtained by averaging the corresponding conditional probabilities $p(W|\overline{W})$ and $p(I_k|\overline{I}_k)$ over the distributions of $\overline{W}$ and $\overline{I}_k$, respectively. Hence,

$$p(W) = \int_0^\infty \frac{1}{\overline{W}} \exp\left(-\frac{W}{\overline{W}}\right) p(\overline{W}) \, d\overline{W} \tag{11.18}$$

$$p(I) = p(I_k) = \int_0^\infty \frac{1}{\overline{I}_k} \exp\left(-\frac{I_k}{\overline{I}_k}\right) p(\overline{I}_k) \, d\overline{I}_k \tag{11.19}$$

With Equations 11.18 and 11.19 in Equation 11.15 and then using Equation 11.15 in Equation 11.16, we obtain (refer to Appendix 11A.1)

$$r_k = \int_0^\infty p(\overline{W}) \left( \int_0^\infty p(\overline{I}_k) \frac{R_0 \overline{I}_k}{R_0 \overline{I}_k + \overline{W}} d\overline{I}_k \right) d\overline{W} \tag{11.20}$$

The mean duration of the fadings is usually on the order of some milliseconds (see Section 4.7) whereas the duration of a time slot in a random access channel is about fractions of a millisecond.* Accordingly, we may assume that all of the received signals remain with the same mean power for the duration of a time slot so that their powers can be added directly. Therefore, the density of the mean interference signal is obtained by convolving the individual densities of the signal as many times ($k$ times) as required (refer to Appendix 11A.2), i.e.,

$$p(\overline{I}_k) = \left[ p(\overline{W}) \right]^{*k} \tag{11.21}$$

where $*k$ signifies a $k$-fold convolution.

### 11.6.3 Rayleigh Fading and Path Loss Combined

Let $\overline{W}$ be the normalized mean power with respect to the power transmitted by the mobile from the border of the cell. Moreover, let $h$ be the ratio between the distances $d/d_{max}$, where $d$ is the distance from the mobile to the base station and $d_{max}$ is the cell radius. We have seen in Chapter 3 that

$$\overline{W} = h^{-\alpha} \tag{11.22}$$

where $\alpha$ is the path loss slope in the range $2 \leq \alpha \leq 4$.

Define $G(h)$ as the number of packets per time slot offered per unit area at a normalized distance $h$. Let the mean number of packets be both Poisson distributed and a function of $h$ only. Then, the total offered traffic is

$$G_t = 2\pi \int_0^\infty hG(h) \, dh \tag{11.23}$$

Assuming a uniform distribution for the mobiles within the cell, the density of $h$ is

$$p(h) = \frac{2\pi hG(h)}{G_t} \tag{11.24}$$

---

*For the GSM system this is 0.577 ms.[10]

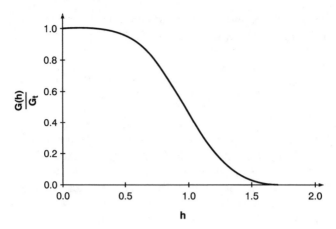

**Figure 11.3.**   Quasiuniform traffic distribution.

Because $\overline{W}$ is a function of $h$ (see Equation 11.22), its density is obtained as

$$p(\overline{W})|d\overline{W}| = p(h)|dh| \tag{11.25}$$

From Equation 11.22, $|d\overline{W}| = \alpha h^{-(\alpha+1)}|dh|$ and $h = \overline{W}^{-1/\alpha}$. Replacing these results and Equation 11.24 in Equation 11.25, we obtain

$$p(\overline{W}) = \frac{2\pi}{\alpha G_t}\overline{W}^{-(1+2/\alpha)}G(\overline{W}^{-1/\alpha}) \tag{11.26}$$

We now have all the necessary tools to evaluate the performance of slotted ALOHA in a mobile radio environment. Once the traffic distribution $G(h)$ is known, the densities $p(\overline{W})$ and $p(\overline{I}_k)$ can be calculated by means of Equations 11.26 and 11.21, respectively. Then these densities are used in Equation 11.20 to obtain $r_k$. Finally, the probability $r_k$ is used in Equation 11.12 to evaluate the system throughput.

As an example, consider the following quasiuniform traffic distribution, as depicted in Figure 11.3,

$$G(h) = \frac{G_t}{\pi}\exp\left(-\frac{\pi}{4}h^4\right) \tag{11.27}$$

This distribution has been "chosen for its analytical convenience".[3]

Setting $\alpha = 4$ and by following the steps previously described, the throughput $S(h)$ as a function of $h$ is

$$S(h) = G(h)\left[1 - \exp(-G_t)\sum_{k=1}^{\infty}\frac{G_t^k}{k!}r_k\right] \tag{11.28}$$

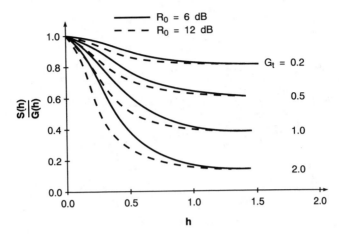

**Figure 11.4.** Probability of successful transmission versus normalized distance for different traffic loads. (*Source:* After J. C. Arnbak and W. Van Blitterswijk, Capacity of slotted ALOHA in Rayleigh fading channels, *J. on Selected Areas in Commun.*, SAC-5(2), 261–269, February 1987.)

where (see Appendix 11A.3)

$$r_k = \sqrt{\pi}\, \gamma_k \exp(\gamma_k^2) \mathrm{erfc}(\gamma_k) \tag{11.29}$$

$$\gamma_k = k \frac{\sqrt{\pi}}{2} \sqrt{R_0}\, h^2 \tag{11.30}$$

The total traffic captured by the receiver is

$$S = 2\pi \int_0^\infty h S(h)\, dh \tag{11.31}$$

yielding

$$S = G_t \exp(-G_t) \sum_{k=1}^\infty \frac{G_t^k}{k!} \frac{1}{k\sqrt{R_0}+1} \tag{11.32}$$

The curves of the normalized conditional throughput $S(h)/G(h)$ versus the normalized distance $h$ with different total traffic $G_t$ and capture threshold $R_0$ are shown in Figure 11.4. The overall throughput $S$ for the various capture thresholds $R_0$ is shown as a function of the total traffic $G_t$ in Figure 11.5. Note from Figure 11.5 that it improves substantially with increasing capture threshold. For a capture threshold tending to infinity, corresponding to the case of no capture, the throughput curve coincides with that of standard slotted ALOHA.

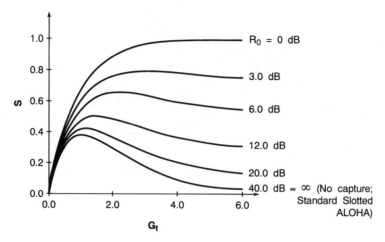

**Figure 11.5.** Throughput curves for different capture thresholds. (*Source:* After J. C. Arnbak and W. Van Blitterswijk, Capacity of slotted ALOHA in Rayleigh fading channels, *J. on Selected Areas in Commun.*, SAC-5(2), 261–269, February 1987.)

## 11.7 SUMMARY AND CONCLUSIONS

This chapter introduced some of the protocols commonly employed in multiple access communications networks. The mobile radio system is an example of multiple access network where the access protocol is used by the mobile station to provide its initial contact with the base station. In particular, the slotted ALOHA is an efficient and easily implemented protocol, chosen to be utilized by the pan-European mobile radio system. The performance of this protocol is affected by the various propagation phenomena occurring in the mobile environment. In our analysis we included the near/far effect (path loss) and Rayleigh fading. The calculations of the throughput are rather laborious but the results are encouraging.

It has been shown that packet destruction and unstability are greatly improved in the mobile radio environment as compared to the standard slotted ALOHA network. Contrary to what is usually expected, performance enhancement is provided by the mobile radio propagation phenomena. Both path loss and Rayleigh fading act independently on the users, naturally splitting them into different closes of access power. This greatly improves the channel efficiency. Because the subscribers are in motion, the randomness of the signal power and the time to initiate a transmission after a collision provide the real mobile channel with a natural priority. This priority changes dynamically at each time instant.

The inclusion of shadowing in the calculations slightly degrades the overall performance but the results are still encouraging.[6]

"The third generation mobile radio system will probably convey information of several types including voice, computer data, facsimile and network control messages".[12] In this sense, new proposals of access protocols able to handle data and speech have emerged and they are under investigation.[11-13]

## APPENDIX 11A   SLOTTED ALOHA IN MOBILE RADIO ENVIRONMENT

### 11A.1   Derivation of Equation 11.20

Given that $I = I_k$ and $W = RI_k$, we use Equations 11.18 and 11.19 in Equation 11.15. We then use Equation 11.15 in Equation 11.16 to obtain $r_k$,

$$r_k = \int_0^{R_0} \int_0^{\infty} \int_0^{\infty} \int_0^{\infty} \frac{I_k}{\overline{W}\overline{I}_k} \exp\left[-I_k\left(\frac{\overline{W} + R\overline{I}_k}{\overline{W}\overline{I}_k}\right)\right] p(\overline{W}) p(\overline{I}_k) \, dR \, dI_k \, d\overline{W} \, d\overline{I}_k$$

By first integrating in $I_k$ and then in $R$ we have

$$r_k = \int_0^{\infty} \int_0^{\infty} \frac{R_0 \overline{I}_k}{\overline{W} + R_0 \overline{I}_k} p(\overline{W}) p(\overline{I}_k) \, d\overline{W} \, d\overline{I}_k \qquad (11A.1)$$

### 11A.2   Probability Density of a Sum of Independent Random Variables

#### 11A.2.1 *Laplace Transform*

The Laplace transform of a function $f(t)$ is $F(s)$, defined as

$$f(t) \Leftrightarrow F(s) \triangleq \int_{-\infty}^{\infty} f(t) e^{-st} \, dt$$

Assuming $f(t) = 0$ for $t < 0$, the one-sided Laplace transform of $f(t)$ is

$$F(s) = \int_0^{\infty} f(t) e^{-st} \, dt \qquad (11A.2)$$

#### 11A.2.2 *Convolution*

Define $*$ as the convolution operator. The convolution of two functions of continuous time, $f(t)$ and $g(t)$, is

$$f(t) * g(t) = \int_{-\infty}^{\infty} f(t - \tau) g(\tau) \, d\tau$$

If $f(t)$ and $g(t)$ take zero values for $t < 0$, then

$$f(t) * g(t) = \int_0^t f(t - \tau) g(\tau) \, d\tau \tag{11A.3}$$

### 11A.2.3 *Laplace Transform of the Convolution*

Equation 11A.3 can be rewritten as

$$f(t) * g(t) = \int_0^t f(t - \tau) g(\tau) \, d\tau + 0 \int_t^\infty f(t - \tau) g(\tau) \, d\tau$$

$$= \int_0^\infty u(t - \tau) f(t - \tau) g(\tau) \, d\tau \tag{11A.4}$$

where

$$u = (t - \tau) = \begin{cases} 1 & \tau \leq t \\ 0 & \tau > t \end{cases}$$

Substituting Equation 11A.4 in Equation 11A.2 and then interchanging the order of integration with respect to the variables $t$ and $\tau$, we have

$$f(t) * g(t) \Leftrightarrow \int_{\tau=0}^\infty g(\tau) \left[ \int_{t=0}^\infty u(t - \tau) f(t - \tau) e^{-st} \, dt \right] d\tau \tag{11A.5}$$

The inner integral in Equation 11A.5 is equal to $e^{-\tau s} F(s)$. Therefore,

$$f(t) * g(t) \Leftrightarrow F(s) \int_{\tau=0}^\infty g(\tau) e^{-s\tau} \, d\tau$$

Thus

$$f(t) * g(t) \Leftrightarrow F(s) G(s) \tag{11A.6}$$

This result applies to the convolution of any number of functions. Accordingly, if $p(x_i)$, $i = 1, \ldots, n$, have Laplace transforms $P_i(s)$, $i = 1, \ldots, n$, respectively, then

$$p(x_i) * p(x_2) * \cdots * p(x_n) \Leftrightarrow \prod_{i=1}^n P_i(s) \tag{11A.7}$$

### 11A.2.4 *Sum of Independent Random Variables*

Let $X_i$, $i = 1, \ldots, n$, be independent random variables with probability density functions equal to $p(x_i)$, $i = 1, \ldots, n$, respectively. Assume that the Laplace transform of $p(x_i)$ is $P_i(s)$. Given that $X = \sum_{i=1}^n X_i$, we want to determine the density of $X$. Let $p(x)$ be this density with Laplace transform

$P(s)$. Thus

$$p(x) \Leftrightarrow P(s) = \int_0^\infty e^{-sx} p(x) \, dx \qquad (11A.8)$$

Using the transformation of random variables,

$$p(x) \, dx = p(x_1, x_2, \ldots, x_n) \, dx_1 \, dx_2 \, \cdots \, dx_n$$

Because $X_i, \; i = 1, \ldots, n$, are independent random variables

$$p(x) \, dx = p(x_1, x_2, \ldots, x_n) \, dx_1 \, dx_2 \, \cdots \, dx_n = \prod_{i=1}^n p(x_i) \, dx_i \qquad (11A.9)$$

Using Equation 11A.9 and Equation 11A.8, we obtain

$$p(x) \Leftrightarrow P(s) = \int_0^\infty e^{-sx_1} p(x_1) \, dx_1 \int_0^\infty e^{-sx_2} p(x_2) \, dx_2 \; \cdots \int_0^\infty e^{-sx_n} P(x_n) \, dx_n$$

Therefore

$$p(x) \Leftrightarrow P(s) = \prod_{i=1}^n P_i(s) \qquad (11A.10)$$

With Equations 11A.7 and 11A.10 we conclude that

$$p(x) = p(x_i) * p(x_2) * \cdots * p(x_n) \Leftrightarrow \prod_{i=1}^n P_i(s) = P(s) \qquad (11A.11)$$

## 11A.3  Distribution $r_k$ of $R$ as a Function of the Normalized Distance $h$ (Equation 11.29)

Assume that at the normalized distance $h$ the mean power $\overline{W}$ is constant and equal to $h^{-\alpha}$. Then, from Equation 11.20

$$r_k = \int_0^\infty p(\overline{W}) \, d\overline{W} \int_0^\infty p(\bar{I}_k) \frac{R_0 \bar{I}_k}{R_0 \bar{I}_k + h^{-\alpha}} \, d\bar{I}_k$$

$$= \int_0^\infty p(\bar{I}_k) \frac{R_0 \bar{I}_k}{R_0 \bar{I}_k + h^{-\alpha}} \, d\bar{I}_k \qquad (11A.12)$$

With Equation 11.27 in Equation 11.26, we obtain

$$p(\overline{W}) = \frac{1}{2}\overline{W}^{-3/2} \exp\left(-\frac{\pi}{4\overline{W}}\right) \qquad (11A.13)$$

The Laplace transform of $p(\overline{W})$ is equal to $\exp(-\sqrt{\pi}\sqrt{s})$.
Therefore, from Equation 11.21 in connection with Equation 11A.11,

$$p(\overline{I}_k) = \left[p(\overline{W})\right]^{*k} \Leftrightarrow \exp(k\sqrt{\pi}\sqrt{s})$$

From the tables of the Laplace transform

$$p(\overline{I}_k) = \frac{k}{2}\overline{I}_k^{-3/2} \exp\left(-\frac{\pi k^2}{4\overline{I}_k}\right) \Leftrightarrow \exp(k\sqrt{\pi}\sqrt{s}) \qquad (11A.14)$$

With Equation 11A.14 in Equation 11A.12 we obtain

$$r_k = k\int_0^\infty \frac{R_0 h^4}{R_0 h^4 + t^2} \exp\left(-\frac{\pi}{4}k^2 t^2\right) dt$$

From tables of integrals

$$r_k = \sqrt{\pi}\,\gamma_k \exp(\gamma_k^2)\mathrm{erfc}(\gamma_k) \qquad (11A.15)$$

where

$$\gamma_k = k\frac{\sqrt{\pi}}{2}\sqrt{R_0}\,h^2$$

## REFERENCES

1. Li, V. O. K., Multiple access communication networks, *IEEE Communications Magazine*, 25(6), 41–48, June 1987.
2. Goodman, D. J. and Saleh, A. A. M., The near/far effect in local ALOHA radio communications, *IEEE Trans. Vehicular Tech.*, VT-36(1), 19–27, February 1987.
3. Arnbak, J. C. and Van Blitterswijk, W., Capacity of slotted ALOHA in Rayleigh fading channels, *IEEE J. Selected Areas Commun.*, SAC-5(2), 261–269, February 1987.
4. Abramson, N., The throughput of packet broadcasting channels, *IEEE Trans. Commun.*, COM-25, 117–128, January 1977.
5. Namislo, C., Analysis of mobile radio slotted ALOHA network, *IEEE J. Selected Areas Commun.*, SAC-2, 583–588, July 1984.

6. Ling, Y. K. and Cattermole, K. W., The random-access control channel of a cellular mobile radio system, in *5th U.K. Teletraffic Symposium*, Birmingham, England, 1988.

7. Capetanakis, J. I., Tree algorithms for packet broadcasting channels, *IEEE Trans. Inform. Theory*, IT-25, 505–515, September 1979.

8. Gallager, R. G., Conflict resolution in random access broadcast networks, in *Proc. AFOSR Workshop on Commun. Theory and Appl.*, Provincetown, MA, September 1978.

9. Kleinrock, L. and Tobagi, F. A., Packet switching in radio channels: Part 1—carrier sense multiple access modes and their throughput delay characteristics, *IEEE Trans. Commun.*, COM-23(12), 1400–1416, December 1975.

10. CEPT/CCH/GSM, Physical Layer on the Radio Path: General Description, GSM recommendation 05.01, Draft 3.1.0, February 1988.

11. Nanda, S., Analysis of packet reservation multiple access: voice data integration for wireless network, in *IEEE Globecom Conf.*, San Diego, pp. 1984–1988, December 1990.

12. Goodman, D. J. and Wei, S. X., Efficiency of packet reservation multiple access, *IEEE Trans. Vehicular Tech.*, 40(1), 170–176, February 1991.

13. Goodman, D. J., Trends in cellular and cordless communications, *IEEE Communications Magazine*, 31–40, June 1991.

# PART VI
# Traffic

# CHAPTER 12

# Traffic Aspects in Mobile Radio Systems

This chapter examines some of the main channel allocation techniques that can be used in a mobile radio system. It starts by reviewing the basic principles of queueing and traffic theory, commonly used in the teletraffic analysis. The allocation techniques are divided into two groups, namely, global and local. The global techniques usually imply a substantial change in the usage pattern of the channels, requiring the involvement of the central processor. In the local techniques, although some small change in the usage pattern of the channels may occur, the channels are used within or closely adjacent to their planned service area. In this case the decisions are taken locally, involving two or three neighboring cells.

The global allocation techniques are investigated by means of simulation, where the main phenomena, such as hand-off, cochannel interference, adjacent-channel interference, and others, are taken into account. Special attention is given to the hybrid channel allocation technique, where it is shown that a small proportion of dynamic channels is enough to provide a substantial gain in traffic capacity.

The local techniques are investigated by means of the Markov process in a two-cell system. The aim is to get an insight into the main phenomena involved and not to provide a quantitative performance measure. In particular, we examine the blocking threshold variation technique using numerical analysis. It is shown that even with the use of a small proportion of the channels for alternative routing, a significant gain in traffic capacity can be achieved. An approximate expression for the mean blocking probability is found to give results very close to those of the exact solution.

## 12.1  INTRODUCTION

Although the traffic characteristics of a mobile radio system is rather distinct from that of a fixed telephone network, the system planning and design are still carried out with the tools of conventional traffic theory. According to the expected traffic and cochannel interference requirements, the geographical region is divided into cells. Given the traffic in each cell and the desired mean blocking probability, the Erlang-B formula is used to determine the number of channels per cell, assuming no hand-off and no roaming. The design is later adjusted to take into account these factors.

All the cells can be built to have the same area but different number of channels per cell, chosen to meet the blocking probability requirements. Alternatively, the cells can present the same number of channels but their sizes must be tailored so that the traffic in each cell can experience the desired grade of service.

The traffic distribution vary in time and in space, but they are commonly bell-shaped. High concentrations are found in the city center during the rush hour, decreasing toward the outskirts. After the rush hour and toward the end of the day, this concentration changes as the users move from the town center to their homes.

Note that, because of the mobility of the users, hand-offs and roaming are always occurring, reducing the channel holding times in the cell where the calls are generated. This increases the traffic in the cell where the mobiles go to. Accordingly, the Erlang-B formula no longer applies. A full investigation on the traffic performance of a mobile radio system requires all the phenomena to be taken into account, rendering intricate any traffic model. We may, however, introduce many simplifications and obtain a qualitative rather than a quantitative result to understand the main phenomena better.

In this chapter we shall examine some channel assignment techniques, particularly analyzing the hybrid assignment algorithm. Moreover, we shall investigate an alternative routing technique using the fuzzy traffic of the cells —the blocking threshold variation algorithm. This chapter starts by presenting some queueing and traffic theory fundamentals.

## 12.2   QUEUEING AND TRAFFIC THEORY FUNDAMENTALS

In the traffic analysis of a mobile radio system it is widely accepted that (1) calls have a Poisson arrival, (2) holding times have a negative exponential distribution, and (3) blocked calls are lost.

In this section we shall examine the basic principles of queueing and traffic theories used to analyze the various channel assignment algorithms in a mobile radio system.

### 12.2.1   Call Arrival Process

Consider the call arrival as a random process. Let a time interval $t$ be divided into $n$ equal subintervals with length $t/n$ each. Choose $n$ to be sufficiently large that the following hold:

1. Only one arrival can occur in any subinterval $t/n$.
2. Call arrivals are independent from each other.
3. The probability $p_1(1)$ that an arrival occurs in one of the subintervals is proportional to the subinterval length. Hence $p_1(1) = \lambda t/n$, where $\lambda > 0$. Accordingly, the probability of no arrivals in $t/n$ is $1 - \lambda t/n$.

The probability of exactly $k$ arrivals in $n$ subintervals can be evaluated using the binomial distribution. Then

$$p_k(n) = \binom{n}{k} \left( \lambda \frac{t}{n} \right)^k \left( 1 - \lambda \frac{t}{n} \right)^{n-k} \tag{12.1}$$

The limit of $p_k(n)$ as $n$ tends to infinity will be the probability $p_k$ of $k$ arrivals in the time interval $t$. Using Newton's binomial expansion* for $(1 - \lambda t/n)^{n-k}$, it is straightforward to show that, as $n$ tends to infinity, Equation 12.1 becomes

$$p_k = \frac{(\lambda t)^k}{k!} \exp(-\lambda t) \tag{12.2}$$

Equation 12.2 is the Poisson probability distribution. Its mean is $E[k] = \sum_{k=0}^{\infty} k p_k = \lambda t$. The parameter $\lambda$ is the mean arrival rate of calls (calls per second). The variance is $E[k^2] - E^2[k] = \lambda t$.

### 12.2.1.1 Mean Interarrival Time

Let $\tau$ be a random variable denoting the time between adjacent arrivals with probability distribution and density given by $A(t)$ and $a(t)$, respectively. The distribution $A(t)$ is the probability that the time between arrivals $\tau$ is less than or equal to $t$. Hence

$$A(t) = \text{prob}(\tau \leq t) = 1 - \text{prob}(\tau > t)$$

But $\text{prob}(\tau > t)$ is exactly the probability that no arrivals occur within $t$, that is, $p_0$. Therefore, with $k = 0$ in Equation 12.2,

$$A(t) = 1 - \exp(-\lambda t) \tag{12.3}$$

---

*$(a + b)^N = \sum_{i=1}^{N} \binom{N}{i} a^i b^{N-i}$.

The density $a(t)$ is obtained by differentiating $A(t)$ with respect to $t$. Thus

$$a(t) = \lambda \exp(-\lambda t) \qquad (12.4)$$

Equation 12.4 is referred to as the negative exponential distribution. The mean interarrival time is

$$E[t] = \int_0^\infty t a(t)\, dt = a/\lambda$$

### 12.2.1.2 *Memoryless Property of the Negative Exponential Distribution*

This property refers to the fact that the past history of an exponentially distributed random variable has no influence in predicting its future. In our particular case, consider the situation where a call has just arrived. The distribution of the time until arrival of the next call is still a negative exponential, as we shall demonstrate.

Let $t_0$ be a time where no arrival occurs. What is the probability that the next arrival occurs within a time $t$ from $t_0$? To answer this question we must calculate the probability $\mathrm{prob}(\tau \le t + t_0 | \tau > t_0)$. Using the conditional probability properties we have

$$\mathrm{prob}(\tau \le t + t_0 | \tau > t_0) = \frac{\mathrm{prob}(t_0 < \tau \le t + t_0)}{\mathrm{prob}(\tau > t_0)}$$

$$= \frac{\mathrm{prob}(\tau \le t + t_0) - \mathrm{prob}(\tau \le t_0)}{\mathrm{prob}(\tau > t_0)} \qquad (12.5)$$

The probability $\mathrm{prob}(\tau \le x)$ is given by $A(x)$. Therefore, using Equation 12.3 in Equation 12.5, we obtain

$$\mathrm{prob}(\tau \le t + t_0 | \tau > t_0) = 1 - \exp(-\lambda t)$$

which is exactly the distribution $A(t)$.

### 12.2.2  Call Holding Time

Let $t$ be a time interval divided into $n$ equal subintervals of length $t/n$ each. Choose $n$ to be sufficiently large that the following hold:

1. The probability that a call terminates within one subinterval is proportional to its length. That is, this probability equals $\mu t/n$, where $\mu > 0$.
2. The call termination occurs independently of which subinterval is considered.

Let $\tau$ be a random variable denoting the call holding time with probability distribution and density given by $H(t)$ and $h(t)$, respectively. The distribution $H(t)$ is the probability that the holding time $\tau$ is less than or equal to $t$. Hence

$$1 - H(t) = 1 - \text{prob}(\tau \le t) = \text{prob}(\tau > t)$$

The probability $\text{prob}(\tau > t)$ that a call originated at time zero will not terminate before $t$ is given by the probability that it will not terminate in any of the $n$ subintervals of length $t/n$, when $n$ tends to infinity. Then

$$1 - H(t) = \lim_{n \to \infty} \left(1 - \mu \frac{t}{n}\right)^n = \exp(-\mu t)$$

Thus

$$H(t) = 1 - \exp(-\mu t) \tag{12.6}$$

The density is obtained by differentiating Equation 12.6 with respect to $t$. Hence

$$h(t) = \mu \exp(-\mu t) \tag{12.7}$$

In the same way, the mean holding time for the interarrival time distribution case is $1/\mu$.

## 12.2.3 Birth–Death Processes

"A Markov process with a discrete state space is referred to as a Markov chain. A set of random variables $\{X_n\}$ forms a Markov chain if the probability that the next value (state) is $x_{n+1}$ depends only upon the current value (state) $x_n$ and not upon any previous values."[1] In analytical form we have

$$\text{prob}\big[X(t_{n+1}) = x_{n+1}|X(t_n) = x_n, \ldots, X(t_1) = x_1\big]$$

$$= \text{prob}\big[X(t_{n+1}) = x_{n+1}|X(t_n) = x_n\big] \tag{12.8}$$

A birth–death process constitutes a special case of Markov process where transitions are allowed to occur only between neighboring states. Let $S_k$ denote the state of the system at a given time when the population of the system (number of busy channels) is $k$. A transition from $S_k$ to $S_{k+1}$ implies a birth within the population, occurring with a rate $\lambda_k$. A transition from $S_k$ to $S_{k-1}$ implies a death with a rate $\mu_k$. Note that $\lambda_k$ and $\mu_k$ are rates and not probabilities. They can, however, be converted into probabilities if they are multiplied by the infinitesimal time $dt$, leading to the probability of such a transition occurring during $dt$.

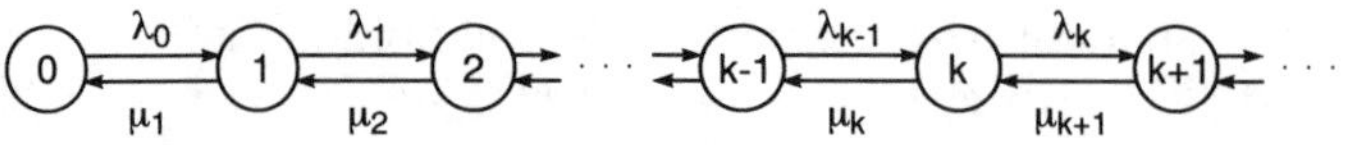

**Figure 12.1.** State-transition diagram for the one-dimensional birth–death process.

A one-dimensional birth–death process can be represented by means of a state transition rate diagram as depicted in Figure 12.1. Let $p_k(t)$ be the probability that the system is in state $S_k$ at the time instant $t$. By simply inspecting Figure 12.1 we may write the following relation:

probability of reaching the state $S_k = \left[\lambda_{k-1} p_{k-1}(t) + \mu_{k+1} p_{k+1}(t)\right] dt$

probability of departing from the state $S_k = (\lambda_k + \mu_k)\, dt\, p_k(t)$

The difference between these two probabilities equals the differential probability $dp_k(t)$, such that

$$\frac{dp_k(t)}{dt} = \lambda_{k-1} p_{k-1}(t) + \mu_{k+1} p_{k+1}(t) - (\lambda_k + \mu_k) p_k(t) \quad (12.9)$$

As $t$ tends to infinity, the system tends to an equilibrium solution where the transient behavior vanishes. Accordingly, $dp_k(t)/dt$ tends to zero and we denote $p_k(t)$ by $p_k$. Therefore, from Equation 12.9,

$$\lambda_{k-1} p_{k-1} + \mu_{k+1} p_{k+1} = (\lambda_k + \mu_k) p_k, \qquad k \geq 0 \qquad (12.10)$$

where $p_{-1} = 0$, $\mu_0 = 0$, and $\lambda_{-1} = 0$. Equation 12.10 shows that, in equilibrium, the flow rate into $S_k$ equals the flow rate out of $S_k$. By writing Equation 12.10 sequentially for $k = 0, 1, 2, 3, \ldots$ and observing that the probabilities $p_k$ must sum to unity,

$$\sum_{k=0}^{\infty} p_k = 1 \qquad (12.11)$$

we may solve the set of equations to obtain

$$p_k = p_0 \prod_{i=0}^{k-1} \lambda_i / \mu_{i+1} \qquad (12.12)$$

where

$$p_0 = \left[ 1 + \sum_{k=1}^{\infty} \prod_{i=0}^{k-1} \lambda_i / \mu_{i+1} \right]^{-1} \qquad (12.13)$$

### 12.2.3.1 *Blocked Calls Held*

Consider a system with an infinite number of servers. Assume that the arrival (birth) rate is constant and equal to $\lambda$ and that the departure (death) rate from each server is equal to $\mu$. Thus

$$\lambda_k = \lambda, \qquad k \geq 0$$

$$\mu_k = k\mu, \qquad k \geq 1$$

Then, using these parameters in Equations 12.12 and 12.13, we obtain

$$p_k = \left[(\lambda/\mu)^k/k!\right]\left[\sum_{i=0}^{\infty}(\lambda/\mu)^i\right]^{-1}$$

Hence

$$p_k = \frac{A^k}{k!}\exp(-A) \tag{12.14}$$

where $A = \lambda/\mu$ is the traffic offered in erlangs (erl). The mean value and the variance of the preceding distribution have already been determined in Section 12.2.1, and they are both equal to $A$. Assume that the system has $N$ channels ($N$ servers) and that arrivals, occurring when all $N$ channels are busy, remain in the system for one holding time. If during this period a channel becomes free, a waiting customer seizes the channel for the remainder of the holding time. The blocking probability in this case is

$$B = \sum_{k=N}^{\infty}\frac{A^k}{k!}\exp(-A) \tag{12.15}$$

This equation is referred to as Molina's formula.[2]

### 12.2.3.2 *Blocked Calls Cleared*

In this case the system is assumed to have $N$ channels, and calls arriving when all the channels are found to be busy are lost. Then, the state transition diagram of Figure 12.1 terminates at the state $S_N$ ($N$ channels) and

$$\lambda_k = \lambda, \qquad k \leq N - 1$$

$$\mu_k = k\mu, \qquad k < N$$

With these transition rates in Equations 12.12 and 12.13 and noticing that the upper limit of the sum in Equation 12.13 is equal to $N$, we obtain

$$p_k = \frac{A^k/k!}{\sum_{i=0}^{N}A^i/i!} \tag{12.16a}$$

Blocking will occur when all the $N$ channels are busy. The probability of this event is

$$E(A, N) = p_N = \frac{A^N/N!}{\sum_{i=0}^{N} A^i/i!} \qquad (12.16b)$$

which is known as the Erlang-B formula. Another useful way of writing the Erlang-B formula is in its recursive form. It is easy to show that

$$E(A, N) = \frac{E(A, N-1)}{N/A + E(A, N-1)} \qquad (12.16c)$$

The Erlang-B formula is well established in the form of table, as shown in Appendix 12C (from Reference 21). An approximation can be obtained for large values of $N$, when the denominator of Equation 12.16b tends to $\exp(A)$ such that

$$E(A, N) \simeq \frac{A^N}{N!} \exp(-A), \quad \text{for large } N \qquad (12.17)$$

### 12.2.3.3 Blocked Calls Delayed

In this model the blocked calls are allowed to queue up and wait to be served. Then

$$\lambda_k = \lambda, \qquad k \geq 0$$

$$\mu_k = \begin{cases} k\mu, & 0 \leq k \leq N \\ N\mu, & k \geq N \end{cases}$$

Using the same procedure as in the previous case we obtain

$$p_k = \begin{cases} \dfrac{A^k}{k!} p_0, & 0 \leq k \leq N \\ \dfrac{A^k}{N!} N^{N-k} p_0, & k \geq N \end{cases}$$

where

$$p_0 = \left[ \sum_{k=0}^{N-1} \frac{A^k}{k!} + \frac{A^N}{N!} \frac{1}{1 - A/N} \right]^{-1} \qquad (12.18)$$

The probability of delaying (no channel is available) is

$$\text{prob[queueing]} \triangleq C(N, A) = \sum_{k=N}^{\infty} p_k$$

Using $p_k$ for $k \geq N$, we obtain

$$C(N, A) = \frac{A^N}{N!} \frac{1}{1 - A/N} p_0 \qquad (12.19)$$

This equation is known as the Erlang-C formula.

### 12.2.3.4 Finite Population

In the previous cases we considered a Poisson traffic generated by an infinite population. If, however, the population is finite, say with $M$ customers, the birth rate becomes dependent on the size of this population, decreasing as each customer seizes a server. Assuming a pure loss system, we have

$$\lambda_k = (M - k)\lambda, \qquad k \leq N - 1$$

$$\mu_k = k\mu, \qquad k < N$$

Accordingly,

$$p_k = \frac{\binom{M}{k} A^k}{\sum_{i=0}^{N} \binom{M}{i} A^i} \qquad (12.20)$$

This is known as the Engset distribution.

### 12.2.4 Alternative Routing (Wilkinson's Theory)

In standard telephone networks a stream of traffic can reach its destination through a direct path or through indirect paths. The trunks in the direct path form a high-usage route constituting the first option routing. If all these trunks are busy, the overflow traffic may use an alternative path to reach its destination. Although the traffic using the direct route has Poisson characteristics, the overflow traffic presents peakednesses and is not considered to be Poisson any longer. It is better described by its variance $V_i$, such that[3]

$$V_i = F_i \left[ 1 - F_i + \frac{A_i}{1 + N_i + F_i - A_i} \right] \qquad (12.21)$$

where $A_i$ is the traffic offered to route $i$,
$F_i$ is the overflow traffic from $i$,
$N_i$ is the number of trunks on route $i$.

The overflow traffic is obtained from

$$F_i = A_i E(A_i, N_i) \tag{12.22}$$

Because an overflow route takes traffic from $n$ other routes, its mean total traffic $A$ and variance $V$ are

$$A = \sum_{i=1}^{n} A_i \quad \text{and} \quad V = \sum_{i=1}^{n} V_i \tag{12.23}$$

The equivalent random (Poisson) traffic $A_e$, when offered to an "imaginary" full-availability group of $N_e$ trunks, results in an overflow traffic with mean and variance equal to $A$. Recall from Section 12.2.1 that the variance-to-mean ratio of a Poisson traffic is the unity. It is possible, to show[4] that these parameters are approximately given by

$$A_e = V + \frac{3V}{A}\left(\frac{V}{A} - 1\right) \tag{12.24}$$

$$N_e = A_e\left(1 - \frac{1}{A + V/A}\right)^{-1} - A - 1 \tag{12.25}$$

Suppose that the stipulated blocking probability for the alternative route is $B$. Hence the overflow traffic from this route is $AB$. Consider that the alternative route has $N_a$ trunks. Accordingly, the overflow traffic produced by the equivalent traffic $A_e$ when offered to a group of $N_e + N_a$ trunks should be equal to $AB$. Therefore,

$$B = \frac{A_e}{A} E(N_e + N_a, A_e) \tag{12.26}$$

## 12.3   TRAFFIC PERFORMANCE ENHANCEMENT TECHNIQUES

Channel allocation plays a very important role in the traffic performance of a mobile radio system. The simplest channel assignment algorithm uses fixed allocation. It has the maximum spatial efficiency in channel reuse, as the channels are always assigned at the minimum reuse distance. Moreover, because each cell has a fixed set of channels, the channel assignment control for the calls can be distributed among the base stations.[5] The main problem of fixed allocation is its inability to deal with the alteration of the traffic pattern. Due to the mobility of the subscribers, some cells may experience a sudden growth in the traffic offered, with a consequent deterioration of the grade of service, while other cells may have free channels that cannot be used by the congested cells.

The steps to assign channels on a fixed basis are as follows:

1. The repeat pattern (cluster size) is determined according to the allowable signal-to-interference ratio.
2. The cluster with the highest traffic is chosen for the initial design. The traffic per cell in this cluster is then determined.
3. Given the traffic per cell and an allowable blocking probability, the number of channels per cell can be determined using the Erlang-B formula.
4. In case the total number of available channels is not enough to provide the required grade of service, the area covered by one cluster should be reduced in order to reduce the traffic per cell.
5. The other clusters can reuse the same channels according to the reuse pattern. However, cells with less traffic may use fewer channels so that not all channels need to be provided by the base stations.

### 12.3.1 Global Assignment Techniques

The main channel allocation techniques such as borrowing, dynamic, and hybrid algorithms are briefly described in Chapter 2 (Section 2.12.1). These techniques have been extensively explored in the literature.[5, 12] In particular, they are thoroughly investigated by Sánchez[5] and Sánchez and Eade[12] with most of the mobile radio phenomena (cochannel interference, adjacent-channel interference, hand-offs, etc.) included.

Dynamic channel allocation comprises a set of algorithms with the common characteristics that all the channels are available for all the cells. The dynamic algorithm that chooses the first dynamic channel available not exceeding the reuse constraints is of particular interest due to its simplicity. In the algorithms where there is a pool of channels to be dynamically used according to the traffic demand, spatial efficiency is not usually considered.

The results show[5] that these techniques require a substantial amount of memory and data interchange between base station and mobile switching centers. For an even traffic distribution and low blocking probability, these global techniques perform better than the fixed allocation algorithm. With the increase of traffic, the spatial inefficiency of dynamic channels cause both the dynamic algorithms and the hybrid techniques with high proportion of dynamic channels to reverse their performance with respect to the fixed assignment. For uneven traffic distribution, where more flexibility is required, the dynamic followed by hybrid and borrowing techniques perform better than the fixed algorithm. In Section 12.4 we shall examine the hybrid technique in more detail.

## 12.3.2   Local Assignment Techniques

These techniques treat the traffic with access to more than one base station as available for alternative routing. This traffic is mostly encountered at the border regions of the cells where base station service areas overlap. Although the application of these techniques implies some change in the usage pattern of the channels, the rearrangement is purely local and does not imply any complex global network management. Moreover, because the flexible channels are within or closely adjacent to their service area, there are no serious implications for frequency reuse pattern. Let us describe the main techniques.[14-20]

- *Mean adaptation* (MAP)—divides the flexible traffic among the cells in a proportion controlled by the estimate of the imbalance of each cell.
- *Adaptive response to blocking* (ARB) (also known as directed retry[13])—diverts the blocked flexible calls to the cell presenting free channels.
- *Adaptation to mean and blocking*—is the combination of MAP and ARB.
- *Instantaneous adaptation* (IAP)—diverts the flexible traffic to the cell with a smaller number of busy channels. When the cells present an equal number of busy channels, the flexible traffic is equally shared by these cells.
- *Instantaneous and mean adaptation* (IMA)—is similar to IAP. However, in case the cells present an equal number of busy channels, the flexible traffic is directed to the least imbalanced cell.
- *Full IMA* (FIMA)—combines the characteristics of IMA and AMB to yield an optimum strategy.
- *Blocking threshold variation* (BTV)—to be described in Section 12.5.

These strategies have been extensively analyzed in References 14–20, showing that a substantial improvement in traffic performance can be achieved. In the results to be given next we use the symbol " $<$ " to mean "better than" (i.e., less mean blocking) and " $\equiv$ " to mean "equivalent to". Hence:

1. For an even traffic distribution, $IAP \equiv IMA < ARB \equiv AMB < MAP$
2. For an uneven traffic distribution, $IMA < IAP < ARB$ and $AMB < ARB$
   a. When traffic balance cannot be accomplished and
      For high imbalance, $IMA < AMB$
      For low imbalance, $AMB < IMA$
      For very high imbalance, $AMB < MAP < IMA < IAP < ARB$

FIMA gives the best performance among all these strategies. However, it is substantially more complicated to implement, not justifying its application. Moreover, the improvement achieved is not significantly greater than that of IMA.[18]

## 12.4   HYBRID CHANNEL ALLOCATION[5, 12]

In hybrid allocation a proportion of the channels is assigned on a fixed basis, while the remaining channels form a common pool for dynamic assignment. The dynamic channels ($N_d$) are requested by the cell when its fixed channels ($N_f$) are all being used. It can be seen that this technique may range from the fixed ($N_d = 0$) to the totally dynamic ($N_f = 0$) allocation. Accordingly, its performance depends on the traffic profile and on the proportion of dynamic-to-fixed ($N_d/N_f$) channels.

### 12.4.1   Initial Performance Assessment

The usefulness of the hybrid technique can be initially assessed using an analytical approach such as that described by the Wilkinson theory (refer to Section 12.2.4). The traffic blocked when all the fixed channels are busy can be treated as an overflow traffic. The dynamic channels are equivalent to the trunks available on the indirect (alternative) route. Therefore, for a given traffic per cell $A_i$, offered to a group of fixed channels $N_i$ ($N_i = N_f$), the overflow traffic $F_i$ can be calculated using Equation 12.22. The variance $V_i$ is then determined by Equation 12.21 and the total traffic $A$ with variance $V$ on the alternative route can be estimated by Equation 12.23, where $n$ is equal to

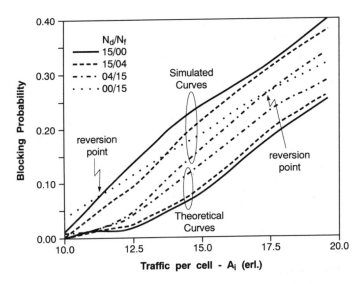

**Figure 12.2.**   Theoretical and simulation results of the hybrid channel assignment technique. (*Source*: After J. H. Sánchez V. and J. P. Eade, A simulation study of cellular and sectorized mobile telephone systems using a hybrid channel allocation technique, in *Proc. 5th UK Teletraffic Symposium*, Aston, U.K., pp. 11/1–10, July 1988.)

the number of cells using the pool of dynamic channels. Then the equivalent random traffic $A_e$ offered to the group of $N_e$ channels is determined by Equations 12.24 and 12.25, respectively. The overall blocking probability is determined by Equation 12.26, where $N_a = nN_d$.

This procedure has been used to assess the performance of a seven-cell cluster with a total of 105 channels (15 channels per cell) and an even traffic distribution. The result is shown in Figure 12.2 (theoretical curves). Note that the curves for the proportion of dynamic-to-fixed channels equal to 00/15 and 15/00 can be obtained directly by the Erlang-B formula by performing $E(A_i, 15)$ and $E(7A_i, 105)$, respectively. In particular, the theoretical and the simulated curves (see Section 12.4.2) for the proportion 00/15 differ by a factor smaller than 1%. Therefore, for practical purposes they are considered to be "coincident". However, this may not be true if the proportion of hand-offs increases.

## 12.4.2   Simulation Results

The results obtained in the previous subsection apply to an ideal situation where the traffic in each cell is considered to be independent from that of its neighbors. The hand-offs of calls imply a mutual traffic dependence between cells, so that these results are not accurate. Moreover, in our analysis the call using a dynamic channel is supposed to continue with this channel until the termination of the call.

It is possible, however, to rearrange the calls so that the dynamic channels can be used only if there are no fixed channels available. This means that a call, using a dynamic channel, would be transferred to a fixed channel, in case this fixed channel is rendered available during that period. Consequently, more dynamic channels can be left free, rendering the system more capable to cope with alterations in the traffic. Call rearrangement can also be carried out among the dynamic channels so that they can be more efficiently used. We may also allow the hand-offs to be requested a certain number of times to decrease their blocking probability.

It can be seen that, if all the phenomena occurring in a real mobile radio system are to be taken into account, then an analytical solution for the channel allocation techniques is not feasible. The use of simulation can overcome the intractability of the analytical solutions, constituting an important tool for the traffic analysis in a mobile radio system.

### 12.4.2.1 *Traffic Model*

In order to simulate an infinite system with a finite number of cells, these cells can be placed on a toroidal surface so that the mobiles can roam indefinitely in any direction over the torus surface without leaving the system. A system with seven-cell clusters (49 cells), as shown in Figure 12.3, can be

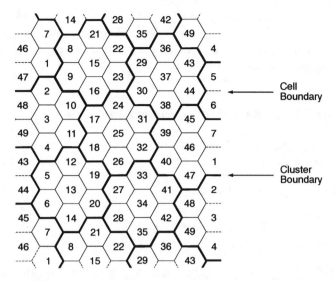

**Figure 12.3.** "Infinite" system with seven seven-cell clusters. (*Source*: After J. H. Sánchez V. and J. P. Eade, A simulation study of cellular and sectorized mobile telephone systems using a hybrid channel allocation technique in *Proc. 5th UK Teletraffic Symposium*, Aston, U.K., pp. 11/1–10, July 1988.)

used for this purpose. Note that the cells at the edges are repeated to form a toroidal surface.

This system was simulated assuming the following: (1) 105 channels per cluster, (2) one guard channel between adjacent channels within the cell, (3) 40% of the calls being handed off, (4) up to 10 attempts of hand-offs during a time-out period of 0.5 min, (5) a negative exponential distribution with mean value equal to 2 min for the call holding time, and (6) Poisson-originated traffic per cell with even distribution. The results are shown in Figure 12.2.

Compare the theoretical and the simulated curves. Note that "as the traffic level increases more channels are needed in each cell up to a point where the Fixed Allocation 15/00 outperforms the other proportions".[5] This is a direct consequence of the spatial inefficiency in the dynamic channels, because they are not always allocated at the minimum reuse distance. Accordingly, there may be reversion points where the hybrid technique become worse than the fixed one.

## 12.5 BLOCKING THRESHOLD VARIATION[20,22]

The blocking threshold variation (BTV) is a channel assignment strategy treating the traffic with access to more than one base station as available for alternative routing. The performance of this strategy has been thoroughly investigated by Mencia[20] and Mencia and Yacoub,[22] where a two-cell system is considered. It is recognized, however, that there are reservations to be made, because in a practical network larger clusters interact. The joint process of two cells gives some guide to the phenomena to be expected, but not a precise quantitative solution.

### 12.5.1   The BTV Strategy Routing Procedure

Consider two adjacent cells in a mobile radio system each of which with $N$ channels. Define flexible (or alternative) traffic as the traffic with access to more than one base station. As far as the flexible traffic is concerned, the BTV strategy divides the cells into two categories as follows:

1. *First-option cell*—the subscriber's own cell
2. *Second-option cell*—the subscriber's alternative cell

Let $T$ $(0 \leq T \leq N + 1)$ be the blocking threshold of each cell. For convenience we shall assume the same threshold for both cells. The blocking threshold $T$ corresponds to the starting point from which the first-option cell with $T$ busy channels should be avoided. In this case the second-option cell is chosen to carry this alternative traffic, if its number of busy channels is smaller than $T$. If the number of busy channels is greater than $T$ in both cells, the flexible traffic is diverted to the least-loaded cell.

Suppose now that the cells present the same number of busy channels, exceeding the threshold $T$. In this case the strategy assumes one out of three possible alternatives:

- Alt1—to keep the flexible traffic within the first-option cell
- Alt2—to split the flexible traffic equally between the cells
- Alt3—to divert the flexible traffic to the second-option cell

It is interesting to note that, for $T = N + 1$, the flexible traffic is not used. For $T = N$ this strategy coincides with ARB (directed retry). For $T = 0$ and with Alt2, BTV is equivalent to IAP. For $T = 0$ and with Alt3, BTV is equivalent to IMA.

### 12.5.2   Traffic Parameters

The traffic offered is defined by three independent variables: fixed traffic of cell 1 ($A_1$), fixed traffic of cell 2 ($A_2$), and flexible traffic ($A_{12}$). It is convenient to analyze the system in terms of a global traffic variable $A$, such that

$$A = A_1 + A_2 + A_{12} \tag{12.27}$$

Define $\beta$ $(-1 \leq \beta \leq 1)$ as the imbalance of fixed traffic between cells and $\gamma$ $(0 \leq \gamma \leq 1)$ as the proportion of flexible traffic. Thus

$$\beta = (A_1 - A_2)/(A_1 + A_2) \tag{12.28}$$

$$\gamma = A_{12}/A \tag{12.29}$$

Assuming the subscribers to be uniformly distributed within the cell, the parameter $\gamma$ may be obtained as the proportion of overlapped area, as estimated in Section 3.7.4. In terms of the parameters $A$, $\beta$, and $\gamma$ we have

$$A_1 = \tfrac{1}{2}A(1 + \beta)(1 - \gamma)$$
$$A_2 = \tfrac{1}{2}A(1 - \beta)(1 - \gamma) \tag{12.30}$$

The number of channels per cell $(N)$ and the blocking threshold $(T)$ also constitute parameters to be considered in our traffic analysis.

### 12.5.3 Traffic Model

The traffic process is modelled by means of a two-dimensional birth–death process. The state number is a two-component vector $(i, j)$, where $i$ and $j$ are the number of busy channels in cell 2 and cell 1, respectively. There are two upward transition densities from each state $(i, j)$ namely, $\lambda1_{ij}$ and $\lambda2_{ij}$, defined by the rate of arrival in cell 1 and cell 2, respectively. Similarly, there are two downward transitions $\mu1_{ij}$ and $\mu2_{ij}$ defined by the rate of departure from cells 1 and 2. The state-transition diagram is shown in Figure 12.4.

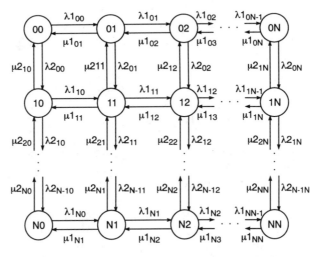

**Figure 12.4.** State-transition diagram of a two-dimensional birth–death process.

The balance equation for the state $(i, j)$ can be written as

$$-(\lambda 1_{ij} + \mu 1_{ij} + \lambda 2_{ij} + \mu 2_{ij})p_{ij} + \mu 1_{ij+1}p_{ij+1} + \lambda 1_{ij-1}p_{ij-1}$$

$$+ \mu 2_{i+1j}p_{i+1j} + \lambda 2_{i-1j}p_{i-1j} = 0 \qquad (12.31)$$

where $p_{xy}$ is the equilibrium state probability. By writing Equation 12.31 for all the $(N + 1)^2$ states, we obtain a system of $(N + 1)^2$ linearly dependent equations. Because $p_{xy}$ are probability distributions, the usual constraint

$$\sum_{i=0}^{N} \sum_{j=0}^{N} p_{ij} = 1 \qquad (12.32)$$

replacing any one of the balance equations eliminates the indeterminacy. Hence, the equilibrium state probabilities can be calculated by solving a system of $(N + 1)^2$ equations.

In our analysis we shall assume that (1) the traffic is Poisson, (2) the holding time has a negative exponential distribution, (3) flexible calls occur with probability $\gamma$, and (4) flexible calls may be routed to either cell as required.

### 12.5.3.1 *Transition Probability Densities*

With the negative exponential distribution assumption, the probability of a reduction in the occupancy of either cell is proportional to the number of busy channels in that cell. For simplicity, we consider the mean holding time to be equal to 1. Hence,

$$\mu 1_{ij} = j$$
$$\qquad (12.33)$$
$$\mu 2_{ij} = i$$

Assuming that each cell takes its own fixed traffic plus some proportion of the flexible traffic, we have

$$\lambda 1_{ij} = A_1 + \tfrac{1}{2}(1 + \alpha_{ij})A_{12}$$
$$\qquad (12.34)$$
$$\lambda 2_{ij} = A_2 + \tfrac{1}{2}(1 - \alpha_{ij})A_{12}$$

where $\alpha_{ij}$ are routing coefficients defining the proportion of flexible traffic to be taken by each cell. These coefficients are state-dependent and are determined according to the routing procedure. Hence:

1. If the number of busy channels $i$ and $j$ in each cell is smaller than the blocking threshold $T$, then the flexible calls of a given cell will stay in their own cell. Therefore, the ratio $A_1/A_2$ must equal that of $(1 + \alpha_{ij})/(1 - \alpha_{ij})$. Accordingly, $\alpha_{ij} = \beta$.

In the same way, the following hold:

2. If $i > T$ and $j \geq i$, then $\alpha_{ij} = -1$.
3. If $j > T$ and $i \geq j$, then $\alpha_{ij} = 1$.
4. If $i = j \geq T$, then $\alpha_{ij}$ depends on the alternative:
   a. for Alt1, $\alpha_{ij} = \beta$;
   b. for Alt2, $\alpha_{ij} = 0$;
   c. for Alt3, $\alpha_{ij} = -\text{sgn}(\beta)$, where $\text{sgn}(x) = -1, 0,$ or $1$ for $x < 0$, $x = 0$, or $x > 0$, respectively.

The routing coefficients $\alpha_{ij}$ can be written as functions of $i$, $j$, and $T$, using a three-level logic, as shown in Appendix 12A.

### 12.5.4  Performance Measures

Define $B_1$, $B_2$, and $B_{12}$ as the blocking probabilities of cell 1, cell 2, and cells 1 and 2, respectively. They can be obtained from the equilibrium state probabilities as follows

$$B_1 = \sum_{i=0}^{N} p_{iN}, \qquad B_2 = \sum_{j=0}^{N} p_{Nj}, \qquad B_{12} = B_1 \cap B_2 = p_{NN} \quad (12.35)$$

The principal measure used to assess the system performance is the mean blocking probability $B_m$ over all bundles of traffic,

$$B_m = (A_1 B_1 + A_2 B_2 + A_{12} B_{12})/A \qquad (12.36)$$

### 12.5.5  Analytical Solutions

As we have seen, the number of states in the two-dimensional birth–death process is a quadratic function of the number of channels per cell. In general, the two-dimensional process does not admit the simple one-dimensional form of solution given in Section 12.2.3. Accordingly, numerical analysis using a computer is the most common tool for solving a set of $(N + 1)^2$ equations. Nevertheless, it is possible to obtain analytical solutions for special cases where not all the parameters are considered or their variations are limited to some specific range. Let us consider these cases.

#### 12.5.5.1  Erlang-B Formula

The Erlang-B formula can be used in three limiting cases as follows.

1. *No flexibility*—In this case the flexible traffic is nil ($\gamma = 0$). Accordingly, the two cells are considered to be isolated from one another so that we have $B_1 = E(A_1, N)$, $B_2 = E(A_2, N)$, and $B_{12} = 0$. The mean blocking is given by Equation 12.36.

2. *Full flexibility*—In this case the fixed streams of traffic are nil and all the traffic is flexible ($\gamma = 1$). Hence, the two cells are equivalent to one cell with $2N$ channels. Therefore, $B_1 = B_2 = 0$ and $B_m = B_{12} = E(A, 2N)$.
3. *Maximum threshold*—The case when $T = N + 1$ is equivalent to that when $\gamma = 0$ (no flexibility).

### 12.5.5.2 *Approximate Formula*

It is shown in Appendix 12B that, by means of low-traffic analysis, an approximate solution is given by

$$B_m = E(A/2, N)(1 + K\beta^2)(1 - \gamma)^{N-T+1}$$

$$+ \gamma^{1/N}\left[1 - (1 - \gamma)^{N-T+1}\right]E(A, 2N) \qquad (12.37)$$

where

$$K = \left[(N + 1)(N - A) + (A/2)^2\right]/2 \qquad (12.38)$$

In fact, this approximate solution yields satisfactory results over a wide range of the parameters, as we shall see in Section 12.5.6.

### 12.5.6  Some Results

The performance of a two-cell system using the blocking threshold variation strategy has been assessed by means of a numerical analysis, as described in the previous subsections. It was verified[20, 22] that there is no significant difference between the three alternatives (Alt1, Alt2, Alt3) of the strategy. Rigorously speaking, Alt3 gives the best performance, followed by Alt2 and Alt1, in this order. Recall from the definition of these alternatives that in Alt3, when both cells present an equal number of busy channels, the flexible traffic is diverted to the second-option cell whereas, in Alt1, it remains in its own cell. In this condition (both cells with an equal number of busy channels), the second-option cell is more likely to be less loaded than the other cell. Therefore, traffic balance can be more quickly reached with Alt3. However, we must emphasize that the performance of these strategies is fairly similar, so that, for practical purposes, they are considered to be equivalent to each other.

In the results to be given here, each cell is assumed to have six channels so that there are up to seven blocking thresholds. Moreover, because the mean blocking is an even function with respect to the imbalance parameter, just half of the range of this parameter needs to be explored.

Figure 12.5 shows the mean blocking versus the imbalance parameter for 50% flexibility ($\gamma = 0.5$) and total traffic $A = 6$ erl. Note that there is a substantial improvement in the mean blocking with the use of the strategy.

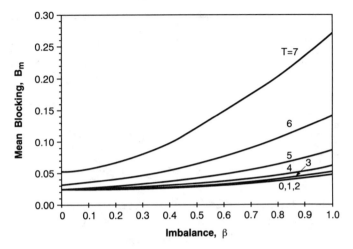

**Figure 12.5.** Mean blocking versus imbalance ($A = 6$ erl, $N = 6$, $\gamma = 0.5$).

The improvement is more significant when the threshold varies from $T = 7$ to $T = 6$. We notice that for $T = 5$ and $T = 4$ the results are even better, but below this ($T \leq 3$) there is not much gain, and the curves practically coincide with each other.

Figure 12.6 shows the mean blocking versus the flexibility parameter for a traffic imbalance of 50% ($\beta = 0.5$) and total traffic $A = 6$ erl. Note that for

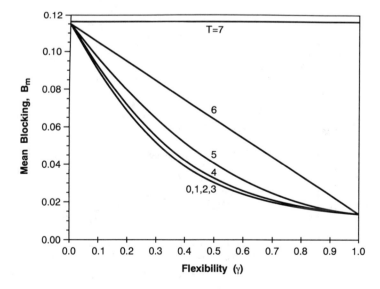

**Figure 12.6.** Mean blocking versus flexibility ($A = 6$ erl, $N = 6$, $\beta = 0.5$).

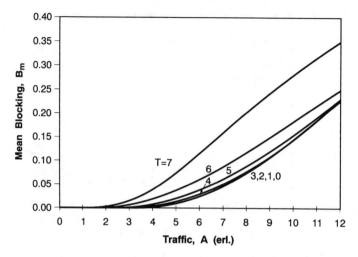

**Figure 12.7.**   Mean blocking versus traffic ($N = 6$, $\beta = 0.5$, $\gamma = 0.5$).

$T = N + 1$ the mean blocking is independent of the flexibility, as expected. For $\gamma = 0.5$ the mean blocking reduces from approximately 12% (for $T = 7$) to less than 3% (for $T \leq 4$) with the use of the BTV strategy.

Figure 12.7 shows the mean blocking versus traffic. Note that within the range 2–20% of the blocking probability the system has the best performance. It can be seen that, for $B_m = 5\%$, the improvement in traffic capacity is greater than 50% when BTV is used with a threshold $T \leq 4$.

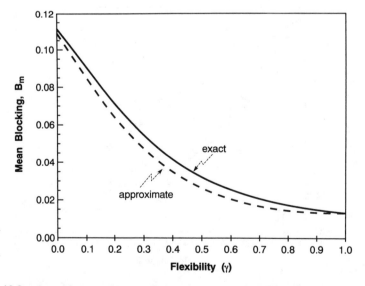

**Figure 12.8.**   Approximate and exact solutions ($A = 6$ erl, $N = 6$, $\beta = 0.5$, $T = 4$) for mean blocking versus flexibility.

The approximate formula given by Equation 12.37 was also checked against the exact solution throughout a wide range of variation of the parameters. The results are quite encouraging not only for low traffic, where the exact and approximate curves are almost coincident, but also for high traffic as shown in Figure 12.8.

### 12.5.7  Extension of the Model for Larger Clusters

The model used here can be extended to deal with larger clusters where more cells interact. In this case we may also consider the flexible traffic with access to three base stations as available for alternative routing. We showed in Chapter 3 (Section 3.6.9) that, if $\delta$ is the proportion of traffic with three or more alternative radio paths and $\gamma$ is this proportion for two or more paths, then, $\delta \simeq 1.25\gamma^2$ is a good approximation in the range $0 \leq \gamma \leq 0.8$. More-over, because $\gamma = 50\%$ is a fairly realistic figure in mobile radio systems (refer to Section 3.7), then $\delta \simeq 31\%$. It has been shown[19] that with the use of the third path option, the increase in traffic capacity may exceed the figure of 20% on top of the improvement achieved when only two paths are used for alternative routing, for a grade of service of 5%.

When clusters including different-capacity cells are considered, the imbal-ance parameter must be modified to include not only the traffic per cell, but also the number of channels per cell. Hence each cell $i$ would be associated with an individual imbalance $\beta_i$ such that

$$\beta_i = \left( B_i - \overline{B} \right) / \overline{B} \tag{12.39}$$

where $B_i$ is the blocking probability of cell $i$ due to its fixed traffic, and $\overline{B}$ is the average blocking probability experienced by all of the fixed streams of traffic.

If cell $i$ has $N_i$ channels and a fixed traffic equal to $A_i$, then

$$B_i = E(A_i, N_i) \tag{12.40}$$

The average blocking $\overline{B}$ is

$$\overline{B} = \frac{\left( \sum_{i=1}^{n} A_i B_i \right)}{\left( \sum_{i=1}^{n} A_i \right)} \tag{12.41}$$

where $n$ is the number of cells in the system.

## 12.6  SUMMARY AND CONCLUSIONS

The traffic process in a mobile radio system greatly departs from that of the fixed telephone network. However, in general, design and planning have not been optimized for the practical situation in which the subscribers are mobile. In this case classical teletraffic theory can be used just to provide an insight into the problem. A better assessment of the system performance can be achieved by means of simulation, where the complex realistic situations can be taken into account without much difficulty.

Because of the reduced availability of frequency spectrum, many techniques to improve the spectrum efficiency have been proposed. One way of increasing the channel utilization and, consequently, the spectrum efficiency is by means of properly assigning the channels in the system. Channel assignment can be carried out in a global or in a local manner. In the global approach a centralized control of the allocation process must be provided. The local approach takes advantage of the fact that mobiles with access to more than one base station can be used for alternative routing. In this case the control for the channel assignment process can be local since only two or three neighbouring cells may be involved.

From the various global assignment algorithms investigated in References 5 and 12, the hybrid channel allocation with rearrangement seemed to give the best results. In particular, the borrowing technique requires a high control complexity without much improvement. The totally dynamic algorithm has low spatial efficiency and may have performance inferior to that of the fixed allocation for high traffic. On the other hand, the performance of the hybrid allocation is very dependent on the traffic characteristics. The right proportion of dynamic-to-fixed channels can only be determined if the traffic distribution, proportion of hand-offs, and other factors are considered to be known. For irregular traffic distributions, more dynamic channels are necessary to cope with the unexpected peaks of traffic. Under normal conditions, a low proportion of dynamic channels can be used with very good results in the system performance.

As for the local approach techniques, they are usually very simple to implement and do not require the involvement of the system central control. Although there may be some change in the usage pattern of channels, all channels are used within or closely adjacent to their planned service area, so that there are no serious implications for frequency reuse pattern.

From the various available techniques we chose to analyze the *blocking threshold variation* (BTV) algorithm because of its "multipurpose" characteristics. In fact, the BTV strategy may coincide with some of the other techniques, depending on the blocking threshold $T$. If $T$ is set to be equal to $N + 1$, where $N$ is the number of channels per cell, then the flexible traffic is not used for alternative routing. If $T = N$, then BTV coincides with the adaptive response to blocking (ARB) strategy (also known as directed retry). In the case where $T = 0$ and the alternative Alt2 is used, BTV is equivalent to the instantaneous adaptation (IAP) technique. If, however, $T = 0$ and Alt3 is used, then BTV coincides with the instantaneous and mean adaptation (IMA) technique.

It has been verified that the difference in performance between the three alternatives (Alt1, Alt2, and Alt3) is not significant, so that any one of them can be used to give a similar gain in traffic. Accordingly, when both cells present an equal number of busy channels, exceeding the blocking threshold $T$, then the flexible traffic can be directed to any cell, indistinguishably.

It has also been shown that it is not necessary to set a very low blocking threshold to obtain a reasonable gain in performance. For example, in a six-channel-per-cell system, a threshold equal to 4 was already enough to yield a considerable gain in traffic. Below 4 the gain is even greater, but not significant for practical purposes. Accordingly, we may reserve a fixed number of channels within each cell and use the others for alternative routing. The fixed channels can be primarily used for some services considered to be essential (e.g., hand-off).

## APPENDIX 12A   THREE-LEVEL LOGIC

### 12.A.1   Three-Level Algebra

Define the following:

- *Input variable*—a variable assuming only the logic levels $-1$, 0, or $+1$
- *Output variable*—the value of an algebraic expression assuming any real value

Define $X_{-1}$, $X_0$, and $X_1$ as *indication functions* assuming the value $+1$ only when the input variable $X$ is equal to $-1$, 0, or $+1$, respectively, and 0 otherwise. Accordingly,

$$X_{-1} = -X(1 - X)/2$$

$$X_0 = 1 - |X| \qquad (12A.1)$$

$$X_1 = X(1 + X)/2$$

Define the following:

- *Term*—the product of indication functions
- *Factor*—the value of the output variable for a given term

The output variable is then given by the sum of the terms, each of which is multiplied by its respective factor. As an example, consider the truth table given in Table 12A.1, where $A$ and $B$ are the input variables and $C$ is the output variable.

From this table the output variable can be written as

$$C = 5A_{-1}B_{-1} + 3A_{-1}B_0 - 2A_0B_0 + 0.3A_1B_{-1}$$

TABLE 12A.1.   An Example of Three-Level Logic Algebra

| A | B | C |
|---|---|---|
| −1 | −1 | 5 |
| −1 | 0 | 3 |
| −1 | 1 | 0 |
| 0 | −1 | 0 |
| 0 | 0 | −2 |
| 0 | 1 | 0 |
| 1 | −1 | 0.3 |
| 1 | 0 | 0 |
| 1 | 1 | 0 |

## 12A.2   Routing Coefficients for BTV Strategy

The input variables can be written as functions of the instantaneous states $(i, j)$ and the threshold $T$. Let $I, J$, and $K$ be these variables such that

$$I = \text{sgn}(i - T)$$

$$J = \text{sgn}(j - T) \tag{12A.2}$$

$$K = \text{sgn}(i - j)$$

where $\text{sgn}(x) = -1, 0,$ or $1$ for $x < 0, x = 0,$ or $x > 0$, respectively. The routing coefficients $\alpha_{ij}$ can be written as functions of $I, J$, and $K$. Following the routing procedures of the BTV strategy and having in mind Equations 12.34, Table 12A.2 is set up.

The factor $A$ is obtained according to the procedure taken by each alternative. It can be easily verified that

$$\text{for Alt1} \quad A = \beta$$

$$\text{for Alt2} \quad A = 0 \tag{12A.3}$$

$$\text{for Alt3} \quad A = -\text{sgn}(\beta)$$

TABLE 12A.2.   Routing Coefficients for BTV Strategy

| $I$ | $J$ | $\alpha_{ij}$ | Routing Procedure Relative to Flexible Traffic |
|---|---|---|---|
| −1 | −1 | $\beta$ | To keep it in its own cell |
| −1 | 0 | −1 | To divert it to cell 2 |
| −1 | 1 | −1 | To divert it to cell 2 |
| 0 | −1 | 1 | To divert it to cell 1 |
| 0 | 0 | $A$ | To be defined; depends on the alternative |
| 0 | 1 | −1 | To divert it to cell 2 |
| 1 | −1 | 1 | To divert it to cell 1 |
| 1 | 0 | 1 | To divert it to cell 1 |
| 1 | 1 | $B$ | To be defined; depends on the values of $i$ and $j$ |

The factor $B$ corresponds to the case $i > T$ and $j > T$. Following the same procedure as for Table 12A.2, we obtain

$$B = -K_{-1} + AK_0 + K_1 \tag{12A.4}$$

Therefore, from Table 12A.2 we have

$$\alpha_{ij} = \beta I_{-1} J_{-1} - I_{-1} J_0 - I_{-1} J_1 + I_0 J_{-1} + A I_0 J_0$$

$$- I_0 J_1 + I_1 J_{-1} + I_1 J_0 + B I_1 J_1 \tag{12A.5}$$

where $A$ and $B$ are given by Equations 12A.3 and 12A.4, respectively.

## APPENDIX 12B   APPROXIMATE MEAN BLOCKING PROBABILITY FOR THE BTV STRATEGY

### 12B.1   Mean Blocking Versus Imbalance

In the absence of flexibility ($\gamma = 0$) we have $B_1 = (A_1, N)$, $B_2 = E(A_2, N)$, and $B_{12} = 0$. Hence with $\gamma = 0$ in Equation 12.30 and then using these results in Equation 12.36, we obtain

$$B_m(A, \beta, 0) = \tfrac{1}{2}(1 + \beta) E\left[\tfrac{1}{2}(1 + \beta) A, N\right] + \tfrac{1}{2}(1 - \beta) E\left[\tfrac{1}{2}(1 - \beta) A, N\right] \tag{12B.1}$$

Suppose that for an arbitrary $f(x)$ expandable as a power series,

$$g(x, \beta) = \tfrac{1}{2}(1 + \beta) f(x + \beta x) + \tfrac{1}{2}(1 - \beta) f(x - \beta x) \tag{12B.2}$$

Define

$$f(x) = \sum_k a_k x^k$$

and

$$g(x) = x^n f(x) = \sum_k a_k x^{k+n} \tag{12B.3}$$

with $k, n \geq 0$.

Then, differentiating both sides of Equation 12B.3 $r$ times

$$g^{(r)}(x) = \sum_k a_k (k + n)^{(r)} x^{k+n-r} = r! \sum_k a_k \binom{k + n}{r} x^{k+n-r}$$

where $(k + n)^{(r)} = (k + n)(k + n - 1) \cdots (k + n - r + 1)$.
Then

$$\sum_k a_k \binom{k + n}{r} x^k = \frac{x^{r-n}}{r!} g^{(r)}(x)$$

Specifically, with $n = 1$

$$\sum_k a_k \binom{k + 1}{r} x^k = \frac{x^{r-1}}{r!} g^{(r)}(x)$$

Now this can be used to evaluate Equation 12B.2. Hence,

$$g(x, \beta) = \tfrac{1}{2} \sum_k a_k x^k \left[ (1 + \beta)^{k+1} + (1 - \beta)^{k+1} \right]$$

$$g(x, \beta) = \sum_k a_k x^k \sum_j \beta^{2j} \binom{k + 1}{2j}$$

$$g(x, \beta) = \sum_j \beta^{2j} \sum_k a_k x^k \binom{k + 1}{2j}$$

Finally,

$$g(x, \beta) = \sum_j \beta^{2j} \frac{x^{2j-1}}{(2j)!} g^{(2j)}(x), \quad \text{where } g(x) = xf(x) \quad \text{(12B.4)}$$

As an example, consider $f(x)$ as the one-channel Erlang blocking probability

$$f(x) = E(x, 1) = \frac{x}{1 + x}$$

Then

$$g(x, \beta) = \frac{x}{1 + x} + \beta^2 \frac{x}{(1 + x)^3} + O(x^3 \beta^4)$$

With large values of $N$, the exact expression is rather complex, but there is a good approximation, such as that given by Equation 12.17, i.e.,

$$f(x) = \frac{1}{N!} x^N e^{-x} \simeq E(x, N)$$

which gives

$$g(x, \beta) = f(x)\{1 + \beta^2[\tfrac{1}{2}N(N + 1) - (N + 1)x + \tfrac{1}{2}x^2]\} + O(\beta^4)$$

$$(12B.5)$$

Note that in our case $x = A/2$ (compare Equations 12B.2 and 12B.1). Therefore, from Equations 12B.1 and 12B.5 we obtain

$$B_m(A, \beta, 0) \simeq E(A/2, N)(1 + K\beta^2) \qquad (12B.6)$$

where

$$K = [N(N + 1) - (N + 1)A + (A^2/2)]/2 \qquad (12B.7)$$

## 12B.2   Low-Traffic Analysis

Consider a low-traffic approximation where the blocking states are dominated by the states $(0, N)$ for cell 1 and $(N, 0)$ for cell 2. As an approximation we may reduce the two-dimensional diagram of Figure 12.4 to a one-dimensional diagram containing the states $(0, 0), (0, 1), \ldots, (0, N)$ and $(0, 0), (1, 0), \ldots, (N, 0)$. Hence, the mean blocking probability is

$$B_m \simeq \frac{A_1 p_{0N} + A_2 p_{N0}}{A} \qquad (12B.8)$$

where $p_{0N}$ and $p_{N0}$ are the equilibrium state probabilities of the states $(0, N)$ and $(N, 0)$, respectively. These probabilities can be calculated by means of the standard one-dimensional form of solution given in Section 12.2.3. Up to the state $(0, T)$ the traffic offered to cell 1 is $A(1 + \beta)/2$, obtained from Equation 12.30 with $\gamma = 0$. From $(0, T)$ up to $(0, N)$ the traffic is exactly that given by Equation 12.30. The same reasoning applies to cell 2. Therefore, from the one-dimensional solution

$$p_{0N} = \left[\frac{A}{2}(1 + \beta)\right]^T \left[\frac{A}{2}(1 + \beta)(1 - \gamma)\right]^{N-T} \frac{p_{00}}{N!}$$

and

$$p_{N0} = \left[ \frac{A}{2}(1 - \beta) \right]^{T} \left[ \frac{A}{2}(1 - \beta)(1 - \gamma) \right]^{N-T} \frac{p_{00}}{N!} \qquad (12B.9)$$

Using Equations 12.30 and 12B.9 in Equation 12B.8, we obtain

$$B_m = \left[ \frac{p_{00}(A/2)^N}{N!} \right] (1 + \beta^2)(1 - \gamma)^{N-T+1} + O(\beta^4) \quad (12B.10)$$

## 12B.3  Approximate Formula

Observe the similarity between Equations 12B.10 and 12B.6. As an attempt to obtain an approximate expression for the overall mean blocking, we may combine both equations and neglect the higher-order terms as follows:

$$B_m \simeq E(A/2, N)(1 + K\beta^2)(1 - \gamma)^{N-T+1} \qquad (12B.11)$$

Note, however, that for $\gamma = 1$ Equation 12B.11 reduces to zero, whereas the true value would be $E(A, 2N)$. We may overcome this limitation by adding a new term such that

$$B_m \simeq E(A/2, N)(1 + K\beta^2)(1 - \gamma)^{N-T+1} + f(\gamma, N)E(A, 2N) \quad (12B.12)$$

where the function $f(\gamma, N)$ must satisfy the following conditions:

$$f(\gamma, N) = \begin{cases} 0 & \text{for } \gamma = 0, & \text{regardless of } T \\ 0 & \text{for } T = N + 1, & \text{regardless of } \gamma \\ 1 & \text{for } \gamma = 1, & \text{regardless of } T \end{cases}$$

One function satisfying these restrictions is

$$f(\gamma, N) = \gamma^a \left[ 1 - (1 - \gamma)^{N-T+1} \right] \qquad (12B.13)$$

where $a$ is a constant. It was found that satisfactory results are obtained for $a = 1/N$. Therefore, the approximate formula for the mean blocking is given by Equation 12B.12, where $f(\gamma, N)$ is defined in Equation 12B.13 with $a = 1/N$ and $K$ is given by Equation 12B.7.

**TABLE  12C.1.   Blocked Calls Cleared — Erlang B**

A (erlangs)

Blocking Probability

| N | 1.0% | 1.2% | 1.5% | 2% | 3% | 5% | 7% | 10% | 15% | 20% | 30% | 40% | 50% |
|---|---|---|---|---|---|---|---|---|---|---|---|---|---|
| 1 | 0.0101 | 0.0121 | 0.0152 | 0.0204 | 0.0309 | 0.0526 | 0.0753 | 0.111 | 0.176 | 0.250 | 0.429 | 0.667 | 1.00 |
| 2 | 0.153 | 0.168 | 0.190 | 0.223 | 0.282 | 0.381 | 0.470 | 0.595 | 0.796 | 1.00 | 1.45 | 2.00 | 2.73 |
| 3 | 0.455 | 0.489 | 0.535 | 0.602 | 0.715 | 0.899 | 1.06 | 1.27 | 1.60 | 1.93 | 2.63 | 3.48 | 4.59 |
| 4 | 0.869 | 0.922 | 0.992 | 1.09 | 1.26 | 1.52 | 1.75 | 2.05 | 2.50 | 2.95 | 3.89 | 5.02 | 6.50 |
| 5 | 1.36 | 1.43 | 1.52 | 1.66 | 1.88 | 2.22 | 2.50 | 2.88 | 3.45 | 4.01 | 5.19 | 6.60 | 8.44 |
| 6 | 1.91 | 2.00 | 2.11 | 2.28 | 2.54 | 2.96 | 3.30 | 3.76 | 4.44 | 5.11 | 6.51 | 8.19 | 10.4 |
| 7 | 2.50 | 2.60 | 2.74 | 2.94 | 3.25 | 3.74 | 4.14 | 4.67 | 5.46 | 6.23 | 7.86 | 9.80 | 12.4 |
| 8 | 3.13 | 3.25 | 3.40 | 3.63 | 3.99 | 4.54 | 5.00 | 5.60 | 6.50 | 7.37 | 9.21 | 11.4 | 14.3 |
| 9 | 3.78 | 3.92 | 4.09 | 4.34 | 4.75 | 5.37 | 5.88 | 6.55 | 7.55 | 8.52 | 10.6 | 13.0 | 16.3 |
| 10 | 4.46 | 4.61 | 4.81 | 5.08 | 5.53 | 6.22 | 6.78 | 7.51 | 8.62 | 9.68 | 12.0 | 14.7 | 18.3 |
| 11 | 5.16 | 5.32 | 5.54 | 5.84 | 6.33 | 7.08 | 7.69 | 8.49 | 9.69 | 10.9 | 13.3 | 16.3 | 20.3 |
| 12 | 5.88 | 6.05 | 6.29 | 6.61 | 7.14 | 7.95 | 8.61 | 9.47 | 10.8 | 12.0 | 14.7 | 18.0 | 22.2 |
| 13 | 6.61 | 6.80 | 7.05 | 7.40 | 7.97 | 8.83 | 9.54 | 10.5 | 11.9 | 13.2 | 16.1 | 19.6 | 24.2 |
| 14 | 7.35 | 7.56 | 7.82 | 8.20 | 8.80 | 9.73 | 10.5 | 11.5 | 13.0 | 14.4 | 17.5 | 21.2 | 26.2 |
| 15 | 8.11 | 8.33 | 8.61 | 9.01 | 9.65 | 10.6 | 11.4 | 12.5 | 14.1 | 15.6 | 18.9 | 22.9 | 28.2 |
| 16 | 8.88 | 9.11 | 9.41 | 9.83 | 10.5 | 11.5 | 12.4 | 13.5 | 15.2 | 16.8 | 20.3 | 24.5 | 30.2 |
| 17 | 9.65 | 9.89 | 10.2 | 10.7 | 11.4 | 12.5 | 13.4 | 14.5 | 16.3 | 18.0 | 21.7 | 26.2 | 32.2 |
| 18 | 10.4 | 10.7 | 11.0 | 11.5 | 12.2 | 13.4 | 14.3 | 15.5 | 17.4 | 19.2 | 23.1 | 27.8 | 34.2 |
| 19 | 11.2 | 11.5 | 11.8 | 12.3 | 13.1 | 14.3 | 15.3 | 16.6 | 18.5 | 20.4 | 24.5 | 29.5 | 36.2 |
| 20 | 12.0 | 12.3 | 12.7 | 13.2 | 14.0 | 15.2 | 16.3 | 17.6 | 19.6 | 21.6 | 25.9 | 31.2 | 38.2 |
| 21 | 12.8 | 13.1 | 13.5 | 14.0 | 14.9 | 16.2 | 17.3 | 18.7 | 20.8 | 22.8 | 27.3 | 32.8 | 40.2 |
| 22 | 13.7 | 14.0 | 14.3 | 14.9 | 15.8 | 17.1 | 18.2 | 19.7 | 21.9 | 24.1 | 28.7 | 34.5 | 42.1 |
| 23 | 14.5 | 14.8 | 15.2 | 15.8 | 16.7 | 18.1 | 19.2 | 20.7 | 23.0 | 25.3 | 30.1 | 36.1 | 44.1 |
| 24 | 15.3 | 15.6 | 16.0 | 16.6 | 17.6 | 19.0 | 20.2 | 21.8 | 24.2 | 26.5 | 31.6 | 37.8 | 46.1 |
| 25 | 16.1 | 16.5 | 16.9 | 17.5 | 18.5 | 20.0 | 21.2 | 22.8 | 25.3 | 27.7 | 33.0 | 39.4 | 48.1 |
| 26 | 17.0 | 17.3 | 17.8 | 18.4 | 19.4 | 20.9 | 22.2 | 23.9 | 26.4 | 28.9 | 34.4 | 41.1 | 50.1 |
| 27 | 17.8 | 18.2 | 18.6 | 19.3 | 20.3 | 21.9 | 23.2 | 24.9 | 27.6 | 30.2 | 35.8 | 42.8 | 52.1 |
| 28 | 18.6 | 19.0 | 19.5 | 20.2 | 21.2 | 22.9 | 24.2 | 26.0 | 28.7 | 31.4 | 37.2 | 44.4 | 54.1 |
| 29 | 19.5 | 19.9 | 20.4 | 21.0 | 22.1 | 23.8 | 25.2 | 27.1 | 29.9 | 32.6 | 38.6 | 46.1 | 56.1 |
| 30 | 20.3 | 20.7 | 21.2 | 21.9 | 23.1 | 24.8 | 26.2 | 28.1 | 31.0 | 33.8 | 40.0 | 47.7 | 58.1 |
| 31 | 21.2 | 21.6 | 22.1 | 22.8 | 24.0 | 25.8 | 27.2 | 29.2 | 32.1 | 35.1 | 41.5 | 49.4 | 60.1 |
| 32 | 22.0 | 22.5 | 23.0 | 23.7 | 24.9 | 26.7 | 28.2 | 30.2 | 33.3 | 36.3 | 42.9 | 51.1 | 62.1 |
| 33 | 22.9 | 23.3 | 23.9 | 24.6 | 25.8 | 27.7 | 29.3 | 31.3 | 34.4 | 37.5 | 44.3 | 52.7 | 64.1 |
| 34 | 23.8 | 24.2 | 24.8 | 25.5 | 26.8 | 28.7 | 30.3 | 32.4 | 35.6 | 38.8 | 45.7 | 54.4 | 66.1 |
| 35 | 24.6 | 25.1 | 25.6 | 26.4 | 27.7 | 29.7 | 31.3 | 33.4 | 36.7 | 40.0 | 47.1 | 56.0 | 68.1 |
| 36 | 25.5 | 26.0 | 26.5 | 27.3 | 28.6 | 30.7 | 32.3 | 34.5 | 37.9 | 41.2 | 48.6 | 57.7 | 70.1 |
| 37 | 26.4 | 26.8 | 27.4 | 28.3 | 29.6 | 31.6 | 33.3 | 35.6 | 39.0 | 42.4 | 50.0 | 59.4 | 72.1 |
| 38 | 27.3 | 27.7 | 28.3 | 29.2 | 30.5 | 32.6 | 34.4 | 36.6 | 40.2 | 43.7 | 51.4 | 61.0 | 74.1 |
| 39 | 28.1 | 28.6 | 29.2 | 30.1 | 31.5 | 33.6 | 35.4 | 37.7 | 41.3 | 44.9 | 52.8 | 62.7 | 76.1 |
| 40 | 29.0 | 29.5 | 30.1 | 31.0 | 32.4 | 34.6 | 36.4 | 38.8 | 42.5 | 46.1 | 54.2 | 64.4 | 78.1 |
| 41 | 29.9 | 30.4 | 31.0 | 31.9 | 33.4 | 35.6 | 37.4 | 39.9 | 43.6 | 47.4 | 55.7 | 66.0 | 80.1 |
| 42 | 30.8 | 31.3 | 31.9 | 32.8 | 34.3 | 36.6 | 38.4 | 40.9 | 44.8 | 48.6 | 57.1 | 67.7 | 82.1 |
| 43 | 31.7 | 32.2 | 32.8 | 33.8 | 35.3 | 37.6 | 39.5 | 42.0 | 45.9 | 49.9 | 58.5 | 69.3 | 84.1 |
| 44 | 32.5 | 33.1 | 33.7 | 34.7 | 36.2 | 38.6 | 40.5 | 43.1 | 47.1 | 51.1 | 59.9 | 71.0 | 86.1 |
| 45 | 33.4 | 34.0 | 34.6 | 35.6 | 37.2 | 39.6 | 41.5 | 44.2 | 48.2 | 52.3 | 61.3 | 72.7 | 88.1 |
| 46 | 34.3 | 34.9 | 35.6 | 36.5 | 38.1 | 40.5 | 42.6 | 45.2 | 49.4 | 53.6 | 62.8 | 74.3 | 90.1 |
| 47 | 35.2 | 35.8 | 36.5 | 37.5 | 39.1 | 41.5 | 43.6 | 46.3 | 50.6 | 54.8 | 64.2 | 76.0 | 92.1 |
| 48 | 36.1 | 36.7 | 37.4 | 38.4 | 40.0 | 42.5 | 44.6 | 47.4 | 51.7 | 56.0 | 65.6 | 77.7 | 94.1 |
| 49 | 37.0 | 37.6 | 38.3 | 39.3 | 41.0 | 43.5 | 45.7 | 48.5 | 52.9 | 57.3 | 67.0 | 79.3 | 96.1 |
| 50 | 37.9 | 38.5 | 39.2 | 40.3 | 41.9 | 44.5 | 46.7 | 49.6 | 54.0 | 58.5 | 68.5 | 81.0 | 98.1 |
| 51 | 38.8 | 39.4 | 40.1 | 41.2 | 42.9 | 45.5 | 47.7 | 50.6 | 55.2 | 59.7 | 69.9 | 82.7 | 100.1 |
| 52 | 39.7 | 40.3 | 41.0 | 42.1 | 43.9 | 46.5 | 48.8 | 51.7 | 56.3 | 61.0 | 71.3 | 84.3 | 102.1 |
| 53 | 40.6 | 41.2 | 42.0 | 43.1 | 44.8 | 47.5 | 49.8 | 52.8 | 57.5 | 62.2 | 72.7 | 86.0 | 104.1 |
| 54 | 41.5 | 42.1 | 42.9 | 44.0 | 45.8 | 48.5 | 50.8 | 53.9 | 58.7 | 63.5 | 74.2 | 87.6 | 106.1 |
| 55 | 42.4 | 43.0 | 43.8 | 44.9 | 46.7 | 49.5 | 51.9 | 55.0 | 59.8 | 64.7 | 75.6 | 89.3 | 108.1 |
| 56 | 43.3 | 43.9 | 44.7 | 45.9 | 47.7 | 50.5 | 52.9 | 56.1 | 61.0 | 65.9 | 77.0 | 91.0 | 110.1 |
| 57 | 44.2 | 44.8 | 45.7 | 46.8 | 48.7 | 51.5 | 53.9 | 57.1 | 62.1 | 67.2 | 78.4 | 92.6 | 112.1 |

TABLE 12C.1. (*Continued*)

| N | \multicolumn{12}{c}{A (erlangs) Blocking Probability} |
|---|------|------|------|------|------|------|------|------|------|------|------|------|------|
| | 1.0% | 1.2% | 1.5% | 2% | 3% | 5% | 7% | 10% | 15% | 20% | 30% | 40% | 50% |
| 58 | 45.1 | 45.8 | 46.6 | 47.8 | 49.6 | 52.6 | 55.0 | 58.2 | 63.3 | 68.4 | 79.8 | 94.3 | 114.1 |
| 59 | 46.0 | 46.7 | 47.5 | 48.7 | 50.6 | 53.6 | 56.0 | 59.3 | 64.5 | 69.7 | 81.3 | 96.0 | 116.1 |
| 60 | 46.9 | 47.6 | 48.4 | 49.6 | 51.6 | 54.6 | 57.1 | 60.4 | 65.6 | 70.9 | 82.7 | 97.6 | 118.1 |
| 61 | 47.9 | 48.5 | 49.4 | 50.6 | 52.5 | 55.6 | 58.1 | 61.5 | 66.8 | 72.1 | 84.1 | 99.3 | 120.1 |
| 62 | 48.8 | 49.4 | 50.3 | 51.5 | 53.5 | 56.6 | 59.1 | 62.6 | 68.0 | 73.4 | 85.5 | 101.0 | 122.1 |
| 63 | 49.7 | 50.4 | 51.2 | 52.5 | 54.5 | 57.6 | 60.2 | 63.7 | 69.1 | 74.6 | 87.0 | 102.6 | 124.1 |
| 64 | 50.6 | 51.3 | 52.2 | 53.4 | 55.4 | 58.6 | 61.2 | 64.8 | 70.3 | 75.9 | 88.4 | 104.3 | 126.1 |
| 65 | 51.5 | 52.2 | 53.1 | 54.4 | 56.4 | 59.6 | 62.3 | 65.8 | 71.4 | 77.1 | 89.8 | 106.0 | 128.1 |
| 66 | 52.4 | 53.1 | 54.0 | 55.3 | 57.4 | 60.6 | 63.3 | 66.9 | 72.6 | 78.3 | 91.2 | 107.6 | 130.1 |
| 67 | 53.4 | 54.1 | 55.0 | 56.3 | 58.4 | 61.6 | 64.4 | 68.0 | 73.8 | 79.6 | 92.7 | 109.3 | 132.1 |
| 68 | 54.3 | 55.0 | 55.9 | 57.2 | 59.3 | 62.6 | 65.4 | 69.1 | 74.9 | 80.8 | 94.1 | 111.0 | 134.1 |
| 69 | 55.2 | 55.9 | 56.9 | 58.2 | 60.3 | 63.7 | 66.4 | 70.2 | 76.1 | 82.1 | 95.5 | 112.6 | 136.1 |
| 70 | 56.1 | 56.8 | 57.8 | 59.1 | 61.3 | 64.7 | 67.5 | 71.3 | 77.3 | 83.3 | 96.9 | 114.3 | 138.1 |
| 71 | 57.0 | 57.8 | 58.7 | 60.1 | 62.3 | 65.7 | 68.5 | 72.4 | 78.4 | 84.6 | 98.4 | 115.9 | 140.1 |
| 72 | 58.0 | 58.7 | 59.7 | 61.0 | 63.2 | 66.7 | 69.6 | 73.5 | 79.6 | 85.8 | 99.8 | 117.6 | 142.1 |
| 73 | 58.9 | 59.6 | 60.6 | 62.0 | 64.2 | 67.7 | 70.6 | 74.6 | 80.8 | 87.0 | 101.2 | 119.3 | 144.1 |
| 74 | 59.8 | 60.6 | 61.6 | 62.9 | 65.2 | 68.7 | 71.7 | 75.6 | 81.9 | 88.3 | 102.7 | 120.9 | 146.1 |
| 75 | 60.7 | 61.5 | 62.5 | 63.9 | 66.2 | 69.7 | 72.7 | 76.7 | 83.1 | 89.5 | 104.1 | 122.6 | 148.0 |
| 76 | 61.7 | 62.4 | 63.4 | 64.9 | 67.2 | 70.8 | 73.8 | 77.8 | 84.2 | 90.8 | 105.5 | 124.3 | 150.0 |
| 77 | 62.6 | 63.4 | 64.4 | 65.8 | 68.1 | 71.8 | 74.8 | 78.9 | 85.4 | 92.0 | 106.9 | 125.9 | 152.0 |
| 78 | 63.5 | 64.3 | 65.3 | 66.8 | 69.1 | 72.8 | 75.9 | 80.0 | 86.6 | 93.3 | 108.4 | 127.6 | 154.0 |
| 79 | 64.4 | 65.2 | 66.3 | 67.7 | 70.1 | 73.8 | 76.9 | 81.1 | 87.7 | 94.5 | 109.8 | 129.3 | 156.0 |
| 80 | 65.4 | 66.2 | 67.2 | 68.7 | 71.1 | 74.8 | 78.0 | 82.2 | 88.9 | 95.7 | 111.2 | 130.9 | 158.0 |
| 81 | 66.3 | 67.1 | 68.2 | 69.6 | 72.1 | 75.8 | 79.0 | 83.3 | 90.1 | 97.0 | 112.6 | 132.6 | 160.0 |
| 82 | 67.2 | 68.0 | 69.1 | 70.6 | 73.0 | 76.9 | 80.1 | 84.4 | 91.2 | 98.2 | 114.1 | 134.3 | 162.0 |
| 83 | 68.2 | 69.0 | 70.1 | 71.6 | 74.0 | 77.9 | 81.1 | 85.5 | 92.4 | 99.5 | 115.5 | 135.9 | 164.0 |
| 84 | 69.1 | 69.9 | 71.0 | 72.5 | 75.0 | 78.9 | 82.2 | 86.6 | 93.6 | 100.7 | 116.9 | 137.6 | 166.0 |
| 85 | 70.0 | 70.9 | 71.9 | 73.5 | 76.0 | 79.9 | 83.2 | 87.7 | 94.7 | 102.0 | 118.3 | 139.3 | 168.0 |
| 86 | 70.9 | 71.8 | 72.9 | 74.5 | 77.0 | 80.9 | 84.3 | 88.8 | 95.9 | 103.2 | 119.8 | 140.9 | 170.0 |
| 87 | 71.9 | 72.7 | 73.8 | 75.4 | 78.0 | 82.0 | 85.3 | 89.9 | 97.1 | 104.5 | 121.2 | 142.6 | 172.0 |
| 88 | 72.8 | 73.7 | 74.8 | 76.4 | 78.9 | 83.0 | 86.4 | 91.0 | 98.2 | 105.7 | 122.6 | 144.6 | 174.0 |
| 89 | 73.7 | 74.6 | 75.7 | 77.3 | 79.9 | 84.0 | 87.4 | 92.1 | 99.4 | 106.9 | 124.0 | 145.9 | 176.0 |
| 90 | 74.7 | 75.6 | 76.7 | 78.3 | 80.9 | 85.0 | 88.5 | 93.1 | 100.6 | 108.2 | 125.5 | 147.6 | 178.0 |
| 91 | 75.6 | 76.5 | 77.6 | 79.3 | 81.9 | 86.0 | 89.5 | 94.2 | 101.7 | 109.4 | 126.9 | 149.3 | 180.0 |
| 92 | 76.6 | 77.4 | 78.6 | 80.2 | 82.9 | 87.1 | 90.6 | 95.3 | 102.9 | 110.7 | 128.3 | 150.9 | 182.0 |
| 93 | 77.5 | 78.4 | 79.6 | 81.2 | 83.9 | 88.1 | 91.6 | 96.4 | 104.1 | 111.9 | 129.7 | 152.6 | 184.0 |
| 94 | 78.4 | 79.3 | 80.5 | 82.2 | 84.9 | 89.1 | 92.7 | 97.5 | 105.3 | 113.2 | 131.2 | 154.3 | 186.0 |
| 95 | 79.4 | 80.3 | 81.5 | 83.1 | 85.8 | 90.1 | 93.7 | 98.6 | 106.4 | 114.4 | 132.6 | 155.9 | 188.0 |
| 96 | 80.3 | 81.2 | 82.4 | 84.1 | 86.8 | 91.1 | 94.8 | 99.7 | 107.6 | 115.7 | 134.0 | 157.6 | 190.0 |
| 97 | 81.2 | 82.2 | 83.4 | 85.1 | 87.8 | 92.2 | 95.8 | 100.8 | 108.8 | 116.9 | 135.5 | 159.3 | 192.0 |
| 98 | 82.2 | 83.1 | 84.3 | 86.0 | 88.8 | 93.2 | 96.9 | 101.9 | 109.9 | 118.2 | 136.9 | 160.9 | 194.0 |
| 99 | 83.1 | 84.1 | 85.3 | 87.0 | 89.8 | 94.2 | 97.9 | 103.0 | 111.1 | 119.4 | 138.3 | 162.6 | 196.0 |
| 100 | 84.1 | 85.0 | 86.2 | 88.0 | 90.8 | 95.2 | 99.0 | 104.1 | 112.3 | 120.6 | 139.7 | 164.3 | 198.0 |
| 102 | 85.9 | 86.9 | 88.1 | 89.9 | 92.8 | 97.3 | 101.1 | 106.3 | 114.6 | 123.1 | 142.6 | 167.6 | 202.0 |
| 104 | 87.8 | 88.8 | 90.1 | 91.9 | 94.8 | 99.3 | 103.2 | 108.5 | 116.9 | 125.6 | 145.4 | 170.9 | 206.0 |
| 106 | 89.7 | 90.7 | 92.0 | 93.8 | 96.7 | 101.4 | 105.3 | 110.7 | 119.3 | 128.1 | 148.3 | 174.2 | 210.0 |
| 108 | 91.6 | 92.6 | 93.9 | 95.7 | 98.7 | 103.4 | 107.4 | 112.9 | 121.6 | 130.6 | 151.1 | 177.6 | 214.0 |
| 110 | 93.5 | 94.5 | 95.8 | 97.7 | 100.7 | 105.5 | 109.5 | 115.1 | 124.0 | 133.1 | 154.0 | 180.9 | 218.0 |
| 112 | 95.4 | 96.4 | 97.7 | 99.6 | 102.7 | 107.5 | 111.7 | 117.3 | 126.3 | 135.6 | 156.9 | 184.2 | 222.0 |
| 114 | 97.3 | 98.3 | 99.7 | 101.6 | 104.7 | 109.6 | 113.8 | 119.5 | 128.6 | 138.1 | 159.7 | 187.6 | 226.0 |
| 116 | 99.2 | 100.2 | 101.6 | 103.5 | 106.7 | 111.7 | 115.9 | 121.7 | 131.0 | 140.6 | 162.6 | 190.9 | 230.0 |
| 118 | 101.1 | 102.1 | 103.5 | 105.5 | 108.7 | 113.7 | 118.0 | 123.9 | 133.3 | 143.1 | 165.4 | 194.2 | 234.0 |
| 120 | 103.0 | 104.0 | 105.4 | 107.4 | 110.7 | 115.8 | 120.1 | 126.1 | 135.7 | 145.6 | 168.3 | 197.6 | 238.0 |
| 122 | 104.9 | 105.9 | 107.4 | 109.4 | 112.6 | 117.8 | 122.2 | 128.3 | 138.0 | 148.1 | 171.1 | 200.9 | 242.0 |
| 124 | 106.8 | 107.9 | 109.3 | 111.3 | 114.6 | 119.9 | 124.4 | 130.5 | 140.3 | 150.6 | 174.0 | 204.2 | 246.0 |
| 126 | 108.7 | 109.8 | 111.2 | 113.3 | 116.6 | 121.9 | 126.5 | 132.7 | 142.7 | 153.0 | 176.8 | 207.6 | 250.0 |
| 128 | 110.6 | 111.7 | 113.2 | 115.2 | 118.6 | 124.0 | 128.6 | 134.9 | 145.0 | 155.5 | 179.7 | 210.9 | 254.0 |
| 130 | 112.5 | 113.6 | 115.1 | 117.2 | 120.6 | 126.1 | 130.7 | 137.1 | 147.4 | 158.0 | 182.5 | 214.2 | 258.0 |
| 132 | 114.4 | 115.5 | 117.0 | 119.1 | 122.6 | 128.1 | 132.8 | 139.3 | 149.7 | 160.5 | 185.4 | 217.6 | 262.0 |
| 134 | 116.3 | 117.4 | 119.0 | 121.1 | 124.6 | 130.2 | 134.9 | 141.5 | 152.0 | 163.0 | 188.3 | 220.9 | 266.0 |
| 136 | 118.2 | 119.4 | 120.9 | 123.1 | 126.6 | 132.3 | 137.1 | 143.7 | 154.4 | 165.5 | 191.1 | 224.2 | 270.0 |

TABLE 12C.1.  (Continued)

| | | | | | | A (erlangs) | | | | | | | |
|---|---|---|---|---|---|---|---|---|---|---|---|---|
| | | | | | | Blocking Probability | | | | | | | |
| N | 1.0% | 1.2% | 1.5% | 2% | 3% | 5% | 7% | 10% | 15% | 20% | 30% | 40% | 50% |
|---|---|---|---|---|---|---|---|---|---|---|---|---|---|
| 138 | 120.1 | 121.3 | 122.8 | 125.0 | 128.6 | 134.3 | 139.2 | 145.9 | 156.7 | 168.0 | 194.0 | 227.6 | 274.0 |
| 140 | 122.0 | 123.2 | 124.8 | 127.0 | 130.6 | 136.4 | 141.3 | 148.1 | 159.1 | 170.5 | 196.8 | 230.9 | 278.0 |
| 142 | 123.9 | 125.1 | 126.7 | 128.9 | 132.6 | 138.4 | 143.4 | 150.3 | 161.4 | 173.0 | 199.7 | 234.2 | 282.0 |
| 144 | 125.8 | 127.0 | 128.6 | 130.9 | 134.6 | 140.5 | 145.6 | 152.5 | 163.8 | 175.5 | 202.5 | 237.6 | 286.0 |
| 146 | 127.7 | 129.0 | 130.6 | 132.9 | 136.6 | 142.6 | 147.7 | 154.7 | 166.1 | 178.0 | 205.4 | 240.9 | 290.0 |
| 148 | 129.7 | 130.9 | 132.5 | 134.8 | 138.6 | 144.6 | 149.8 | 156.9 | 168.5 | 180.5 | 208.2 | 244.2 | 294.0 |
| 150 | 131.6 | 132.8 | 134.5 | 136.8 | 140.6 | 146.7 | 151.9 | 159.1 | 170.8 | 183.0 | 211.1 | 247.6 | 298.0 |
| 152 | 133.5 | 134.8 | 136.4 | 138.8 | 142.6 | 148.8 | 154.0 | 161.3 | 173.1 | 185.5 | 214.0 | 250.9 | 302.0 |
| 154 | 135.4 | 136.7 | 138.4 | 140.7 | 144.6 | 150.8 | 156.2 | 163.5 | 175.5 | 188.0 | 216.8 | 254.2 | 306.0 |
| 156 | 137.3 | 138.6 | 140.3 | 142.7 | 146.6 | 152.9 | 158.3 | 165.7 | 177.8 | 190.5 | 219.7 | 257.6 | 310.0 |
| 158 | 139.2 | 140.5 | 142.3 | 144.7 | 148.6 | 155.0 | 160.4 | 167.9 | 180.2 | 193.0 | 222.5 | 260.9 | 314.0 |
| 160 | 141.2 | 142.5 | 144.2 | 146.6 | 150.6 | 157.0 | 162.5 | 170.2 | 182.5 | 195.5 | 225.4 | 264.2 | 318.0 |
| 162 | 143.1 | 144.4 | 146.1 | 148.6 | 152.7 | 159.1 | 164.7 | 172.4 | 184.9 | 198.0 | 228.2 | 267.6 | 322.0 |
| 164 | 145.0 | 146.3 | 148.1 | 150.6 | 154.7 | 161.2 | 166.8 | 174.6 | 187.2 | 200.4 | 231.1 | 270.9 | 326.0 |
| 166 | 146.9 | 148.3 | 150.0 | 152.6 | 156.7 | 163.3 | 168.9 | 176.8 | 189.6 | 202.9 | 233.9 | 274.2 | 330.0 |
| 168 | 148.9 | 150.2 | 152.0 | 154.5 | 158.7 | 165.3 | 171.0 | 179.0 | 191.9 | 205.4 | 236.8 | 277.6 | 334.0 |
| 170 | 150.8 | 152.1 | 153.9 | 156.5 | 160.7 | 167.4 | 173.2 | 181.2 | 194.2 | 207.9 | 239.7 | 280.9 | 338.0 |
| 172 | 152.7 | 154.1 | 155.9 | 158.5 | 162.7 | 169.5 | 175.3 | 183.4 | 196.6 | 210.4 | 242.5 | 284.2 | 342.0 |
| 174 | 154.6 | 156.0 | 157.8 | 160.4 | 164.7 | 171.5 | 177.4 | 185.6 | 198.9 | 212.9 | 245.4 | 287.6 | 346.0 |
| 176 | 156.6 | 158.0 | 159.8 | 162.4 | 166.7 | 173.6 | 179.6 | 187.8 | 201.3 | 215.4 | 248.2 | 290.9 | 350.0 |
| 178 | 158.5 | 159.9 | 161.8 | 164.4 | 168.7 | 175.7 | 181.7 | 190.0 | 203.6 | 217.9 | 251.1 | 294.2 | 354.0 |
| 180 | 160.4 | 161.8 | 163.7 | 166.4 | 170.7 | 177.8 | 183.8 | 192.2 | 206.0 | 220.4 | 253.9 | 297.5 | 358.0 |
| 182 | 162.3 | 163.8 | 165.7 | 168.3 | 172.8 | 179.8 | 185.9 | 194.4 | 208.3 | 222.9 | 256.8 | 300.9 | 362.0 |
| 184 | 164.3 | 165.7 | 167.6 | 170.3 | 174.8 | 181.9 | 188.1 | 196.6 | 210.7 | 225.4 | 259.6 | 304.2 | 366.0 |
| 186 | 166.2 | 167.7 | 169.6 | 172.3 | 176.8 | 184.0 | 190.2 | 198.9 | 213.0 | 227.9 | 262.5 | 307.5 | 370.0 |
| 188 | 168.1 | 169.6 | 171.5 | 174.3 | 178.8 | 186.1 | 192.3 | 201.1 | 215.4 | 230.4 | 265.4 | 310.9 | 374.0 |
| 190 | 170.1 | 171.5 | 173.5 | 176.3 | 180.8 | 188.1 | 194.5 | 203.3 | 217.7 | 232.9 | 268.2 | 314.2 | 378.0 |
| 192 | 172.0 | 173.5 | 175.4 | 178.2 | 182.8 | 190.2 | 196.6 | 205.5 | 220.1 | 235.4 | 271.1 | 317.5 | 382.0 |
| 194 | 173.9 | 175.4 | 177.4 | 180.2 | 184.8 | 192.3 | 198.7 | 207.7 | 222.4 | 237.9 | 273.9 | 320.9 | 386.0 |
| 196 | 175.9 | 177.4 | 179.4 | 182.2 | 186.9 | 194.4 | 200.8 | 209.9 | 224.8 | 240.4 | 276.8 | 324.2 | 390.0 |
| 198 | 177.8 | 179.3 | 181.3 | 184.2 | 188.9 | 196.4 | 203.0 | 212.1 | 227.1 | 242.9 | 279.6 | 327.5 | 394.0 |
| 200 | 179.7 | 181.3 | 183.3 | 186.2 | 190.9 | 198.5 | 205.1 | 214.3 | 229.4 | 245.4 | 282.5 | 330.9 | 398.0 |
| 202 | 181.7 | 183.2 | 185.2 | 188.1 | 192.9 | 200.6 | 207.2 | 216.5 | 231.8 | 247.9 | 285.4 | 334.2 | 402.0 |
| 204 | 183.6 | 185.2 | 187.2 | 190.1 | 194.9 | 202.7 | 209.4 | 218.7 | 234.1 | 250.4 | 288.2 | 337.5 | 406.0 |
| 206 | 185.5 | 187.1 | 189.2 | 192.1 | 196.9 | 204.7 | 211.5 | 221.0 | 236.5 | 252.9 | 291.1 | 340.9 | 410.0 |
| 208 | 187.5 | 189.1 | 191.1 | 194.1 | 199.0 | 206.8 | 213.6 | 223.2 | 238.8 | 255.4 | 293.9 | 344.2 | 414.0 |
| 210 | 189.4 | 191.0 | 193.1 | 196.1 | 201.0 | 208.9 | 215.8 | 225.4 | 241.2 | 257.9 | 296.8 | 347.5 | 418.0 |
| 212 | 191.4 | 193.0 | 195.1 | 198.1 | 203.0 | 211.0 | 217.9 | 227.6 | 243.5 | 260.4 | 299.6 | 350.9 | 422.0 |
| 214 | 193.3 | 194.9 | 197.0 | 200.0 | 205.0 | 213.0 | 220.0 | 229.8 | 245.9 | 262.9 | 302.5 | 354.2 | 426.0 |
| 216 | 195.2 | 196.9 | 199.0 | 202.0 | 207.0 | 215.1 | 222.2 | 232.0 | 248.2 | 265.4 | 305.3 | 357.5 | 430.0 |
| 218 | 197.2 | 198.8 | 201.0 | 204.0 | 209.1 | 217.2 | 224.3 | 234.2 | 250.6 | 267.9 | 308.2 | 360.9 | 434.0 |
| 220 | 199.1 | 200.8 | 202.9 | 206.0 | 211.1 | 219.3 | 226.4 | 236.4 | 252.9 | 270.4 | 311.1 | 364.2 | 438.0 |
| 222 | 201.1 | 202.7 | 204.9 | 208.0 | 213.1 | 221.4 | 228.6 | 238.6 | 255.3 | 272.9 | 313.9 | 367.5 | 442.0 |
| 224 | 203.0 | 204.7 | 206.8 | 210.0 | 215.1 | 223.4 | 230.7 | 240.9 | 257.6 | 275.4 | 316.8 | 370.9 | 446.0 |
| 226 | 204.9 | 206.6 | 208.8 | 212.0 | 217.1 | 225.5 | 232.8 | 243.1 | 260.0 | 277.8 | 319.6 | 374.2 | 450.0 |
| 228 | 206.9 | 208.6 | 210.8 | 213.9 | 219.2 | 227.6 | 235.0 | 245.3 | 262.3 | 280.3 | 322.5 | 377.5 | 454.0 |
| 230 | 208.8 | 210.5 | 212.8 | 215.9 | 221.2 | 229.7 | 237.1 | 247.5 | 264.7 | 282.8 | 325.3 | 380.9 | 458.0 |
| 232 | 210.8 | 212.5 | 214.7 | 217.9 | 223.2 | 231.8 | 239.2 | 249.7 | 267.0 | 285.3 | 328.2 | 384.2 | 462.0 |
| 234 | 212.7 | 214.4 | 216.7 | 219.9 | 225.2 | 233.8 | 241.4 | 251.9 | 269.4 | 287.8 | 331.1 | 387.5 | 466.0 |
| 236 | 214.7 | 216.4 | 218.7 | 221.9 | 227.2 | 235.9 | 243.5 | 254.1 | 271.7 | 290.3 | 333.9 | 390.9 | 470.0 |
| 238 | 216.6 | 218.3 | 220.6 | 223.9 | 229.3 | 238.0 | 245.6 | 256.3 | 274.1 | 292.8 | 336.8 | 394.2 | 474.0 |
| 240 | 218.6 | 220.3 | 222.6 | 225.9 | 231.3 | 240.1 | 247.8 | 258.6 | 276.4 | 295.3 | 339.6 | 397.5 | 478.0 |
| 242 | 220.5 | 222.3 | 224.6 | 227.9 | 233.3 | 242.2 | 249.9 | 260.8 | 278.8 | 297.8 | 342.5 | 400.9 | 482.0 |
| 244 | 222.5 | 224.2 | 226.5 | 229.9 | 235.3 | 244.3 | 252.0 | 263.0 | 281.1 | 300.3 | 345.3 | 404.2 | 486.0 |
| 246 | 224.4 | 226.2 | 228.5 | 231.8 | 237.4 | 246.3 | 254.2 | 265.2 | 283.4 | 302.8 | 348.2 | 407.5 | 490.0 |
| 248 | 226.3 | 228.1 | 230.5 | 233.8 | 239.4 | 248.4 | 256.3 | 267.4 | 285.8 | 305.3 | 351.0 | 410.9 | 494.0 |
| 250 | 228.3 | 230.1 | 232.5 | 235.8 | 241.4 | 250.5 | 258.4 | 269.6 | 288.1 | 307.8 | 353.9 | 414.2 | 498.0 |
| | 0.976 | 0.982 | 0.988 | 0.988 | 1.014 | 1.042 | 1.070 | 1.108 | 1.176 | 1.250 | 1.428 | 1.666 | 2.000 |
| 300 | 277.1 | 279.2 | 281.9 | 285.7 | 292.1 | 302.6 | 311.9 | 325.0 | 346.9 | 370.3 | 425.3 | 497.5 | 698.0 |
| | 0.982 | 0.984 | 0.990 | 1.000 | 1.016 | 1.044 | 1.070 | 1.108 | 1.174 | 1.248 | 1.428 | 1.668 | 2.000 |

**TABLE 12C.1.** *(Continued)*

| N | 1.0% | 1.2% | 1.5% | 2% | 3% | 5% | 7% | 10% | 15% | 20% | 30% | 40% | 50% |
|---|------|------|------|-----|-----|-----|-----|------|------|------|------|------|------|
|   | | | | **A (erlangs)** | | | | | | | | | |
|   | | | | **Blocking Probability** | | | | | | | | | |
| 350 | 326.2 | 328.4 | 331.4 | 335.7 | 342.9 | 354.8 | 365.4 | 380.4 | 405.6 | 432.7 | 496.7 | 580.9 | 698.0 |
|     | 0.982 | 0.988 | 0.994 | 1.004 | 1.020 | 1.046 | 1.070 | 1.108 | 1.176 | 1.250 | 1.430 | 1.666 | 2.000 |
| 400 | 375.3 | 377.8 | 381.1 | 385.9 | 393.9 | 407.1 | 418.9 | 435.8 | 464.4 | 495.2 | 568.2 | 664.2 | 798.0 |
|     | 0.986 | 0.990 | 0.996 | 1.004 | 1.018 | 1.046 | 1.072 | 1.110 | 1.176 | 1.250 | 1.428 | 1.666 | 2.000 |
| 450 | 424.6 | 427.3 | 430.9 | 436.1 | 444.8 | 459.4 | 472.5 | 491.3 | 523.2 | 557.7 | 639.6 | 747.5 | 898.0 |
|     | 0.988 | 0.994 | 0.998 | 1.006 | 1.022 | 1.048 | 1.070 | 1.108 | 1.176 | 1.250 | 1.428 | 1.668 | 2.000 |
| 500 | 474.0 | 477.0 | 480.8 | 486.4 | 495.9 | 511.8 | 526.0 | 546.7 | 582.0 | 620.2 | 711.0 | 830.9 | 998.0 |
|     | 0.991 | 0.994 | 1.000 | 1.008 | 1.022 | 1.047 | 1.073 | 1.110 | 1.176 | 1.249 | 1.429 | 1.666 | 2.000 |
| 600 | 573.1 | 576.4 | 580.8 | 587.2 | 598.1 | 616.5 | 633.3 | 657.7 | 699.6 | 745.1 | 853.9 | 997.5 | 1198. |
|     | 0.993 | 0.997 | 1.002 | 1.010 | 1.024 | 1.049 | 1.073 | 1.110 | 1.176 | 1.250 | 1.428 | 1.665 | 2.00 |
| 700 | 672.4 | 676.1 | 681.0 | 688.2 | 700.5 | 721.4 | 740.6 | 768.7 | 817.2 | 870.1 | 996.7 | 1164. | 1398. |
|     | 0.994 | 0.998 | 1.004 | 1.011 | 1.025 | 1.050 | 1.073 | 1.110 | 1.176 | 1.250 | 1.433 | 1.67 | 2.00 |
| 800 | 771.8 | 775.9 | 781.4 | 789.3 | 803.0 | 826.4 | 847.9 | 879.7 | 934.8 | 995.1 | 1140. | 1331. | 1598. |
|     | 0.997 | 1.000 | 1.004 | 1.013 | 1.025 | 1.050 | 1.074 | 1.111 | 1.172 | 1.249 | 1.42 | 1.67 | 2.00 |
| 900 | 871.5 | 875.9 | 881.8 | 890.6 | 905.5 | 931.4 | 955.3 | 990.8 | 1052. | 1120. | 1282. | 1498. | 1798. |
|     | 0.997 | 1.001 | 1.006 | 1.013 | 1.025 | 1.046 | 1.077 | 1.112 | 1.18 | 1.25 | 1.43 | 1.66 | 2.00 |
| 1000 | 971.2 | 976.0 | 982.4 | 991.9 | 1008. | 1036. | 1063. | 1102. | 1170. | 1245. | 1425. | 1664. | 1998. |
|      | 0.998 | 1.000 | 1.006 | 1.011 | 1.03 | 1.05 | 1.07 | 1.11 | 1.18 | 1.25 | 1.43 | 1.67 | 2.00 |
| 1100 | 1071. | 1076. | 1083. | 1093. | 1111. | 1141. | 1170. | 1213. | 1288. | 1370. | 1568. | 1831. | 2198. |

*Source*: From *Telephone Traffic Theory Tables and Charts*, Siemens Aktiengesellschaft, Munich, 1970.

## APPENDIX 12C

## REFERENCES

1. Kleinrock, L., *Queueing Systems*, vol. I: *Theory*, John Wiley & Sons, New York, 1975.
2. Molina, E. C., The theory of probabilities applied to telephone trunking problems, *Bell Syst. Tech. J.*, 6, 69–81, 1922.
3. Wilkinson, R. I., Theories for toll traffic engineering in the USA, *Bell Syst. Tech. J.*, 35, 421–514, 1956.
4. Rapp, Y., Planning of junction network in a multi-exchange area, *Ericsson Technics.*, 20, 77, 1964.
5. Sánchez V., J. H., Traffic performance of cellular mobile radio systems, Ph.D. thesis, University of Essex, June 1988.
6. Elnoubi, S. M., Singh, R., and Gupta, S. C., A new frequency channel assignment algorithm in high capacity communication systems, *IEEE J. Trans. Vehicular Tech.* VT-31(3), 125–131, August 1982.
7. Sekiguchi, H., Ishikawa, H., Koyama, M., and Sawada, H., Techniques for increasing frequency spectrum utilization in mobile radio communication systems, *IEEE* CH2037-0/85/0000-0026, pp. 26–31, 1985.
8. Cox, D. C. and Reudnik, D. O., Dynamic channel assignment in two-dimensional large-scale mobile radio systems, *Bell Syst. Tech. J.*, 51(7), 1611–1629, September 1972.
9. Furuya, Y., and Alaiwa, Y., Channel segregation, a distributed channel allocation scheme for mobile communication systems, in *Proc. 2nd Nordic Seminar on Digital Land Mobile Communications*, 311–315, October 1986.

10. Everitt, D. and Manfield, D., Performance analysis of cellular mobile communication systems with dynamic channel assignment, *IEEE Selected Areas Commun.*, SAC-7(8), 1172–1180, October 1989.

11. Arazi, B., New channel assignment strategy in cellular mobile radiocommunication systems, *IEE Proc.*, 133(6), Part F, 569–575, October 1986.

12. Sánchex V., J. H. and Eade, J. P., A simulation study of cellular and sectorized mobile telephone systems using a hybrid channel allocation technique, in *Proc. 5th UK Teletraffic Symp.*, Aston U.K. pp. 11/1–10, July 1988.

13. Eklundh, B., Channel utilization and blocking probability in a cellular mobile telephone system with directed retry, *11th ITC*, 1985.

14. Yacoub, M. D., Cattermole, K. W., and Rodriguez, D. M., Alternative routing in cellular mobile radio, Third U.K. Teletraffic Symp., Colchester, June 1986.

15. Yacoub, M. D. and Cattermole, K. W., Cellular mobile radio with fuzzy cell boundaries, Fourth U.K. Teletraffic Symp., Bristol, May 1987.

16. Yacoub, M. D., Mobile radio with fuzzy cell boundaries, Ph.D. thesis, University of Essex, April 1988.

17. Yacoub, M. D. and Cattermole, K. W., Improving traffic capacity with the use of the fuzzy cell boundary traffic in a mobile radio system, ISSSE'89, Erlangen, West Germany, September 1989.

18. Yacoub, M. D. and Cattermole, K. W., Novel technique for efficient channel utilization in a mobile radio system, IEEE Global Telecommunications Conference, San Diego, Calif., December 2–5, 1990.

19. Yacoub, M. D. and Cattermole, K. W., Use of the third path option as a means of improving capacity of mobile radio system, 13th ITC, Copenhagen, Denmark, June 1991.

20. Mencia, J. C. E., Performance of a mobile radio system with the blocking threshold variation (in Portuguese), M.Sc. Dissertation, UNICAMP, Campinas, Brazil, January 1991.

21. Telephone Traffic Theory, Tables and Charts, Siemens Aktiengesellschaft, Munich, 1970.

22. Mencia, J. C. E. and Yacoub, M. D., Blocking threshold variation in a mobile radio system, submitted for publication.

# Index